T0212556

Lecture Notes in Computer Science 9703

Commenced Publication in 1973
Founding and Former Series Editors:
Gerhard Goos, Juris Hartmanis, and Jan van Leeuwen

More information about this series at http://www.springer.com/series/7412

José Francisco Martínez-Trinidad
Jesús Ariel Carrasco-Ochoa
Víctor Ayala-Ramírez
José Arturo Olvera-López
Xiaoyi Jiang (Eds.)

Pattern Recognition

8th Mexican Conference, MCPR 2016
Guanajuato, Mexico, June 22–25, 2016
Proceedings

 Springer

Editors
José Francisco Martínez-Trinidad
INAOE
Sta. Maria Tonantzintla
Mexico

Jesús Ariel Carrasco-Ochoa
INAOE
Sta. Maria Tonantzintla, Puebla
Mexico

Víctor Ayala-Ramírez
University of Guanajuato
Salamanca
Mexico

José Arturo Olvera-López
Autonomous University of Puebla
Puebla
Mexico

Xiaoyi Jiang
University of Münster
Münster
Germany

ISSN 0302-9743   ISSN 1611-3349   (electronic)
Lecture Notes in Computer Science
ISBN 978-3-319-39392-6   ISBN 978-3-319-39393-3   (eBook)
DOI 10.1007/978-3-319-39393-3

Library of Congress Control Number: 2016939375

LNCS Sublibrary: SL6 – Image Processing, Computer Vision, Pattern Recognition, and Graphics

Printed on acid-free paper

This Springer imprint is published by Springer Nature
The registered company is Springer International Publishing AG Switzerland

Preface

The Mexican Conference on Pattern Recognition 2016 (MCPR 2016) was the eighth event in the series organized by the Computer Science Department of the National Institute for Astrophysics Optics and Electronics (INAOE) of Mexico. This year the conference was jointly organized with the University of Guanajuato, under the auspices of the Mexican Association for Computer Vision, Neurocomputing and Robotics (MACVNR), which is a member society of the International Association for Pattern Recognition (IAPR). MCPR 2016 was held in Guanajuato, Mexico, during June 22–25, 2016.

This conference aims to provide a forum for the exchange of scientific results, practice, and new knowledge, as well as to promote collaboration among research groups in pattern recognition and related areas in Mexico and around the world.

As in previous years, MCPR 2016 attracted not only Mexican researchers but also worldwide participation. We received contributions from 13 countries. In total, 60 manuscripts were submitted, out of which 34 were accepted for publication in these proceedings and for presentation at the conference. Each of these submissions was strictly peer-reviewed by at least two members of the Program Committee, all of them experts in their respective fields of pattern recognition, which resulted in these excellent conference proceedings.

We were very honored to have as invited speakers the following internationally recognized researchers:

- Prof. Michel Devy, Laboratoire d'Analyse et d'Architecture des Systèmes, LAAS-CNRS, France
- Prof. Theo Gevers, Informatics Institute, University of Amsterdam, The Netherlands
- Prof. Balakrishnan Prabhakaran, Department of Computer Science, University of Texas at Dallas, USA

These distinguished researchers gave keynote addresses on various pattern recognition topics and also presented enlightening tutorials during the conference. To all of them, we express our appreciation for these presentations.

This conference has been possible thanks to the efforts of many people. We express our gratitude to them all. In particular, we extend our gratitude to all authors who submitted their papers to the conference and our regrets to those we turned down. We would like to thank to all Program Committee members and additional reviewers for their efforts and the quality of their reviews. Their work allowed us to maintain the high quality of the conference and provided a conference program of high standard. We would also like to thank Springer for giving us the opportunity of continuing to publish MCPR proceedings in the LNCS series. Finally, but not less important, our thanks go to the University of Guanajuato for providing a key support to this event.

The authors of selected papers were invited to submit extended versions of their papers for a special issue of the *International Journal of Pattern Recognition and Artificial Intelligence* published by World Scientific.

We are sure that MCPR 2016 provided a fruitful forum for the Mexican pattern recognition researchers and the broader international pattern recognition community.

June 2016

José Francisco Martínez-Trinidad
Jesús Ariel Carrasco-Ochoa
Víctor Ayala-Ramírez
José Arturo Olvera-López
Xiaoyi Jiang

Organization

MCPR 2016 was sponsored by the University of Guanajuato and the Computer Science Department of the National Institute of Astrophysics, Optics and Electronics (INAOE).

General Conference Co-chairs

Xiaoyi Jiang	University of Münster, Germany
Víctor Ayala-Ramírez	University of Guanajuato, Mexico
Jesús Ariel Carrasco-Ochoa	National Institute of Astrophysics, Optics and Electronics (INAOE), Mexico
José Francisco Martínez-Trinidad	National Institute of Astrophysics, Optics and Electronics (INAOE), Mexico
José Arturo Olvera-López	Autonomous University of Puebla (BUAP), Mexico

Local Arrangements Committee

Cepeda Negrete Jonathan
Cerón Benítez Gorgonio
Cervantes Cuahuey Brenda Alicia
Correa Tomé Fernando Enrique
Hernández Belmonte Uriel Haile
Hernández Gómez Geovanni
Lizárraga Morales Rocío Alfonsina
López Lucio Gabriela
López Pérez José Jesús
Martínez Rodriguez Diana Eréndira
Rojas Laguna Roberto
Sánchez Yáñez Raúl Enrique

Program Committee

Asano, A.	Kansai University, Japan
Batyrshin, I.	Mexican Petroleum Institute, Mexico
Benedi, J.M.	Universidad Politécnica de Valencia, Spain
Castelan, M.	CINVESTAV, Mexico
Chen, Chia-Yen	National University of Kaohsiung, Taiwan
Escalante-Balderas, H.J.	INAOE, Mexico
Facon, J.	Pontifícia Universidade Católica do Paraná, Brazil
Gelbukh, A.	CIC-IPN, Mexico
Goldfarb, L.	University of New Brunswick, Canada
Gomes, H.	Universidade Federal de Campina Grande, Brazil
Graña, M.	University of the Basque Country, Spain

Heutte, L.	Université de Rouen, France
Kampel, M.	Vienna University of Technology, Austria
Klette, R.	University of Auckland, New Zealand
Kober, V.	CICESE, Mexico
Koster, W.	Universiteit Leiden, The Netherlands
Laurendeau, D.	Université Laval, Canada
Lazo-Cortés, M.S.	Universidad de las Ciencias Informaticas, Cuba
Lopez-de-Ipiña-Peña, M.K.	Universidad del País Vasco, Spain
Lorenzo-Ginori, J.V.	Universidad Central de Las Villas, Cuba
Mayol-Cuevas, W.	University of Bristol, UK
Menezes, P.	University of Coimbra-Polo II, Brazil
Montes-Y-Gomez, M.	INAOE, Mexico
Mora, M.	Catholic University of Maule, Chile
Morales, E.	INAOE, Mexico
Pina, P.	Instituto Superior Técnico, Portugal
Pinho, A.	University of Aveiro, Portugal
Pinto, J.	Instituto Superior Técnico, Portugal
Pistori, H.	Dom Bosco Catholic University, Brazil
Raducanu, B.	Universitat Autònoma de Barcelona, Spain
Real, P.	University of Seville, Spain
Roman-Rangel, E.F.	University of Geneva, Switzerland
Ross, A.	West Virginia University, USA
Rueda, L.	University of Windsor, Canada
Ruiz-Shulcloper, J.	CENATAV, Cuba
Sanchez-Cortes, D.	Idiap Research Institute, Switzerland
Sang-Woon, K.	Myongji University, South Korea
Sanniti di Baja, G.	Istituto di Cibernetica, CNR, Italy
Sansone, C.	Università di Napoli, Italy
Sappa, A.	Universitat Autònoma de Barcelona, Spain
Spyridonos, P.	University of Loannina, Greece
Sucar, L.E.	INAOE, Mexico
Valev, V.	University of North Florida, USA
Vaudrey, T.	University of Auckland, New Zealand
Vitria, J.	University of Barcelona, Spain
Zhi-Hua, Z.	Nanjing University, China

Additional Reviewers

Carbajal-Hernández, J.J.	Morales-Reyes, A.
Chien, H.J.	Qian, C.
Feregrino-Uribe, C.	Reyes-García, C.A.
Gómez-Gil, P.	Rodríguez-González, A.Y.
González-Bernal, J.A.	Saleem, N.
Hasan, M.R.	Zhang, T.
Martínez-Carranza, J.	

Sponsoring Institutions

University of Guanajuato (UGTO)
National Institute of Astrophysics, Optics and Electronics (INAOE)
Mexican Association for Computer Vision, Neurocomputing
and Robotics (MACVNR)
National Council of Science and Technology of Mexico (CONACYT)

Contents

Pattern Recognition and Artificial Intelligent Techniques

Signal Processing and Analysis

Applications of Pattern Recognition

Computer Vision and Image Analysis

Text Detection in Digital Images Captured with Low Resolution Under Nonuniform Illumination Conditions

Julia Diaz-Escobar[1(✉)] and Vitaly Kober[1,2]

[1] Department of Computer Science, CICESE, B.C. 22860 Ensenada, Mexico
jdiaz@cicese.edu.mx, vkober@cicese.mx
[2] Department of Mathematics, Chelyabinsk State University,
Chelyabinsk, Russian Federation

Abstract. The text detection task becomes difficult when the image content is complex. Nonuniform illumination, camera perspective, low resolution, complex backgrounds and others, are some of new challenges. Nowadays, most of digital information is obtained using mobile devices. In particular, digital images with textual content bring us useful information which leads to the development of helpful applications such as document classification, augmented reality, language translator, text to voice converter, multimedia retrieval, and so on. However, most of existing text recognition methods are not invariant to illumination, low resolution or geometric distortions. In this work, a method for text detection using adaptive synthetic discriminant functions and a synthetic hit-miss transform is proposed. The suggested method is based on threshold decomposition and a bank of adaptive filters. Finally the performance of the proposed system is tested in terms of miss detections and false alarms with help of computer simulations.

Keywords: Text detection · OCR · Nonuniform illumination

1 Introduction

Nowadays, Optical Character Recognition (well-known as OCR) is considered by many researches as a solved problem when digital images are obtained from scanners [1]. However, in the last years new imaging devices have been developed, including smartphones, digital cameras, web cams, and so on. As a result, digital images are the most important source of information and millions of images are shared every day. In particular, digital images with textual content bring us useful information obtained from everywhere: documents, street signs, books, signboards and so on, which leads to development of helpful applications such as document classification, augmented reality, language translator, text to voice converter, industrial automation, multimedia retrieval and much more.

Unfortunately, traditional OCR engines often fail due to complexity of the imagery, becoming more complicated recognition tasks. Nonuniform illumination, camera perspective, resolution, CCD noise, complex backgrounds and others, are some of new challenges.

© Springer International Publishing Switzerland 2016
J.F. Martínez-Trinidad et al. (Eds.): MCPR 2016, LNCS 9703, pp. 3–12, 2016.
DOI: 10.1007/978-3-319-39393-3_1

Text detection is one of the first stages in character recognition task. OCR techniques consider simple backgrounds without geometric distortions or illumination variations and text detection is usually obtained by only image binarization. However, known binarization and segmentation techniques often fail in nonuniform illumination or low resolution conditions affecting the overall system performance.

Many techniques have been explored to solve the text detection problem. The fundamental goal is to determine whether or not there is text in a given image. Connected Component Analysis (CCA), sliding window classification, Stroke Width Transform (SWT), Maximally Stable Extremal Regions (MSER), and others are some of the state-of-the-art approaches for the extraction of textual information from imagery. For a deep explanation we refer to the following survey [1–3].

The local operator SWT [4] computes the stroke width for each image pixel, then places with similar stroke width can be grouped together into bigger components that are likely to be words. More recently, MSER approach [5] have become one of the basic methods for detection of text in imagery. Newmann and Matas proposed to use all extremal regions whereupon classification is improved using more computationally expensive features [6,7]. More recently, the same authors developed the FASText algorithm based on the well-known FAST corner detector to obtain character strokes as features for AdaBoost classifier [8]. On the other hand, Yin et al. use the MSER method to extract character candidates and then they are grouped in text candidates using single-link clustering [9].

However, most of existing text recognition methods are not invariant to nonuniform illumination, low resolution or geometric distortions. In this work, a method for text detection using adaptive Synthetic Discriminant Functions (SDF) [10] and Synthetic Hit-Miss Transform (SHMT) [11] is proposed. The suggested method is based on threshold decomposition and a bank of adaptive SDF filters. The filters are designed by incorporating information from a set of training images. Finally, the performance of the proposed method is tested in terms of miss and false detections with the help of computer simulation.

The paper is organized as follows. In Sect. 2, threshold decomposition, SDF filters and SHMT are recalled. In Sect. 3, the proposed text detection method is described. In Sect. 4, computer simulation results are presented and discussed. Section 5, summarizes our conclusions.

2   Background

In this section we briefly describe some of techniques used for the proposed text detection method.

2.1   Threshold Decomposition

In accordance with the concept of threshold decomposition, a halftone image $S(x, y)$ with Q quantization levels can be represented as a sum of binary slices $\{S_q(x, y), q = 1, ..., Q - 1\}$ as follows [12]:

$$S(x,y) = \sum_{q=1}^{Q-1} S_q(x,y), \tag{1}$$

with

$$S_q(x,y) = \begin{cases} 1, & \text{if } S(x,y) \geq q \\ 0, & \text{otherwise} \end{cases}. \tag{2}$$

2.2 Synthetic Discriminant Functions

The SDF filter is designed to yield a specific value at the origin of the correlation plane in response to each training image [10]. A SDF filter can be composed as a linear combination of the images of training set $T = \{t_i(m,n), i = 1, ..., N\}$, where N is the number of available views of the target. Let u_i be the value at the origin of the correlation plane $c_i(m,n)$, produced by the filter $h(m,n)$ in response to a training pattern $t_i(m,n)$, as follows:

$$u_i = c_i = t_i \otimes h , \tag{3}$$

with $\otimes$ the correlation operator and

$$h(m,n) = \sum_{i=1}^{N} w_i t_i(m,n), \tag{4}$$

where the coefficients $\{w_i, i = 1, ..., N\}$ are chosen to satisfy the prespecified output u_i for each pattern in T.

Using vector-matrix notation, we denote by $\mathbf{R}$ a matrix with N columns and d rows (number of pixels in each image) where each column is given by the vector version of $t_i(m,n)$. Let $\mathbf{u} = [u_1, ..., u_N]^T$ the desired responses to the training patterns, and $\mathbf{S}$ the matrix whose columns are the elements. Equations (3) and (4) can be rewritten as follows:

$$\mathbf{u} = \mathbf{R}^+\mathbf{h}, \tag{5}$$

$$\mathbf{h} = \mathbf{R}\mathbf{a}, \tag{6}$$

with $\mathbf{a} = [w_1, ..., w_N]^T$ a vector of coefficients, where the superscripts ^{T} and $^+$ denotes transpose and conjugate transpose, respectively. By substituting (6) into (5) we obtain,

$$\mathbf{u} = (\mathbf{R}^+\mathbf{R})\mathbf{a}. \tag{7}$$

The $(i,j)'th$ element of the matrix $\mathbf{S} = \mathbf{R}^+\mathbf{R}$ is the value at the origin of cross-correlation between the training patterns $t_i(m,n)$ and $t_j(m,n)$. If the matrix $\mathbf{S}$ is nonsingular, the solution of the equation system is given by:

$$\mathbf{a} = \mathbf{S}^{-1}\mathbf{u} \tag{8}$$

and the filter vector is:

$$\mathbf{h} = \mathbf{R}\mathbf{S}^{-1}\mathbf{u}. \tag{9}$$

The SDF filter with equal output correlation peaks can be used for intraclass distortion-invariant pattern recognition. This can be done by setting all elements of $\mathbf{u}$ to unity.

2.3   Hit-Miss Transform

Consider a composite Structural Element (SE) $B = (B_1, B_2)$ with $B_1 \cap B_2 = \emptyset$. The set of points at which the shifted pair (B_1, B_2) fits inside the image I is the hit-miss transformation ($\odot$) of X by (B_1, B_2):

$$I \odot B = (I \ominus B_1) \cap (I \ominus B_2), \tag{10}$$

where $\ominus$ is the erosion operator.

Doh et al. [11] proposed a SHMT for the recognition of distorted objects. The algorithm uses SDF filters (see Sect. 2.2) as Structural Elements (SE) for distortion-invariant recognition.

Using the synthetic hit SE, H_{SDF}, as the linear combination of the hit reference images $\{H_i, i = 1, .., k\}$ and the synthetic miss SE, M_{SDF}, as the linear combination of the miss reference images $\{M_i, i = 1, .., k\}$, the proposed synthetic SEs are defined as follows:

$$H_{\mathrm{SDF}} = \sum_{i=1}^{k} a_i H_i \quad and \quad M_{\mathrm{SDF}} = \sum_{i=1}^{k} b_i M_i. \tag{11}$$

Let I be a binary image and I^c be the complement of I, using the synthetic hit SE, H_{SDF} and the synthetic miss SE, M_{SDF}, the proposed SHMT is given as follows:

$$X \odot (H_{SDF}, M_{SDF}) \cong (I \otimes H_{SDF})_{T_{\mathrm{H}}} \cap (I^c \otimes M_{SDF})_{T_{\mathrm{M}}}, \tag{12}$$

where T_{H} is the hit threshold, T_{M} is the miss threshold, and $\cap$ is the intersection operator.

3   Proposed Text Detection Method

To solve the text detection problem, we propose to use the threshold decomposition approach and the SHMT to obtain invariance to nonuniform illumination, noise and slight geometric distortions.

3.1   Adaptive SDF Filters

Based on the work of Aguilar-Gonzalez et al. [13], we design a bank of adaptive SDF filters to obtain distortion invariance. Each filter is created using a modification of the adaptive algorithm proposed by Gonzalez-Fraga et al. [14]. In contrast to the Gonzalez-Fraga's algorithm, we want to recognize a set of characters with the help of SDF filters. The adaptive algorithm for the design of SDF filters is presented in Fig. 1. The algorithm steps can be summarized as follows:

1. Compose a basic SDF filter using the training image of prior known views of a character using (9).

2. Correlate the resulting filter with an image containing all the remaining characters in the scene and find the maximum in correlation plane.
3. Synthesize a pattern to be accepted at the location of the highest value in the correlation plane and include it in the training set of true objects.
4. If the number of images is greater or equal to a prescribed value, the algorithm is finished, else go to step 2.

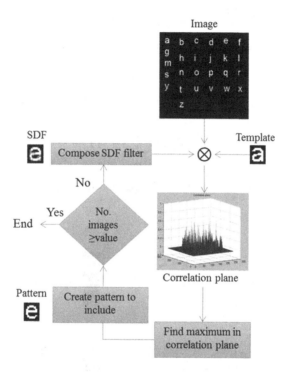

Fig. 1. Block-diagram of the adaptive algorithm for SDF filter design.

As a result we obtain a bank of composite filters. The number of filters depends of the complexity of the geometric distortions. There exists a trade-off between the number of filters and the time consuming.

3.2 Detection Method

In the first stage, using the threshold decomposition concept (described in Sect. 2.1), the image I and it complement I^c are decomposed into binary slices, $\{I_q(x,y),\ q = 1, ..., Q - 1\}$ and $\{I_q^c(x,y), q = 1, ..., Q - 1\}$, respectively. Each binary image is correlated with each filter of the bank, as follows:

$$C_q(x,y) = I_q(x,y) \otimes H_{\text{SDF}} \tag{13}$$

and

$$C_q^c(x, y) = I_q^c(x, y) \otimes M_{\mathrm{SDF}}. \tag{14}$$

Then all correlation planes $C_q(x, y)$ are thresholded by a predefined value T_{H} and T_{M},

$$(C_q(x, y))_{T_{\mathrm{H}}} = \begin{cases} 1, & \text{if } C_q(x, y) \geq T_{\mathrm{H}}, \\ 0, & \text{otherwise} \end{cases} \tag{15}$$

and

$$(C_q^c(x, y))_{T_{\mathrm{M}}} = \begin{cases} 1, & \text{if } C_q^c(x, y) \geq T_{\mathrm{M}} \\ 0, & \text{otherwise} \end{cases}, \tag{16}$$

respectively. Then $SHMT_q$ is obtained by intersection of each pair of binary slices,

$$SHMT_q(x, y) = (C_q(x, y))_{T_{\mathrm{H}}} \cap (C_q^c(x, y))_{T_{\mathrm{M}}}, \tag{17}$$

and, finally, the detection is carried out as union of all $SHMT_q$ results,

$$SHMT_{\mathrm{u}} = \bigcup_{q=1}^{Q-1} SHMT_q. \tag{18}$$

Fig. 2 shows the block-diagram of the proposed method.

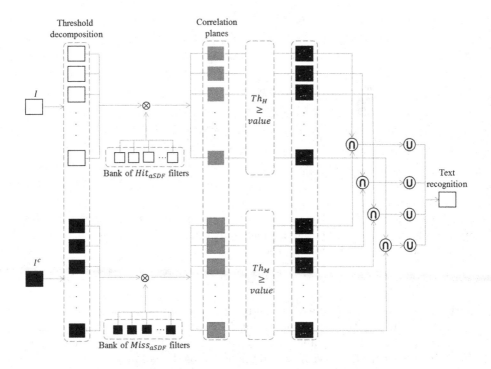

Fig. 2. Diagram of the proposed text detection method.

4 Computer Simulations

In this section we present the results of computer simulations. The performance of the proposed filters is evaluated in terms of false alarms and miss detections.

The size of all synthetic grayscale images used in experiments is 256×256 pixels and the size of character templates is 15×14, using Arial font with size of 16.

In order to analyze the tolerance of the proposed method to geometric distortions (rotation, scaling and shearing) and degradations (noise and nonuniform illumination) we perform experiments using synthetic images, Fig. 3 shows some examples.

We perform 30 experiments for each geometric distortion and degradation changing the character position randomly. The simulation results yield detection errors below than 2 % except for additive noise degradation, where false detections occur. Since false detections could be eliminated in the recognition stage, we do not worry about them. Tables 1 and 2 show the results of geometric distortions and degradations, respectively, in terms of False Positives (FP) and False Negatives (FN).

Inhomogeneous illumination is simulated using a Lambertian model [15],

$$d(x, y) = \cos\{\frac{\pi}{2} - \text{atan}[\frac{\rho}{\cos(\phi)}[(\rho\tan(\phi)\cos(\varphi) - x)^2 \qquad (19)$$
$$+ (\rho\tan(\phi)\sin(\varphi) - y)^2]^{-1/2}]\},$$

Table 1. Tolerance of the proposed method to geometric distortions.

Geometric distortions	Interval	Step size	FP (%)	FN(%)
Rotation	[-30,30]	3	0.00	0.00
Scaling	[0.8,1.5]	0.1	0.00	1.15
Shearing	[-0.5,0.5]	0.1	0.00	1.67

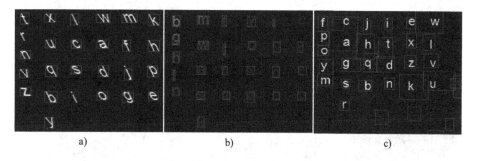

a)   b)   c)

Fig. 3. Example of synthetic images: (a) shearing by factor of 0.5, (b) nonuniform illumination with $\rho = 20$, (c) additive noise with $\sigma = 10$.

where $d(x, y)$ is a multiplicative function which depends on the parameters ρ that is the distance between a point in the surface and the light source, and ϕ, φ that are tilt and slang angles, respectively. For our experiments we use the following parameters: $\phi = 65$, $\varphi = 60$ and varying the parameter ρ in the range of $[5, 50]$ (see Table 2).

Table 2. Tolerance of the proposed method to degradations.

Degradations	Interval	Step size	FP (%)	FN(%)
Illumination	[5,50]	5	0.00	0.64
Additive noise	[0,10]	1	11.41	1.03
Impulsive noise	[-0,0.5]	0.05	0.13	0.90

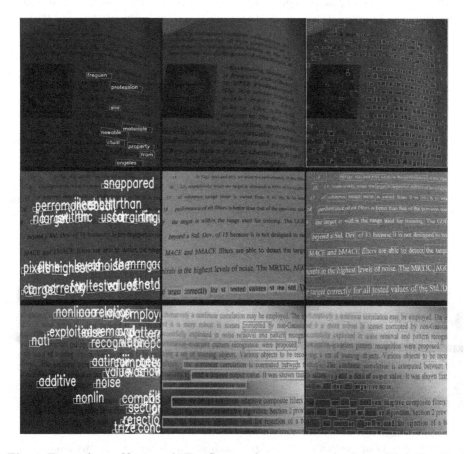

Fig. 4. First column: Neumann's TextSpotter detector, second column: Yin's TextDetector, third column: proposed text detector.

4.1   Real Images

Finally, some preliminary experiments were performed with real images to compare the proposed method with TextSpotter[1] by Neumann et al. [8] and TextDetector[2] by Yin et al. [9] (both described in Sect. 1), using real images. Figure 3 shows the results.

The same image for the three detectors was used. The resulting images look a little different due to the processing of each detector, the best results are obtained with the proposed method.

5   Conclusion

In this work we proposed a new method for text detection in degraded images using the threshold decomposition and adaptive synthetic hit-miss transform. The suggested text detector is robust to slight geometric distortions and degradations such as nonuniform illumination, noise and low resolution. In future we continue to improve the text detection and recognition algorithms to create a real-time OCR system, which is able to reliably recognize characters in low quality images.

Acknowledgments. This work was supported by the Ministry of Education and Science of Russian Federation (grant 2.1766.2014K).

References

1. Qixiang, Y., Doermann, D.: Text detection and recognition in imagery: a survey. Pattern Anal. Mach. Intell. **37**(7), 1480–1500 (2015)
2. Zhu, Y., Yao, C., Bai, X.: Scene text detection and recognition: recent advances and future trends. Front. Comput. Sci. **10**(1), 19–36 (2015)
3. Zhang, H., Zhao, K., Song, Y., Guo, J.: Text extraction from natural scene image: a survey. Neurocomputing **122**, 310–323 (2013)
4. Epshtein, B., Eyal, O., Yonatan, W.: Detecting text in natural scenes with stroke width transform. In: Computer Vision and Pattern Recognition, pp. 2963–2970 (2010)
5. Matas, J., Chum, O., Urban, M., Pajdla, T.: Robust wide-baseline stereo from maximally stable extremal regions. Image Vis. Comput. **22**(10), 761–767 (2004)
6. Neumann, L., Matas, J.: A method for text localization and recognition in real-world images. In: Kimmel, R., Klette, R., Sugimoto, A. (eds.) ACCV 2010. LNCS, vol. 6494, pp. 770–783. Springer, Heidelberg (2011)
7. Neumann, L., Matas, J.: Real-time scene text localization and recognition. Computer Vision and Pattern Recognition, pp. 3538–3545 (2012)
8. Busta, M., Neumann, L., Matas, J.: FASText: Efficient Unconstrained Scene Text Detector. Computer Vision, pp. 1206–1214 (2015)

---

[1] http://www.textspotter.org/.
[2] http://kems.ustb.edu.cn/learning/yin/dtext/.

9. Yin, X.C., Yin, X., Huang, K., Hao, H.W.: Robust text detection in natural scene images. Pattern Anal. Mach. Intell. **36**(5), 970–983 (2014)
10. Casasent, D.: Unified synthetic discriminant function computational formulation. Appl. Opt. **23**(10), 1620–1627 (1984)
11. Doh, Y., Kim, J., Kim, J., Choi, K., Kim, S., Alam, M.: Distortion-invariant pattern recognition based on a synthetic hit-miss transform. Opt. Eng. **43**(8), 1798–1803 (2004)
12. Fitch, J., Coyle, E., Gallagher Jr., N.: Median filtering by threshold decomposition. Acoust. Speech Sig. Proc. **32**(6), 1183–1188 (1984)
13. Aguilar-Gonzalez, P., Kober, V., Diaz-Ramirez, V.: Adaptive composite filters for pattern recognition in nonoverlapping scenes using noisy training images. Pattern Recogn. Lett. **41**, 83–92 (2014)
14. Gonzalez-Fraga, J., Kober, V., Alvarez-Borrego, J.: Adaptive synthetic discriminant function filters for pattern recognition. Opt. Eng. **45**(5), 057005 (2006)
15. Diaz-Ramirez, V., Picos, K., Kober, V.: Target tracking in nonuniform illumination conditions using locally adaptive correlation filters. Opt. Comm. **323**(1), 32–43 (2014)

Rotation Invariant Local Shape Descriptors for Classification of Archaeological 3D Models

Edgar Roman-Rangel[1]([⊠]), Diego Jimenez-Badillo[2],
and Stephane Marchand-Maillet[1]

[1] CVMLab, University of Geneva, Geneva, Switzerland
{edgar.romanrangel,stephane.marchand-maillet}@unige.ch
[2] National Institute of Anthropology and History of Mexico (INAH),
Mexico City, Mexico
diego_jimenez@inah.gob.mx

Abstract. We introduce a method for estimation of rotation invariant local shape descriptors for 3D models. This method follows a successful idea commonly used to obtain rotation invariant descriptors in 2D images, and improves it by tackling the difficulty of the 3 degrees of freedom that exists in 3D models. Our method is simple, yet it achieves high levels of invariance after rotation transformations, and it produces short descriptors that can be efficiently used in several tasks. Such is the case of automatic classification of 3D surfaces with archaeological value, in which the proposed method attains state-of-the-art results in shorter times when compared with previous methods.

Keywords: 3D models · Shape descriptor · Rotation invariant

1 Introduction

Achieving invariance to rotation is among the principal challenges when designing local shape descriptors [1,2]. This problem becomes specially difficult when dealing with 3D data, where an extra degree of freedom comes into play, such that it results unclear how to define an oriented reference frame [3].

Many previous works have proposed approaches of different nature to address the problem of rotation invariance for local 3D shape descriptors. From approximating Gaussian convolutions of point cloud vertices [4], which roughly resembles the traditional Gaussian scale space used for 2D images [5]; to computing local descriptors several times, each one from a different viewpoint, and then selecting one of these descriptors according to the specific needs of the problem of interest [6]; to implementing Singular Value Decomposition (SVD) or Eigenvalue Decomposition (EVD) for the definition of oriented reference frames [3,7]. However, these methods either fail in providing unambiguous reference orientations for the local descriptors, or result in very time-consuming approaches.

We introduce a method that is of fast computation and that produces rotation invariant local shape descriptors for 3D models. We combined our method for

© Springer International Publishing Switzerland 2016
J.F. Martínez-Trinidad et al. (Eds.): MCPR 2016, LNCS 9703, pp. 13–22, 2016.
DOI: 10.1007/978-3-319-39393-3_2

rotation invariance with a method for local description that roughly follows the well known Shape Context used in 2D images [8]. This is, it describes a point of interest using the spatial distribution of its nearby points. Namely, our method aligns each point of interest using spherical coordinates that are computed with respect to the centroid of the cloud. This allows to produce local descriptors that are consistent across rotations changes, even without knowing the actual orientation of the 3D model itself, i.e., free positioning of the model in the 3D space. Note that in general, our method is independent from the local descriptor, as it can be plugged to different descriptive functions.

We evaluated the proposed method on a set of classification experiments, using 3D surfaces that represent archaeological potsherds from the Teotihuacan culture. Classification of potsherds is a very challenging problem because: (1) of the large amount of them that are found in a common excavation site, e.g., tens of thousands of fragments; and (2) the fact that often potsherds from different ceramic types are discovered together and scrambled. Also, it is of special interest in Archaeology, as classifying newly discovered potsherds brings most of the insights regarding the daily life of ancient civilizations. Our results show that the proposed method is both robust to rotations transformations, as it is fast.

The remaining of this papers is organized as follows. Section 2 introduces previous works that have addressed the problem of rotation invariance for 3D shape descriptors. Section 3 presents the framework we propose in this paper to achieve rotation invariant local shape descriptors of 3D models. Section 4 details the experimental setup we followed to evaluate the proposed method. Section 5 discusses our results. And Sect. 6 presents our conclusions.

2   Related Work

Several method for local description of 3D models have been developed [1, 2] focusing on different types of information: such as geometry or depth.

Spin Images [9] are one of the earliest methods to show promise for description of 3D models. It consists in counting the number of neighboring points that fall inside the regions of a volumetric space, which is centered at a point of interest, and that is orientated according to the normal of the tangent plane at such point of interest. In turn, the number of neighbor points in each region is assigned to bins in a histogram. This method was latter improved by using a 3D approximation of a Gaussian scale space that enables invariance to scale transformations: Scale Invariant Spin Images (SISI) [4]. Although this approach works well for oriented 3D models, it fails in the presence of rotated instances due to the lack of definition of an oriented reference frame.

The 3D Shape Context (3DSC) [6] is an extension of the Shape Context descriptor [8]. This method defines a spherical space around a point of interest, and divides it into regions. Then, the number of nearby points, with respect to the point of interest, that are in each of the d-th regions is assigned as the value of the d-th dimension of a descriptive vector. Given the lack of an oriented reference frame, 3DSC aligns the north pole of the sphere with the normal of the 3D surface estimated at the point of interest, then it computes L discrete rotations and a local descriptor for each of them. Finally, during a given comparison task, e.g., classification, a permutation step is conducted to compare the L versions of the descriptor, thus impacting its computational cost.

As improvement to 3DSC, the Unique Shape Context for 3D data descriptor (USC) [7] implements a SVD or EVD approach to estimate an oriented frame. Namely, it approximates a Total Least Square estimation with the three principal eigenvectors of a weighted covariance matrix, followed by a sign disambiguation that defines the orientation of two axis in the 3D space.

Similar to USC, the Unique Signature of Histograms for Local Surface Description (SHOT) [3] also implements SVD or EVD. However, it computes a local reference rather than a global one, as USC does. Namely, SHOT has two main features that differ from USC: (1) it only considers points within the local sphere, and (2) it performs reorientation of local eigenvectors such that their signs are coherent with the majority of vectors that each of them represents.

Besides the similarity in construction that all these methods share, i.e., they rely on the counting of nearby points within spherical regions, they behave differently under rotation variations, i.e., not all of them are robust enough. Also, 3DSC, USC, and SHOT might be very time consuming.

3   Local Descriptor

As previously mentioned, the base approach we use for local descriptions is a Shape Context-like method [8], which builds an histogram of the relative position of the nearby points for a given point of interest. More precisely, we define a spherical local context around a point of interest. Such a sphere is subdivided into 2 distance intervals, 8 azimuth intervals, and 4 zenith intervals, which are equally spaced across their corresponding axis, thus forming 64 spherical regions $R_{d=1,...,64}$, each of which corresponds to one of the dimensions d of the local descriptor $V(d)$.

Mathematically, the value of the d-th dimension of the descriptor is give by,

$$V_i(d) = |\{p_j \neq p_i : (p_i - p_j) \in R_d\}|, \tag{1}$$

where, $|\cdot|$ indicates cardinality operator, $\{\cdot\}$ denotes a set, p_i is the point of interest, p_j is a neighbor point of p_i, and the term $(p_i - p_j)$ in Eq. (1) denotes vector difference.

In practice, we are interested in knowing which of the neighboring points p_j are located within the d-th spherical region defined for the point of interest p_i. This can be easily estimated by knowing the spherical relative coordinates of all points p_j with respect to p_i, i.e., redefining p_j as $p_j = \left(\rho_j^i, \theta_j^i, \phi_j^i\right)$, which can be computed by,

$$\rho_j^i = \sqrt{d_x^2 + d_y^2 + d_z^2}, \, 0 \le \rho \le \infty,$$

$$\theta_j^i = \tan^{-1}\left(\frac{d_y}{d_x}\right), \quad 0 \le \theta \le 2\pi, \tag{2}$$

$$\phi_j^i = \cos^{-1}\left(\frac{d_z}{\rho_j^i}\right), \quad 0 \le \phi \le \pi,$$

where, (d_x, d_y, d_z) is the component-wise distance between points p_i and p_j. Note that the boundaries in Eq. (2) indicate that the azimuth intervals cover a whole round along the sphere, whereas the zenith intervals cover half a round along it.

3.1   Rotation Invariance

To account for rotation invariance, we can realign the point of interest with respect to a reference point, which must be the same for all points of interest. In the case of 2D images, this alignment is computed by subtracting the value of the local orientation of the point of interest from itself and from the local orientation of all its neighboring points, such that p_i would be at $0°$ and all its nearby points would be shifted accordingly [8].

This procedure is rather straight forward in the 2D case, where the local orientation of a point is defined as the counter-clock wise inclination with respect to the horizontal axis. In the case of 3D models, this definition is less clear given the three degrees of freedom that exist. Therefore, we propose the use of the centroid of the point cloud as a reference point to account for rotation invariance. More specifically, we realign the point of interest (and all its nearby points) by subtracting from it its spherical orientations computed with respect to the centroid c of the 3D model, such that p_i would rest at $0°$ both in the azimuth and the zenith axis, and its neighboring points would be shifted accordingly. Mathematically, this is computed as,

$$\bar{\theta}_j^i = \left(\theta_j^i - \theta_i^c\right) \bmod 2\pi,$$

$$\bar{\phi}_j^i = \left(\phi_j^i - \phi_i^c\right) \bmod \pi. \tag{3}$$

where, θ_i^c and ϕ_i^c indicate, respectively, the azimuth and zenith orientations of p_i computed with respect to the centroid c, which in turn, is estimated as the mean of the (x, y, z) coordinates of the points in the 3D model,

$$x_c = \frac{1}{N} \cdot \sum_{i=1}^{N} x_i, \qquad y_c = \frac{1}{N} \cdot \sum_{i=1}^{N} y_i, \qquad z_c = \frac{1}{N} \cdot \sum_{i=1}^{N} z_i, \tag{4}$$

note the modulus normalization in Eq. (3), which is required to comply with the boundaries defined in Eq. (2).

By using $\left(\rho_j^i, \bar{\theta}_j^i, \bar{\phi}_j^i\right)$, from Eqs. (2) and (3), as the arguments to identify the spherical region for each neighboring point p_j (Eq. (1)), it is possible to construct a rotation invariant descriptor for p_i.

Finally, the visual similarity between two points can be estimated as the distance between their respective local descriptors. To account for sets of different size (3D models with different amount of points), we normalize the descriptors such that summing up their dimensions equals 1,

$$H_i\left(d\right) = \frac{V_i\left(d\right)}{\sum_{d=1}^{64} V_i\left(d\right)},\tag{5}$$

where V_i is defined as in Eq. (1).

4   Experimental Setup

In this section we first provide a description of the data used to evaluate the proposed method. Then we present two types of evaluation that we conducted: one for measuring the level of invariance against rotation, and one to evaluate the impact of the method in the task of classification of 3D data.

4.1   Data

As previously mentioned, we used a set of 3D surfaces, which correspond to pot-sherds from the Teotiuhacan culture that developed in ancient Mesoamerica. An important fact about this specific collection is that it has been already cataloged manually by archaeologist, thus it was possible to annotate all its instances with the name of the ceramic type of their provenience, e.g., plate, pot, crater, bowl, vase, amphora, etc. Note however, that these labels are recognizable only for the so-called diagnostic potsherds, this is, potsherds containing specific parts of the ceramic type, e.g., the neck of a pot, the border of a jar, the supports of a base, etc.

Besides those diagnostic potsherds, there also exist potsherds that are very difficult to recognize. Most of these examples come from sections of the main body of the ceramics, and they often correspond to simple curved or flat sections with no much of discriminative visual information. Figure 1 shows examples of both diagnostic and non-diagnostic potsherds.

As shown in Fig. 1, the annotations for the diagnostic potsherds consist in the name of their respective ceramic type: plate, pot, bowl, crater, censer, vase, and vase with support. Regarding the non-diagnostic potsherds (or regular potsherds), their annotations are as follows:

- Curved: potsherds that have moderately curved shapes.
- Highly (curved): potsherds with highly visible curved shapes.
- Slight (border): sections towards the ceramics border, that are barely visible.
- Border: potsherds with clearly visible border.
- Convex: different from the previous concave curved potsherds, this class consists of potsherds whose curvature is convex.

Overall, this dataset is composed of 148 surfaces and 12 visual classes. Figure 2 shows the distribution of instances over them. Note that although the classes are not well balanced, they remain within the same order of magnitude.

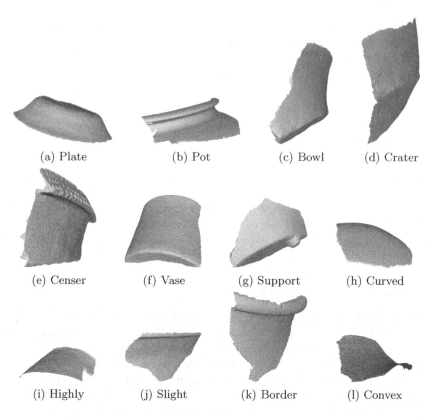

(a) Plate   (b) Pot   (c) Bowl   (d) Crater

(e) Censer   (f) Vase   (g) Support   (h) Curved

(i) Highly   (j) Slight   (k) Border   (l) Convex

Fig. 1. Visual examples of potsherds. Examples 1a to 1g correspond to diagnostic potsherds, and examples 1h to 1l are of non-diagnostic potsherds.

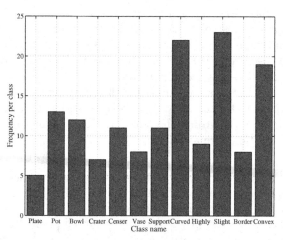

Fig. 2. Frequency of potsherds in each class.

4.2   Consistency and Efficiency

We first compared the time that different methods require for computing local descriptors, including Scale-Invariant Spin Images (SISI) [4], 3D Shape Context (3DSC) [6] with $L = 8$, Unique Shape Context for 3D data (USC) [7], and Unique Signatures of Histograms for Local Surface Description (SHOT) [3].

To check for consistency of the proposed method, we also evaluated the level of degradation that a local descriptor suffers after a rotation change. To this end, we randomly selected 1000 points from all the dataset (different 3D surfaces randomly selected), and computed their local descriptors before and after rotating the 3D surface. Then, we computed the Euclidean distance between the two instances of the same descriptor (before and after rotation).

4.3   Classification Performance

Finally, we compared the classification performance of each method using the dataset of potsherds. To this end, we relied on bag-of-words representations. Namely, we repeated the following protocol independently for each method:

1. Compute a vocabulary of local descriptors using a subset of 20,000 of them randomly selected. In practice, we computed vocabularies of different sizes using the k-means clustering algorithm [10].
2. Represent each 3D surface with a bag representation using the vocabulary previously estimated.
3. Use each 3D surface as query to be classified using a k-NN approach (k=1). This is, a leave-one-out full-cross validation.
4. Compute the average classification accuracy.

5   Results

This section presents the results of our evaluations. First the evaluation of efficiency and consistency, and later the evaluation on classification performance.

5.1   Consistency

Table 1 shows the average time that each method takes to compute a local descriptor alongside its dimensionality. As one can see, both SISI and the proposed method, which we refer to as RI, are the fastest methods for computation of local shape descriptors, as they are also the shortest. Note that the use of the covariance matrix and computation of principal eigenvectors (USC and SHOT) requires more time than the proposed subtraction of local orientations (RI). Special attention is worth to 3DSC, which do not relies on the computation of principal eigenvectors, however it does repeat the description process 8 times.

We compared the degradation induced by our proposed method with respect to that induced by the different previous methods. Figure 3 shows these results. As shown in Fig. 3, almost all descriptors induce comparable levels of distortion

Table 1. Vector size and computational time in seconds for each method.

Method	Vector size	Time(s)
SISI [4]	64	0.91 ± 1.29
3DSC [6]	160 (x8)	33.06 ± 6.29
USC [7]	160	6.03 ± 8.08
SHOT [3]	160	3.12 ± 4.28
RI (ours)	64	0.93 ± 1.20

after rotation transformations, with SISI been the method with highest degradation, as it lacks of a proper definition of orientation frame.

Although the use of eigenvectors induces relative low degradation (USC, SHOT), it is not as low as that induced by the proposed RI. This suggests that the computation of principal eigenvectors is not as strong approach as expected for addressing rotation variations. Also, the results of Fig. 3 shows that the proposed subtraction technique works well in practice.

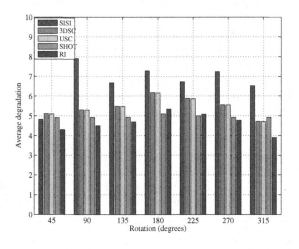

Fig. 3. Euclidean distance computed between two descriptors corresponding to the same point localized at rotated instances of the same 3D model. Rotation in steps of 45 degrees

5.2 Classification Performance

We considered two cases for the classifications experiments. The first one, when the complete dataset is used, and the second case when only the subset of diagnostic potsherds is considered. Table 2 shows the average classification accuracy obtained by the different methods for local description.

Table 2. Average classification accuracy achieved by different 3D local descriptor. These results are for two cases: (1) using the complete dataset, and (2) using only diagnostic potsherds. Best results are highlighted in bold.

Method	Vocabulary size					
	100	250	500	1000	2500	5000
Complete dataset						
SISI [4]	0.295	0.285	0.266	0.261	0.217	0.193
3DSC [6]	0.301	0.304	0.278	0.255	0.246	0.244
USC [7]	**0.481**	0.480	0.458	0.426	0.406	0.367
SHOT [3]	0.466	0.469	0.465	0.460	0.425	0.405
RI (ours)	0.470	0.472	0.447	0.385	0.386	0.374
Only diagnostic potsherds						
SISI [4]	0.362	0.404	0.362	0.319	0.255	0.255
3DSC [6]	0.448	0.492	0.480	0.441	0.416	0.412
USC [7]	0.647	0.652	0.641	0.606	0.588	0.581
SHOT [3]	0.658	**0.668**	0.615	0.639	0.592	0.577
RI (ours)	0.647	0.653	0.622	0.591	0.598	0.579

From Table 2, one can see that higher classification accuracy is achieved by using only diagnostic potsherds. This is expected as it corresponds to a more controlled scenario with respect to using both diagnostic and generic potsherds. Furthermore, although the ceramic pieces from which generic potsherds come are virtually unknown, in practice they will correspond to fragments coming from the same ceramic pieces than diagnostic potsherds. Therefore, they might have sections that visually resemble to one another, which in turn, might lead to confusion during classification. We can also see that short vocabularies suffice for accurate description. In particular, 250 visual words are enough.

Regarding the comparison of methods, SISI and 3DSC achieved the lowest performance. Whereas USC and RI, which are very similar descriptors by construction, performed alike each other. Although USC and RI are partially similar to 3DSC by construction, 3DSC requires the computation of 8 several variants of the local descriptor, and it performs poorer that USC and RI. Note that SHOT achieved a slightly better performance that RI. However, it produces a larger vector, and it takes as much as three times to compute it compared wit RI.

6   Conclusions

We presented a new methodology for achieving rotation invariance on 3D local shape descriptor. Besides its simplicity, the proposed method has high levels of invariance against rotation transformations. Also, it produces a short vector that achieves state-of-the-art performance in classification results, with respect to several previous methods for local shape description of 3D models.

One main feature of the proposed model is that it is independent from the actual function used as descriptor, such that it can be plugged to many local descriptors to boost their robustness against rotation transformations.

We evaluated our method on a set of 3D surfaces that represent archaeological potsherds, and obtained high performance in the classification task. Namely, classifying potsherds is of great importance to archaeologist, and one of the many tasks where pattern analysis could provide with new tools. In particular, this is the case of our research effort, which seeks to develop a machinery for presenting classification suggestions for newly discovered potsherds, such that it could assist archaeologists in deciding classes for potsherds.

Currently, we continue working on the collection of more data for further testing of our method, and the eventual implementation of such system that could handle not only potsherds, but different archaeological artifacts, e.g., masks, ceramics, jewelry, etc.

Acknowledgments. This work was supported by Swiss National Science Foundation through the Tepalcatl project P2ELP2_152166.

References

1. Li, B., Lu, Y., Li, C., Godil, A., Schreck, T., Aono, M., Burtscher, M., Chen, Q., Chowdhury, N.K., Fang, B., Fu, H., Furuya, T., Li, H., Liu, J., Johan, H., Kosaka, R., Koyanagi, H., Ohbuchi, R., Tatsuma, A., Wan, Y., Zhang, C., Zou, C.: A comparison of 3D shape retrieval methods based on a large-scale benchmark supporting multimodal queries. Comp. Vis. Image Underst. **131**, 1–27 (2015)
2. Li, B., Lu, Y., Li, C., Godil, A., Schreck, T., Aono, M., Chen, Q., Chowdhury, N.K., Fang, B., Furuya, T., Johan, H., Kosaka, R., Koyanagi, H., Ohbuchi, R., Tatsuma, A.: SHREC'14 track: large scale comprehensive 3D shape retrieval. In: Eurographics Workshop on 3D Object Retrieval 2014 (3DOR) (2014)
3. Tombari, F., Salti, S., Di Stefano, L.: Unique signatures of histograms for local surface description. In: Maragos, P., Paragios, N., Daniilidis, K. (eds.) ECCV 2010, Part III. LNCS, vol. 6313, pp. 356–369. Springer, Heidelberg (2010)
4. Darom, T., Keller, Y.: Scale-invariant features for 3-D mesh models. IEEE Trans. Image Proc. **21**(5), 2758–2769 (2012)
5. Lowe, D.G.: Distinctive image features from scale-invariant keypoints. Int. J. Comp. Vis. **60**(2), 91–110 (2004)
6. Frome, A., Huber, D., Kolluri, R., Bülow, T., Malik, J.: Recognizing objects in range data using regional point descriptors. In: Pajdla, T., Matas, J.G. (eds.) ECCV 2004. LNCS, vol. 3023, pp. 224–237. Springer, Heidelberg (2004)
7. Tombari, F., Salti, S., Di Stefano, L.: Unique shape context for 3D data description. In: Proceedings of the ACM Workshop on 3D Object Retrieval (2010)
8. Belongie, S., Malik, J., Puzicha, J.: Shape matching and object recognition using shape contexts. IEEE Trans. Pattern Anal. Mach. Intell. **24**(4), 509–522 (2002)
9. Johnson, A.E., Hebert, M.: Using spin images for efficient object recognition in cluttered 3D scenes. IEEE Trans. Pattern Anal. Mach. Intell. **21**(5), 433–449 (1999)
10. Lloyd, S.: Least squares quantization in PCM. IEEE Trans. Inf. Theory **28**(2), 129–137 (1982)

Feature Extraction as Ellipse of Wild-Life Images

Karina Figueroa[1]([✉]), Ana Castro[1], Antonio Camarena-Ibarrola[2], and Héctor Tejeda[1]

[1] Facultad de Ciencias Físico-Matemáticas, Universidad Michoacana, Morelia, Mexico
{karina,acastro,htejeda}@fismat.umich.mx
[2] Facultad de Ing. Eléctrica, Universidad Michoacana, Morelia, Mexico
camarena@umich.mx

Abstract. In this work we propose a new feature for automatic identification of individuals that belong to the species with spot patterns in their fur such as leopards, giraffes, cheetahs, and jaguars. We specifically worked with *ocelots*, kind of leopards found in America, using a collection of photographs taken with trap-cameras installed in the forest in Michoacan, Mexico. Our proposed feature extraction technique obtains for each spot automatically found in the fur the smallest ellipse that encloses the spot. Once the spots are replaced by ellipses, we only have to store for each ellipse, its center and a vector in the direction of its major axis. As a result, we end up with a collection of points and vectors which is exactly the same kind of features that are extracted in fingerprints after locating the minutiae which was precisely what motivated our work. We use two methods to determine the ellipses, one is based on geometric construction and the other one is based on the image moments. The preliminary experiments show that a better representation of the elliptical pattern is obtained when the method of moments is used.

1 Introduction

Identification of human individuals by their faces, fingerprints, or iris are problems that have been of interest to researchers for many years. However, identification of wild animals in a non intrusive fashion is of rather recent interest, it is also a more challenging problem since images are more deformed by position, scalability, rotation and skin folding, after all you cannot ask a wild animal to stand still for a picture.

We in a previous work [4] managed to automatically detect and segment the ocelot for identifying the region of interest in the image. These images were taken by trap-cameras which were triggered by infrared sensors when an animal crossed in front of them. In this work we propose a method to characterize the spots of the ocelot in a way that is very convenient for the next step, the identification of the individual ocelot inside the image.

J.F. Martínez-Trinidad et al. (Eds.): MCPR 2016, LNCS 9703, pp. 23–32, 2016.
DOI: 10.1007/978-3-319-39393-3_3

2   Previous Work

Some recognition systems of wildlife are shown in this section. These works use intrusive techniques (i.e. marks, chips, etc.), the images need to meet certain requirements or the implementation is not fully automatic since it requires the work of an operator to process each image manually.

Jurgen den Hartog et al. [5] implemented a software known as I^3S (Interactive Individual Identification System), for a pattern recognition system over a body of animals, however, it is a supervised technique that needs the work of a human operator which makes it of little use when thousands of pictures have to be processed. The supervisor must mark where the animal is inside the image and I^3S makes a model for each image; after that, this model is searched in the database sequentially. The main disadvantage is that all images must satisfy certain conditions (they must be taken at 30 degrees), which means discarding most of the photos or they have to be taken intrusively which in many cases is not a valid solution.

Charre-Mendel et al. [1] showed with 6 photographs captured from trap-cameras the presence of jaguars (Panthera onca) in Michoacan, Mexico in 2010. The recognition was performed manually by trained employees. All these records belongs to semi-tropical forest located between south of Sierra Madre and the Pacific Coast, they have around 300 records of ocelots that also, have been identified manually.

In [7,11] the authors showed an implementation of a photo-identification system of wildlife with natural marks. In particular in [7], the authors have worked with cheetahs from the Serengeti National Park of Tanzania. However these cheetahs were recognized in a semi-automatic approach, they extract manually every area: shoulder blade, hip joint, the abdomen line and the vertebral column, after a sample of the pattern of the fur. The drawback is of course the need of a human operator to process each image.

A pattern recognition system needs to solve three problems at least:

1. Identify the region of interest, this problem is known as segmentation;
2. Extract features from the pattern;
3. Decide to which class the pattern belongs to, this problem is known as classification.

Segmentation: In [4] the authors proposed a new method to segment the animal from the image. The proposed method consists of getting the difference of two images taken from the same trap-camera and the same position (trap-cameras are placed on non-moving objects like trees or stones).

Feature extraction: There are several ways to model closed-shapes. A convenient way to model closed-shapes is to represent them as ellipses. In [8], the authors used elliptical properties of the Fourier coefficients for characterization purposes. In [10] the authors identified diatoms and extract some features of leaflets using geometric descriptors to approximate the shapes as ellipses. In [3] an algorithm called tangent method rope detect ellipses on an image.

A photo-recognition system for wildlife using non-intrusive techniques is an open problem in this world. Images taken by trap-cameras are a big challenge because they can be in a different position, or side (animals are not symmetric on the distribution of patterns).

This paper is organized as follows: at Sect. 3, our proposal for feature extraction is presented, in Sect. 4 results are showed using the proposed method, and finally, in Sect. 6 we present some conclusions and future work.

3   Our Proposal

Our goal is to extract features from the images to model the spots in the fur of the animal in the pictures taken, in particular, we are interested in those captured with trap-cameras. First, an edge detection algorithm was used in order to emphasize edges, this makes easier to identify the main feature we use to identify the animal, that is, the closed shapes or spots in the fur, then we replace them with ellipses.

3.1   Filters

Digital filters are normally used to enhance contrast or reduce noise of pictures, this is a normal preprocessing phase in image recognition systems. We used two filters: Sobel and Canny (details are not given because both filters are widely known in the literature).

3.2   Feature Extraction Using Ellipses

Firstly, we present two methods to obtain a set of ellipses from the animal's spots in the fur. The first one is a geometric construction and the second one uses the image moments. An ellipse is the trajectory of a point that surrounds two foci such that the sum of the distances to the two foci is constant for every location of the point on the curve [2]. For example, Fig. 1(a) shows it has two fixed points (F and F') called foci of the ellipse. The straight line l through the focus is called the *focal axis*. The ellipse intersects the focal axis at two points V and V' called the *vertices*. The portion of the focal axis between the vertices, or the segment VV' is called the *major axis*, the longest diameters of an ellipse. The point C of the major axis, midpoint of the line segment linking the two foci, is called the *center*. The line l' pass through C and it is perpendicular to the focal axis l and it is called the *normal axis*. The normal axis cuts the ellipse at two points A and A', the segment AA' is called the *minor axis*, the shortest diameters. A segment as BB', that links any two different points of the ellipse, is called the *chord*. In particular a *chord* passing through one focus as EE' is called the *focal chord*. Two diameters are conjugated if and only if they are orthogonal.

In order to find the center of a given ellipse we have to draw two chords EE' and BB', by definition a line passing through at midpoints of these EE' and BB', is a diameter of the ellipse, and therefore passes through the center C.

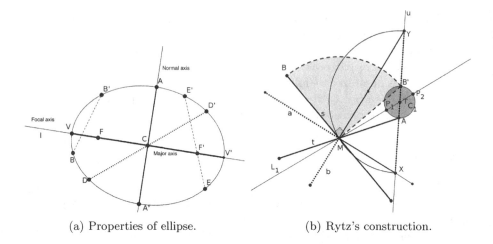

(a) Properties of ellipse.   (b) Rytz's construction.

Fig. 1. Ellipse properties and construction.

Geometric Construction. It is possible to construct an ellipse by knowing a pair of conjugate diameters, in position and magnitude. The Rytz's construction determines both the larger axis and the smaller axis of the ellipse in location and magnitude. Figure 1(b) shows an example. First, we need two conjugate diameters s and t, segment BM is rotated $90°$, the result is segment $B'M$; then B', circle C_1 is drawn with diameter AB', its midpoint is T. The magnitud of the axis is given by the distance of the intersection points of line L_1 with C_1, i.e. (p_1, p_2). In order to find the directions of the axis, we draw a line u through A and B'. The Tales circle with center in the midpoint T that pass at center M cuts u in X and Y lines. The directions of the axis are given by YM and XM.

Using Image Moments. Another way to describe shapes is based on the use of statistical properties called *moments*, the first three *moments* of a discrete one-dimensional function $f(x)$ are related to its mean (μ); its variance(σ) and its *skewness*, which is a measure of the symmetry of the function and it is defined as:

$$skew = \frac{\sum_{x=1}^{N}(x - \mu)^3 f(x)}{\sum_{x=1}^{N} f(x)} \tag{1}$$

Since we are interested in computing descriptors that are invariant under translation, what we really need to compute are the *central moments* of an image. The general two-dimensional $(p + q)$th order moments of a grey-level image $f(x, y)$ are defined as [6]:

$$m_{pq} = \int_{-\infty}^{\infty} \int_{-\infty}^{\infty} x^p y^q f(x, y) dx dy \tag{2}$$

$$p, q = 0, 1, 2 \ldots \tag{3}$$

m_{00} is the moment of degree zero and represents the area of a grey-level image. In order to compute a centroid (or center of mass), it can be determined by combining m_{00} with the image moments of the first degree (m_{01} and m_{10}). Using the moments of second degree (m_{11}, m_{02} and m_{20}), the orientation of the object in the image can be estimated.

$$m_{00} = \sum_{x=1}^{N}\sum_{y=1}^{N} f(x,y) \qquad (4)$$

$$m_{10} = \sum_{x=1}^{N} x f(x,y)$$

$$m_{01} = \sum_{y=1}^{N} y f(x,y)$$

The centroid is computed as follows:

$$\bar{x} = \frac{m_{10}}{m_{00}} \qquad (5)$$

$$\bar{y} = \frac{m_{01}}{m_{00}} \qquad (6)$$

In order to make descriptors invariant to translation, we compute the central moments which are the moments about the image centroid (see Eq. 7).

$$\mu_{p,q} = \sum_{x=1}^{N}\sum_{y=1}^{N} (x-\bar{x})^{p}(y-\bar{y})^{q} f(x,y) \qquad (7)$$

Other descriptors are the size of the principal axis, it can be computed from the *eigenvalues* of the inertia matrix l (covariance matrix [9]), it is composed by central moments μ_{11}, μ_{20} y μ_{02}.

$$I = \begin{pmatrix} \mu_{20} & \mu_{11} \\ \mu_{11} & \mu_{02} \end{pmatrix}$$

This eigenvalues are the principal moments and are given by

$$\lambda_{1,2} = \frac{(\mu_{20}+\mu_{02}) \pm \sqrt{\mu_{20}^{2}+\mu_{02}^{2}-2\mu_{20}\mu_{02}+4\mu_{11}^{2}}}{2} \qquad (8)$$

4   Experiments

We tested our proposal with real images taken in Michoacan, Mexico using a trap-camera, in Fig. 2 we show an original image 2(a) and the segmented one is in Fig. 2(b). These images were segmented partially using the technique described in [4], after the images were delimited by hand since it is not the aim of this paper.

(a) Original image of the ocelot 1.   (b) Ocelot 1 segmented.

Fig. 2. Ocelot used for our tested

Table 1 show both photograph of the ocelots used (the first row corresponds to ocelot 1 and the second row to ocelot 2), each column show these images after a filter was applied. The first column is for the Canny filter, and the second column is for the Sobel filter. For both filters we use the MATLAB's implementation with threshold value between 0.07 and 0.08. Notice that images with Canny filter have more edges detected than Sobel filter. For the next step we use images with Canny filter.

Table 1. Comparison between the use of Canny filters and Sobel filters. MATLAB functions with threshold between 0.07 and 0.08.

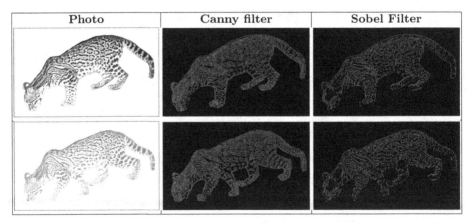

4.1 Selected Patterns

We designed a controlled experiment in order to see if the proposal introduced here was working properly, in our experiment we use some spots from both ocelots, the spots were automatically chosen on the basis that the selected spots were those whose contour were completely closed by the edge detection algorithms that we used, we tried both Canny filters and Sobel filters for this purpose. See Table 2, these images were processed with the MATLAB (function

Table 2. Spot selected for testing our proposal. The spot enclosed in yellow appears in both images, spots in red are only in one of the ocelots, and spots in blue seem to be closed but they are not.

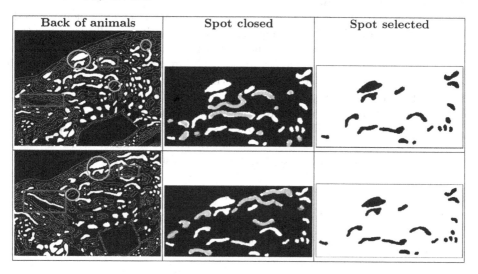

Back of animals	Spot closed	Spot selected

imfill), the first row is for ocelot 1, and second row is for ocelot 2. The spot enclosed in yellow appears in both images, spots in red are only in one of the ocelots, and spots in blue seem to be closed but they are not. Finally, the second column shows the spot selected to test our proposal.

Some examples of feature extraction of spots selected can be seen in figures of Table 3. Notice how well those ellipses correspond to the closed shapes.

Table 3. Spot 1 for ocelot 1 (top) and 2 (bottom), that is column 1, spot 2 for ocelot 1(top) and 2(bottom) at column 3.

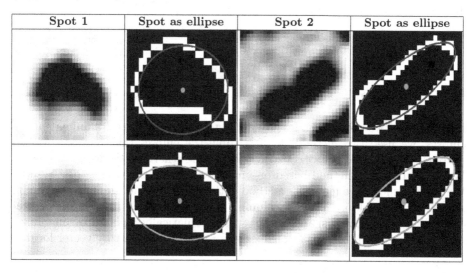

Spot 1	Spot as ellipse	Spot 2	Spot as ellipse

5   Discussion

A comparison between the methods we used for finding the ellipses are shown in figures of Table 4, the first row is for the Geometric construction method and the second row is for the Image moments method. The ellipses from the spots of the Ocelot in Photo 1 should match the corresponding ellipses of the Ocelot in Photo 2 since the Ocelot in both photographs are in fact the same ocelot and in this controlled experiment the spots of Photo 1 were manually compared with the corresponding spots in Photo 2, therefore the ellipses should be very similar both in size (i.e. Major and minor axis length) and in orientation angle. We conclude that the Image moments method outperform the Geometric construction method, particularly the orientation angle is almost identical when using the Image moments method (second row of Table 4).

Table 4. Comparison of some details of spots as ellipses, Geometric method (first row), Moments method (second row).

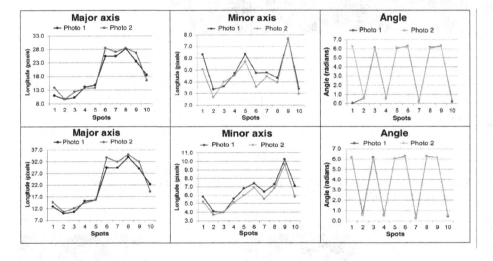

6   Conclusions

In this paper, we proposed a feature extraction of fur patterns of animals using ellipses, we used two methods to determine the ellipses that represent closed shapes, one of them is Rytz's construction and the other uses the image moments. We found that the method based on image moments produces the best results. We applied the image moments method with two ocelot's images and the results are shown in Table 5. The first column shows our ocelot's images, in the second column each closed shape or spot was automatically replaced by an ellipse, finally, in the third column each ellipse was replaced by a unitary vector located in the center of the ellipse. These vectors will be used in our future work where we will compare a set of vectors with another set of vectors to decide if both sets correspond to the same ocelot.

Table 5. Feature extraction process.

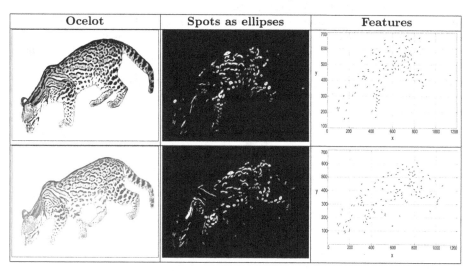

Ocelot	Spots as ellipses	Features

Acknowledgements. We are thankful with Dr. Tiberio Monterrubio, from Universidad Michoacana de San Nicolás de Hidalgo, for sharing his collection of Ocelot's photos with us.

References

1. Charre, J., Monterrubio, T., Botello, F., León, L., Núñez, R.: First records of jaguar (panthera onca) from the state of Michoacán, México. BioOne Res. Evolved **58**, 264–268 (2013)
2. Coxeter, H.: Introduction to Geometry, 2nd edn. Wiley, New York (1969)
3. Espinosa Ceniceros, J.C.: Detección de elipses (2013). http://juankenny.blogspot.mx/2013/04/lab-vc-actividad-6-deteccion-de-elipses.html
4. Figueroa, K., Camarena-Ibarrola, A., García, J., Tejeda, H.V.: Fast automatic detection of wildlife in images from trap cameras. In: Bayro-Corrochano, E., Hancock, E. (eds.) CIARP 2014. LNCS, vol. 8827, pp. 940–947. Springer, Heidelberg (2014)
5. Hartog, J., Reijns, R.: I^3S Pattern Manual. Intractive Individual Identification System, Julio 2014
6. Hu, M.K.: Visual pattern recognition by moment invariants. IEEE Trans. Inf. Theory **8**, 179–187 (1962)

7. Kelly, M.J.: Computer-aided photograph matching in studies using individual identification: an example from serengeti cheetahs. J. Mammal. **82**(2), 440–449 (2001). http://jmammal.oxfordjournals.org/content/82/2/440
8. Kuhl, F.: Elliptic fourier features of a closed contour. Comput. Graph. Image Process. **18**, 236–258 (1982)
9. Lira, J.: Tratamiento digital de imágenes multiespectrales, 2 edn., pp. 185–237. UNAM, ISBN: 978-607-00-3403-9 (2010)
10. Pacheco, E.: Identificación de diatomeas del lago de Pátzcuaro mediante filtros compuestos no lineales y caracterización de las valvas utilizando descriptores geométricos. Master's thesis, Universidad Michoacana de San Nicolás de Hidalgo (2015)
11. Whitehead, H., Christal, J., Tyack, P.: Studying cetacean social structure in space and time. Cetacean Societies: Field Studies of Dolphins and Whales (2000)

Comparing Threshold-Selection Methods for Image Segmentation: Application to Defect Detection in Automated Visual Inspection Systems

Rafael López-Leyva, Alfonso Rojas-Domínguez[(✉)],
Juan Pablo Flores-Mendoza, Miguel Ángel Casillas-Araiza,
and Raúl Santiago-Montero

Tecnológico Nacional de México - Instituto Tecnológico de León,
Av. Tecnológico S/N - Frac. Industrial Julián de Obregón,
37290 Leóng Gto., Mexico
alfonso.rojas@gmail.com

Abstract. Automatic detection of defects on the surface of products or raw material is an important task in the field of automated visual inspection. Thresholding is a method for image segmentation that is often used for the detection of said defects. Several methods to select the optimal thresholding values automatically on a per-image base have been described in the literature. Some of these are particularly designed to deal with mostly homogeneous images such as those of product surfaces with some defects, but have not been tested sufficiently and not in the context of automated visual inspection. In this work we present a comparison based on such experimental conditions by means of an automated visual inspection station and a set of images specially acquired for this purpose. The methods that were compared are: the Otsu's method, the Valley-Emphasis method, the Valley Emphasis with Neighborhood method, the Kittler-Illingworth's method, and the Maximum Similarity Thresholding method. The highest performance, with a statistically significant difference, was obtained by Maximum Similarity Thresholding.

Keywords: Automated thresholding · Otsu · Valley-Emphasis · Maximum Similarity Thresholding · Kittler-Illingworth

1 Introduction

Thresholding is a segmentation technique widely used in industrial applications. It is used when there is a clear difference between the objects and the scene background. Therefore, the scene should be characterized by a uniform background, similarity between the objects to be segmented and homogeneous appearance of these; otherwise all the pixels that compose the scene could be assigned to a single class. In industrial applications

---

A. Rojas-Domínguez—CONACYT Research Fellow

J.F. Martínez-Trinidad et al. (Eds.): MCPR 2016, LNCS 9703, pp. 33–43, 2016.
DOI: 10.1007/978-3-319-39393-3_4

usually the illumination and other scene properties can be easily controlled. Thus, the main problem for automated segmentation via thresholding lies in determining the threshold value that segments the image with optimal results.

Several methods designed to find the optimal thresholding value have been described in the literature; the most popular is Otsu's method [1]. This method selects the threshold value that maximizes the variance between classes of an image histogram and for this reason it is most successful when the images show a bimodal or close to bimodal histogram (in the case of two output classes) or multi modal (in the cases of multiple output classes), but the method usually fails when the histogram is unimodal or near to be unimodal. The latter situation occurs frequently in visual inspection of surfaces, where the test images are mostly uniform and the defects are small and may be represented by grey levels that are similar to those of the image background. Several methods have been proposed to reduce the limitations of Otsu's method, such as the Valley-Emphasis method [2], the Valley-Emphasis using Neighbourhood method [3], the Kittler-Illingworth's method [4] and the MST method [5]. However, although these studies have proposed valuable improvements, there is a lack of experimental results from real-life scenarios to validate the proposals. The need to objectively assess the capabilities of different thresholding methods in order to use them in Automated Visual Inspection (AVI) systems, motivated this study.

For this work we have implement an Automated Visual Inspection System (AVI) [9] consisting of an inspection point equipped with industrial-grade lighting and image acquisition equipment on a conveyor belt. Using this equipment, a set of images designed to test the different thresholding methods were produced, including annotations made by manually selecting a threshold value. For the comparative evaluation two similarity measures were employed: the Tanimoto Coefficient [6, 7] and the Pearson Correlation Coefficient (PCC) [5, 8]. Although our test conditions are not those of an industrial environment, they are sufficiently similar to real-life scenarios as to provide the required evidence in favour of a thresholding method.

The rest of this article is organized as follows: Sect. 2 summarized the thresholding methods under comparison; Sect. 3 describes the implementation of our AVI system and experimental design. The experimental results are contained in Sect. 4. The analysis and discussion of the results are found in Sect. 5 and finally the conclusions and directions of future work are provided in Sect. 6.

2 Theoretical Background

In this section we review a number of methods for automatic selection of threshold values; the methods that we will discuss are: the Otsu's method, the Valley-Emphasis method, the Valley-Emphasis method using Neighborhood, the Kittler-Illingworth's method and the MST method. For a more general discussion regarding thresholding techniques *cf.* [10].

2.1   The Otsu's Method

An image is a bidimensional matrix of gray intensity levels that contains N pixels with gray level values between 0 and 255 (1, L). The number of pixels with a gray level i is denoted as f_i, and the probability of occurrence of gray level i is given by:

$$P_i = \frac{f_i}{N} \tag{1}$$

The total average of gray level in the image can be calculated in this way:

$$\mu_T = \sum_{i=0}^{L-1} i p_i \tag{2}$$

By segmenting the image using a single threshold we get two disjoint regions C_1 and C_2, which are formed by the area of pixels with gray levels $[1,\ldots,t]$ and $[t+1,\ldots,L]$ respectively: Normally these classes correspond to the object of interest and the background of the image. Then the probability distributions of the gray level for these two classes are:

$$C_1 = \frac{P_i}{\omega_1(t)},\ldots,\frac{P_t}{\omega_1(t)} \quad \text{and} \quad C_2 = \frac{P_{t+1}}{\omega_2(t)},\frac{P_{t+2}}{\omega_2(t)},\ldots,\frac{P_L}{\omega_2(t)} \tag{3}$$

Where the probabilities for the two classes are calculated as follows:

$$\omega_1(t) = \sum_{i=1}^{t} P_i \quad \text{and} \quad \omega_2(t) = \sum_{i=t+1}^{L} P_i \tag{4}$$

The mean values of the gray levels for the two classes are calculated as follows:

$$\mu_1 = \sum_{i=1}^{t} \frac{i \cdot P_i}{\omega_1(t)} \quad \text{and} \quad \mu_2 = \sum_{i=t+1}^{L} \frac{i \cdot P_i}{\omega_2(t)} \tag{5}$$

In Otsu's method [1] the optimal threshold t^* is determined by maximizing the variance between classes, which is denoted in the following objective function:

$$t^* = \frac{ArgMax}{0 \leq t < L} \{\sigma_B^2(t)\} \tag{6}$$

Where the variance between classes σ_B^2 is defined by:

$$\sigma_B^2 = \omega_1 \cdot (\mu_1 - \mu_T)^2 + \omega_2 \cdot (\mu_2 - \mu_T)^2 \tag{7}$$

Otsu's method is easy to calculate, but it only works properly for images containing bimodal distribution histograms, which means that the object and background have comparable gray level variances, but fails to find the optimal threshold when the histogram contains a unimodal distribution or close to unimodal. Following this observation, the Valley-Emphasis method [2] was proposed. This is discussed below.

2.2 Valley-Emphasis Method

The Valley-Emphasis method [2] is the result of observing that the Otsu's method does not behave as desired with (near to) unimodal distribution histograms. The emphasis of valley method selects threshold values that have a small probability of occurrence, while at the same time maximizes the variance between groups as the Otsu's method does. This modification means changing the target function of the Otsu's method (Eq. 5) by applying the weight function $(1 - P_t)$:

$$t^* = \frac{ArgMax}{0 \leq t < L} \left\{ (1 - P_t) \left(\omega_1(t) \mu_1^2(t) + \omega_2(t) \mu_2^2(t) \right) \right\} \tag{8}$$

The smaller the value P_t (probability of the threshold value t), the greater the result of (8) since it is multiplied by the complement of this probability. This weight function ensures that the result will always be a threshold value that is in a "valley" or bottom edge of the gray level distribution. The results reported in [2] show that the Valley-Emphasis method produced a misclassification two orders of magnitude below that of the Otsu's method, but these results were not based on sufficient experimentation.

2.3 Valley-Emphasis Using Neighborhood Method

It has been argued that the Valley-Emphasis method cannot improve the quality of segmentation in some cases because, being based on a single gray level value, is not robust enough. A variation of the Valley Emphasis method, called Valley-Emphasis using Neighborhood has been proposed that attempts to alleviate this issue by considering the information contained in the neighborhood around the valley point. The improvement consists in using a modified weight function that is applied on $\sigma_B^2(t)$. This weight function is defined as follows. Using the histogram $\{h(i)\}$ of an image, the neighborhood $\bar{h}(i)$ of gray level i is:

$$\bar{h}(i) = [h(i - m) + \ldots + h(i - 1) + h(i) + h(i + 1) + \ldots + h(i + m)] \tag{9}$$

where m is the size of the neighborhood. The Valley-Emphasis method obtains an optimal threshold by modifying the objective function of the Otsu's method (Eq. 5) in the following way:

$$\xi(t) = \left(1 - \bar{h}(i) \right) \left(\omega_0(i) \mu_1^2(i) + \omega_1(i) \mu_2^2(i) \right) \tag{10}$$

The optimal threshold t^* is the gray level value that maximizes (11):

$$t^* = \frac{ArgMax}{0 \leq t < L} \xi(t) \tag{11}$$

The contribution of the Valley-Emphasis method with Neighborhood is that it considers the neighbors of each gray value in its search for the optimal threshold (this provides robustness) and that it preserves the idea that those with the smallest occurrence must have the largest influence (the proposal of the Valley-Emphasis method). The results obtained in [3] suggest that a neighborhood of size 11 or slightly larger produces the best segmentation results.

2.4   The Kittler–Illingworth's Method

The main idea behind the method called Kittler-Illingworth thresholding is to directly optimize (minimize) the average rate of misclassification. The method obtains the optimal threshold using the minimum thresholding error in calculations based on their objective function:

$$J(t) = 1 + 2[P_1(t)log\sigma_1(t) + P_2(t)log\sigma_2(t)] - 2[P_1(t)logP_1(t) + P_2(t)logP_2(t)] \tag{12}$$

where $\sigma_i^2(t) = \left[\sum_{i=1}^{L-1} \{i - \mu_i(t)\}^2 f_i\right]/P_i(t)$. After obtaining the target values for the gray levels of the image, the method search the minimum value of the function, so that which is defined:

$$t^* = Min\{J(t)\} \tag{13}$$

The results obtained in [4] show that the method achieves a better result than the Otsu's and Ridler's [11] methods when the images are homogeneous or, in cases of bimodal histograms, when one of the peaks is large and the other one is small, meaning that the optimal threshold tends to fall where there is less likelihood of occurrence between the two modes.

2.5   Maximum Similarity Thresholding (MST) Method

The MST method selects the optimal threshold as the value that maximizes the similarity between the edge content of the thresholded image and the edge content of the original image computed as follows:

$$T(t) = \prod_{i=1}^{k} \|\nabla G(x, y, \sigma_i) * t\| \tag{14}$$

where $G(x, y, \sigma_i) = \frac{1}{\sqrt{2\pi}\sigma_i} e^{-\frac{(x^2+y^2)}{2\sigma_i^2}}$ and the symbols ∇ and $*$ represent the derivate and the convolution, respectively. The results obtained in [5] show that, using the Pearson's Correlation Coefficient as a similarity metric, the MST method on average achieves similarities two orders of magnitude larger than the Otsu's method and the Valley-Emphasis method. The disadvantage of this method, compared to other methods, is its greater complexity, which could affect its performance in real time applications.

3 Development

In an industrial environment, control of image capture and the lighting is extremely important because these produce uniformity in the images, thus simplifying inspection tasks. In this work, the following parameters were considered to guarantee said uniformity in the acquisition: exposure time, signal amplification, automatic brightness control, external control to manage lighting (strobe mode), and synchronizing image capture with the movement of the conveyor belt. A CCD digital camera for machine vision applications, the Grey Point Flea3 model [12], equipped with an 8 mm focal-length lens was used in the acquisition. The choice of the lens is related to the working distance (between the object on the conveyor belt and the camera). A ring-type LED lamp with white light was used in the experiments. Synchronization between frame acquisition and light strobe is achieved through a trigger signal produced by the camera (Fig. 1).

Fig. 1. AVI installed on the conveyor band, which has a view camera and white lighting of ring type.

After synchronization of the camera with the strobe light and the movement of the conveyor belt, a number of images of a paper strip moving on the conveyor were obtained. The paper strip contained several defects, designed to simulate those defects that can be observed in the production of different types of fabrics and other homogeneous materials. The size of the defects range from 11 to 27133 pixels; the average intensity value of the defects is 161.8 ± 12.2 (using 8 bpp gray scale images) while the average intensity of the background is 225.7 ± 1.3; the mean intensity difference between defect and background is approximately 64. A sample of the images produced is shown in Fig. 2.

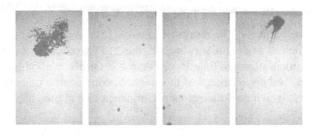

Fig. 2. Different defects found in paper strips moving on the conveyor belt.

After selecting the frames that were deemed most useful to our purposes, 97 images were saved that most clearly show the defects on the paper strips. These constitute the set of test images on which the different threshold selection algorithms summarized in Sect. 2 are tested. An example of applying each of the methods under comparison on a test image is illustrated in Fig. 3.

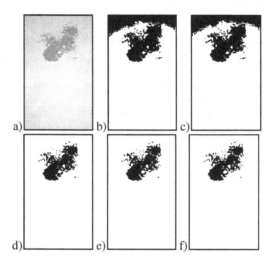

Fig. 3. Example of thresholding of an image with a defect to the paper, where (a) is the original image, (b) is the thresholding with Otsu ($t^* = 204$), (c) Valley-Emphasis ($t^* = 205$), (d) Valley-Emphasis using neighborhood ($t^* = 180$), (e) Kittler Illingworth ($t^* = 179$), (f) and the MST method ($t^* = 174$).

In order to quantify the quality of the results obtained by means of applying each of the threshold selection algorithms on our test images, the similarity between the original image and the thresholded images was computed. For this purpose, a set of binary images was produced by manually selecting the best threshold value in each case. The similarity between these reference images, referred to as Ground Truth masks, and each binary image produced by the thresholding algorithms, was measured using two measures; the Tanimoto's Coefficient (TC) [6, 7] and Pearson's Correlation Coefficient (PCC) [5, 8] which are defined as follows:

$$TC = \frac{N(A \cap B)}{N(A \cup B)} \tag{18}$$

$$PCC = \frac{\sum_{k=1}^{L}(i_k - \mu_A)(i_k - \mu_B)}{\sqrt{\sum_{k=1}^{L}(i_k - \mu_A^2)\sum_{k=1}^{L}(i_k - \mu_B^2)}} \tag{19}$$

where A and B denote the images under comparison, in this case the Ground Truth masks against the automatically segmented images.

4 Experimental Results

Using the Tanimoto's Coefficient [6, 7] as similarity measure, the results of the comparison performed with base on our test set are summarized in Table 1. The corresponding boxplots and the Wilcoxon's rank sum test of these results are shown in Fig. 4 and Table 3 respectively.

Table 1. Tanimoto coefficient between Ground Truth and automatic segmentation

	Otsu	Valley-Emphasis	Valley-Emphasis neighborhood	Kittler-Illingworth	MST
Percent similarity range	0.003–0.932	0.003–0.940	0.005–0.995	0.000–1.000	0.000–1.000
Average	0.158	0.160	0.331	0.614	0.686
Standard deviation	0.232	0.234	0.381	0.406	0.304

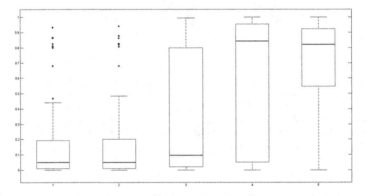

Fig. 4. Boxplots of the Tanimoto's Coefficient between thresholded test images against the corresponding Ground Truth, from left to right: Otsu's method; Valley-Emphasis method; Valley-Emphasis using neighborhood; the Kittler-Illingworth's method and MST method.

Using instead the Pearson's Correlation Coefficient [8, 9] as similarity measure, the results summarized in Table 3 were obtained. The corresponding boxplots are shown in Fig. 5 and the Wilcoxon's rank sum test in Table 3 (Table 2).

Table 2. Pearson's coefficient between Ground Truth and automatic segmentation.

	Otsu	Valley-Emphasis	Valley-Emphasis neighborhood	Kittler-Illingworth	MST
Percent similarity range	0.015–0.961	0.016–0.966	0.021–0.997	0.001–1.000	0.001–1.000
Average	0.281	0.284	0.442	0.679	0.775
Standard deviation	0.249	0.250	0.355	0.382	0.275

Figure 5 displays the results for each of the thresholding methods that were used:

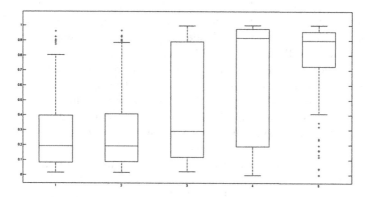

Fig. 5. Boxplot corresponding to the values resulting of similarity of Pearson's Correlation Coefficient between all thresholded images presented defects against the corresponding masks to each them, where (1) refers to Otsu's, (2) the Valley-Emphasis, (3) the Valley-Emphasis using Neighborhood, (4) the Kittler-Illingworth's and (5) the MST methods.

Table 3. Wilcoxon's rank sum test between TC of the methods and PCC of the methods.

Otsu VE	Otsu VEN	Otsu KI	Otsu MST	VE VEN	VE KI	VE MST	VEN KI	VEN MST	KI MST
0.895	0.005	0.000	0.000	0.006	0.000	0.000	0.000	0.000	0.967

Wilcoxon's rank sum test between Tanimoto's Coefficient of the methods

Wilcoxon's rank sum test between PCC of the methods

Otsu VE	Otsu VEN	Otsu KI	Otsu MST	VE VEN	VE KI	VE MST	VEN KI	VEN MST	KI MST
0.891	0.002	0.000	0.000	0.003	0.000	0.000	0.000	0.000	0.904

5 Discussion

The results obtained with the Tanimoto's Coefficient as the similarity measure show that Otsu's method, the Valley-Emphasis method and the Valley-Emphasis method using Neighborhood obtained the lowest values (based on the corresponding median values) with an average similarity against the Ground Truth between 0.15 to 0.33 and a standard deviation between 0.20 and 0.40 all of these methods appear to be inferior to the rest. The Kittler-Illingworth's method and the MST method obtained the highest performance values; however, with average values of 0.61 and 0.68 respectively and a standard deviation of approximately 0.4 and 0.3 respectively, the MST method appears to be superior. The statistical significance of these observations was validated by means of the Wilcoxon's rank sum test, with which we compared all methods against each other to verify that the differences between them is statistically supported. In this test, the null hypothesis is that there is no statistical difference between the data samples being compared. When comparing the MST method against the other methods, the test indicates that the null hypothesis can be rejected (at a significance level of 0.05) in the case of Otsu's, Valley-Emphasis, and Valley-Emphasis using Neighborhood, since all of these produced a $P_{value} = 0.000$. Meanwhile, in the comparison against the Kittler-Illingworth's method, with a $P_{value} = 0.967$ the test shows that there is no evidence to reject the null hypothesis; in other words, there is no statistical difference between the results of the MST and the Kittler-Illingworth's method.

Regarding the results based on the Pearson's Correlation Coefficient as the similarity measure, we can observe that although the numerical values vary slightly, a very similar behavior as that when using the Tanimoto Coefficient is obtained. This observation was also verified by means of the Wilcoxon's rank sum test, with which we compared the methods against each other under the same hypotheses and significance level. With a $P_{value} = 0.904$ the null hypothesis cannot be rejected in the comparison between the MST method and the Kittler-Illingworth's, while in the comparison against the other methods there is evidence to reject it ($P_{value} = 0.000$).

6 Conclusion

In this work, several methods for automatic determination of the optimal threshold value for image segmentation were compared using a set of test images specially designed to test the utility of the algorithms to be employed in AVI systems. The test images depict varying defects in paper strips and were acquired under industrial-standard conditions. Tested under two different similarity measures (the Tanimoto's Coefficient and the Pearson's Correlation Coefficient), the MST method obtained the highest similarity against the other methods, with an average measure of TC = 0.68 and PCC = 0.77. As future work, we plan to employ the MST method in our AVI tasks and particularly in the automated detection of defects on homogeneous surfaces. The experiments presented provide us with the necessary knowledge to make this informed decision, which was not possible prior to the realization of this work.

References

1. Otsu, N.: A threshold selection method from gray-level histograms. IEEE Trans. Syst. Man Cybern. **9**(1), 62–66 (1979)
2. Hui-Fuang, Ng: Automatic thresholding for defect detection. Pattern Recogn. Lett. **27**, 1644–1649 (2006)
3. Fan, J., Lei, B.: A modified valley-emphasis method for automatic thresholding. Pattern Recogn. Lett. **33**, 703–708 (2012)
4. Kittler, J., Illingworth, J.: Minimum error thresholding. Pattern Recogn. **19**, 41–47 (1986)
5. Yaobin, Z., Shuifa, S., Fangmin, D.: Maximum similarity thresholding. Digital Sig. Process. **28**, 120–135 (2014)

6. Crum, W., Camara, O., Hill, D.: Generalized overlap measures for evaluation and validation in medical image analysis. IEEE Trans. Med. Imaging **25**(11), 1451–1461 (2006)
7. Duda, R., Hart, P.: Pattern Classification and Scene Analysis. Wiley, New York (1973)
8. Rodgers, J., Nicewander, W.: Thirteen ways to look at the correlation coefficient. Am. Stat. **42**, 59–66 (1988)
9. Christian, D., Bernd, A., Carsten, G.: Industrial Image Processing. Springer, Heidelberg (2013)
10. Davies, D.: Machine Vision – Theory, Algorithms, Practicalities. Elsevier, Philadelphia (2005)
11. Ridler, T., Calvard, S.: Picture thresholding using an iterative selection method. IEEE Trans. Syst. Man Cybern. **8**, 630–632 (1978)
12. Point Grey: Register Reference for Point Grey Digital Cameras (2015). www.ptgrey.com

Training a Multilayered Perceptron to Compute the Euler Number of a 2-D Binary Image

Humberto Sossa[1(✉)], Ángel Carreón[1], and Raúl Santiago[2]

[1] Instituto Politécnico Nacional-CIC,
Av. Juan de Dios Bátiz S/N, Gustavo a. Madero,
07738 Mexico City, Mexico
`humbertosossa@gmail.com`,
`angelcarreon01@hotmail.com`
[2] Instituto Tecnológico de León,
Av. Tecnológico S/N, Frac. Julián de Obregón,
León, Guanajuato, Mexico
`rsantiago66@gmail.com`

Abstract. In this short communication, we explain how a Multilayered Perceptron (MLP) can be used to compute the Euler number or Genus of a 2-D binary image. We take as basis the results provided by a mathematical formulation that is known providing exact results in the computation of this important topological image feature to derive two MLP-based architectures, one useful for the 4-connected case and one useful for 8-connected case. We present results with a set of realistic images and compare our proposals in terms of processing with other approaches reported in literature.

1 Introduction

The Euler number or Euler characteristic is a feature that allows describing the topological structure of an image or an specific object in an image. As it is known, the Euler number has been used in many applications: industrial part recognition [1], real-time thresholding [2], object number calculation, [3], and real-time Malayan license plate recognition [4], to mention a few.

Mathematically speaking, the Euler number, e, of a digital binary image $I(x, y)$ can be obtained as follows:

$$e = o - h \tag{1}$$

In this case, o is the number of objects or (binary regions) in the image and h is the number of holes (i.e., isolated regions of the image's background).

Many methods have been developed to obtain the Euler number of a digital binary image. Some of these methods, can be found in [5–25].

The algorithm outlined in [5] was one of the first reported in literature. The most popular algorithm of this method is used by the MATLAB image processing tool. It calculates the Euler number of a binary image as:

© Springer International Publishing Switzerland 2016
J.F. Martínez-Trinidad et al. (Eds.): MCPR 2016, LNCS 9703, pp. 44–53, 2016.
DOI: 10.1007/978-3-319-39393-3_5

$$e = \frac{s1 - s3 - 2 \cdot x}{4}. \tag{2}$$

In this case:

1. $s1$ is the number of matrices $\left\{ \begin{bmatrix} 0 & 0 \\ 1 & 0 \end{bmatrix}, \begin{bmatrix} 0 & 0 \\ 0 & 1 \end{bmatrix}, \begin{bmatrix} 0 & 1 \\ 0 & 0 \end{bmatrix}, \begin{bmatrix} 1 & 0 \\ 0 & 0 \end{bmatrix} \right\}$;

2. $s3$ is the number of matrices $\left\{ \begin{bmatrix} 0 & 1 \\ 1 & 1 \end{bmatrix}, \begin{bmatrix} 1 & 0 \\ 1 & 1 \end{bmatrix}, \begin{bmatrix} 1 & 1 \\ 1 & 0 \end{bmatrix}, \begin{bmatrix} 1 & 1 \\ 0 & 1 \end{bmatrix} \right\}$, and

3. x is the number of matrices $\left\{ \begin{bmatrix} 0 & 1 \\ 1 & 0 \end{bmatrix}, \begin{bmatrix} 1 & 0 \\ 0 & 1 \end{bmatrix} \right\}$.

As we can see, before using (2), the MATLAB algorithm needs to perform up to 10 comparisons on each image pixel. Time complexity for this method is of $O(N^2)$ for a $N \times N$ image, which is linearly dependent on the number of pixels. For image processing tasks, where the data could be huge the constant term that is so often hidden in the *big-Oh* notation becomes important.

Artificial Neural Networks (ANN), on the other hand, have been successfully used in many tasks including signal analysis, noise cancellation, model identification, process control, object detection, and pattern recognition, and so on. Many ANN models have been reported in literature, since the very simple Threshold Logic Unit (TLU), introduced by McCulloch-Pitts [26] at the beginning of the 40's, passing by the well-known Perceptron, presented to the world by Rosenblatt in the 50's [27, 28] until the so called Morphological Neural Models with and without Dendritic Processing introduced by Ritter et al. in [29, 30, 31, 32], to mention a few.

In this paper, we show how an MLP can be used to compute the Euler number of a 2-D binary image. By making an analysis of the local results provided by a known formulation to compute the image Euler number we arrive at the specialized ANN architecture. We decided to use a MLP for its versatility since many years ago in a multitude of situations. This is the first, to our knowledge, that a MLP-based architecture is used to compute the Euler number of a 2-D binary image. It constitutes and original an interesting option to compute this topological describing feature.

The rest of this paper is organized as follows. In Sect. 2 we describe our proposed methodology to derive at the end to the specialized MLP based architecture to compute the Euler number of a 2-D binary image. We devote Sect. 3 to report the experimental results that validate the applicability of the derived specialized MLP architecture as well as a comparison with other approaches reported in literature. In short, in Sect. 4 we reach our conclusions and directions for present and future research.

2 Our Proposal

In this section we describe how an MLP can be used to obtain the Euler number of a 2-D digital binary image. We decided it to do so, because as it is known MLPs have shown to be an excellent options to solve many problems in multiple areas where pattern classification is required.

To train a MLP to obtain the Euler number of a 2-D binary image we can proceed as usual by firstly selecting a set of P training samples, for example: $M \times N$ 2-D binary images: $\mathbf{I} = \{I_1, I_2, \ldots, I_P\}$. Before training the ANN, suppose we divide set $\mathbf{I}$ into q sub-sets of images such that each sub-set has the same Euler number, according to (1). With this in mind, we could proceed, for example, as illustrated in Fig. 1(a) by presenting, on the one hand, as input to a known Euler number computation method, that in turn outputs a correct value of e for each image: $I_k, k = 1, 2, \ldots, P$. As can be appreciated from this same figure, each image is also presented at the input of the untrained MLP, that produces a value: $\hat{e}$, as an estimate of e. The resulting error: E, could be then used, in an iterative way, to adjust the MLP weights until it is ready to compute the Euler number of an unknown input image as illustrated in Fig. 1(b).

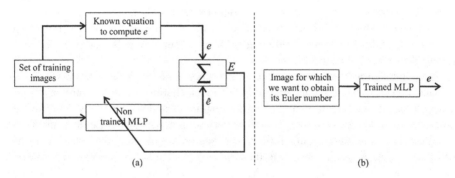

Fig. 1. A first alternative to train a MLP to compute the Euler number of a 2-D binary image. (a) Training of the MLP. (b) Testing of the MLP.

It is clear that if we apply the above described strategy, we would have several inconveniences. A first inconvenience would be the following: If we use a three layer MLP, the number of input neurons would be $M \times N$ (the image size); the number of output neurons would be directly proportional to the number of values NV of e $\{\ldots, -3, -2, -1, 0, 1, 2, 3, \ldots\}$ needed to be computed for an input image. In short, the correct number of hidden neurons to obtain the desired values for e would be certainly very big and rather difficult to find.

A second inconvenience would be that because the Euler number of an image is a function of the number of its objects and its holes, lots of training images would be required to reach good training results, this is because we have too many possibilities. Instead of using an architecture like this, we propose to derive a specialized one as follows. Let us first consider the following two expressions, introduced in [33] to compute the Euler number of a 2-D digital binary image in terms of only three comparisons. For the case of 4-connected regions (regions where their pixels are allowed to be connected only by their sides), the authors propose computing the image Euler number as:

$$e = \#\begin{pmatrix} 1 & 0 \\ 0 & 0 \end{pmatrix} - \#\begin{pmatrix} 1 & 1 \\ 1 & 0 \end{pmatrix} + \#\begin{pmatrix} 1 & 0 \\ 0 & 1 \end{pmatrix} \qquad (3)$$

On the other hand, for the case of 8-connected regions (regions where the pixels are allowed to be connected by their sides and corners), the authors propose to compute e by means of the following equation:

$$e = \#\begin{pmatrix} 1 & 0 \\ 0 & 0 \end{pmatrix} - \#\begin{pmatrix} 1 & 1 \\ 1 & 0 \end{pmatrix} - \#\begin{pmatrix} 0 & 1 \\ 1 & 0 \end{pmatrix} \qquad (4)$$

As referred in [33], these two equations seem to be the smallest expressions (in terms of the necessary operations) that allow computing the Euler number of a 2-D digital binary image, providing exact values as if (1) was used. To appreciate the validity of (3) and (4), let us consider the four academic examples shown in Fig. 2.

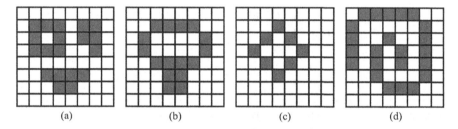

| | (a) | (b) | (c) | (d) |

Fig. 2. Examples to numerically validate the functioning of (3) and (4).

Table 1. Application of (3) and (4) to the four example of Fig. 2.

	Image (a)	Image (b)	Image (c)	Image (d)
(3)	2	4	8	7
(4)	2	0	0	1

Table 1 summarizes the results obtained for these four binary images by means of (3) and (4). In the case of the image (a), all three binary regions, as can be appreciated, are 4-connected, thus the computed results by (3) and (4) are the same. In the case of image (b), some of the pixels are only 8-connected and some others are 4-connected, the reader can see that the obtained results are different by the application of both equations is different. In the case of image (c), when (3) is applied, all pixels are considered as disconnected, that is why an "8" is obtained, however when (4) is used a "0" is obtained due to for this equation the eight pixels as considered as a connected object with a hole. Finally, in the case of image (d), if 4-connectivity is considered, we see that we have seven connected regions; that is why by means of (3) we obtain a "7". On the other side, if 8-connectivity is considered, as can be appreciated from Fig. 2(d), all the pixels are taken as connected forming a spiral, thus the value for e in terms of (4) is "1", as expected.

Suppose now we want to design two specialized MLP architectures that allow computing the Euler number of a 2-D binary image, one based on (3) and the other based on (4). To accomplish this goal, let us represent the four numbers of each of each of the three terms of (3) and (4) by the four variables: v_1, v_2, v_3 and v_4. It is not difficult to see that the three arrangements used by (3) and (4) are three of the sixteen possibilities depicted in Table 2.

Table 2. Values for (3) and (4).

	v_1	v_2	v_3	v_4	Results for (3)	Results for (4)
1	0	0	0	0	0	0
2	0	0	0	1	0	0
3	0	0	1	0	0	0
4	0	0	1	1	0	0
5	0	1	0	0	0	0
6	0	1	0	1	0	0
7	0	1	1	0	0	-1
8	0	1	1	1	0	0
9	1	0	0	0	1	1
10	1	0	0	1	1	0
11	1	0	1	0	0	0
12	1	0	1	1	0	0
13	1	1	0	0	0	0
14	1	1	0	1	0	0
15	1	1	1	0	-1	-1
16	1	1	1	1	0	0

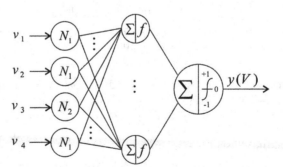

Fig. 3. Sketch of the specialized MLP architecture to compute the Euler number of a 2-D binary image.

From (3) we can also see that the first and third resulting values are both positive (rows 9 and 10), while the second term is negative (row 15). The remaining combinations, according to Table 2, sixth column, are zero. However, from (4) we can appreciate that the first term is positive as shown in row 9, while the second and third terms are both negative as depicted in rows 7 and 15, respectively. Column 6 summarizes the results for (3), while column 7 resumes the results for (4).

In both cases, we propose to use these 16 values to train a MLP that allows producing as output one of the three values: $\{-1, 0, 1\}$, that in turn allow us determining the Euler number of a 2-D binary image in both cases of 4 and 8 connectivity. Figure 3 depicts a sketch of the specialized MLP architecture. It has four input neurons, nine hidden neurons and one output neuron. All activation functions of hidden neurons were chosen as sigmoidal, however the activation function for the output neuron is a hyperbolic tangent to approach at the end of training the three desired values: $\{-1, 0, 1\}$. For adjusting the connections weights among neurons, we have used standard Backpropagation rule with a learning rate of $\alpha = 0.1$. All the synaptic weights were initialized to random values between 0.0 and 1.0. At 5000 iterations good output values were obtained, but because these three values never correspond to the values of $-1, 0, 1$, we took output y of the MLP and if $y > 0.5$ then $r = 1$, else if $y \geq -0.5$ and $y \leq 0.5$ then $r = 0$ else if $y \leq -0.5$ then $r = -1$. Training was attained at 0.7442 s.

The reader can easily verify that all 16 values shown in Table 2 are correctly classified into their corresponding three classes; $-1, 0, 1$, in both cases of 4 and 8 connectivity. This guaranties that at the moment of computing the image Euler number by means of this specialized ANN, we will obtain a correct value as if (1) was used in both cases of 4 and 8-connected images.

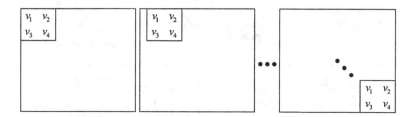

Fig. 4. Sequential way to apply (5) over an image.

To locally compute the Euler number of a 2-D binary image $I(x, y)$ we proceed sequentially as illustrated in Fig. 4. As can be seen from this figure, the generated MLP is displaced from the upper left position down to the right position inside the image, obtaining each time one of the three values: 1, 0 or 1. At the end of the application of this very simple procedure, we should obtain the Euler number of any 2-D binary image $I(x, y)$ as follows:

$$e = \sum_{k=1}^{T} r(k) \tag{5}$$

In this case, T is the number of times the trained MLP is applied to image $I(x, y)$; $r(k) = -1, 0, 1$ is the local result output by the MLP applied to a sector of $I(x, y)$.

3 Experimental Results

To numerically validate the correct functioning of the two derived ANNs, we first took 100 binary images of 256×256 pixels. In the case of the first the 50 images, objects are 4-connected; in the remaining 50 images objects are 8-connected. Due to space limitations, results for only 16 of these images are shown in Fig. 5. Table 3 depicts the values of e for these 16 images by the application of (5) as illustrated in Fig. 4. Only the results with the MLP for the 4-connected cases are shown in this table. As expected, in all cases, the correct Euler number for the 16 images was correctly calculated. The reader can demonstrate that for any other 2-D binary image, the desired e should be correctly computed.

All the experiments were run on a desktop computer with an Intel (R) Core(TM) i7 950 CPU 3.07 GHZ $\times$ 8 cores and 18 GB of RAM; operating system: Windows 7 Ultimate $\times$64.

Fig. 5. 16 of the binary images used to validate the functioning of the derived MLPs.

Table 3. Values of e for the 16 images shown in Fig. 5.

Image 1	Image 2	Image 3	Image 4	Image 5	Image 6	Image 7	Image 8
$e = 1$	$e = -5$	$e = -2$	$e = -6$	$e = -3$	$e = -2$	$e = -1$	$e = 3$
Image 9	Image 10	Image 11	Image 12	Image 13	Image 14	Image 15	Image 16
$e = 1$	$e = -2$	$e = 3$	$e = 0$	$e = -1$	$e = 6$	$e = -5$	$e = 1$

Table 4. Cpmparisson with other proposals.

Method	Average time in seconds over the 100 images
Ref. [14]	14.00
Ref. [22]	18.00
Ref. [5]	20.00
Ref. [17]	73.00
MLP 4-connected case	78.08
MLP 8-connected case	73.52
Ref. [9]	151.00
Ref. [11]	1647.00

The average time to process an image of 256×256 pixels was of 0.7808 (4-connected case) and 0.7352 (8-connected case) seconds, respectively if the MLP is sequentially applied as depicted in Fig. 4. If (3) or (4) are applied over an image the average time reduces to 0.1557 and 0.1456 s, respectively. As can be seen more time is needed to obtain the Euler number of a binary image if the ANN based method is applied. This is normal due to more processing time is required to attain the same goal. What we want to show in this paper is that it is possible to compute the binary image Euler number by means of a Multi-layered Perceptron.

Compared with other standard formulations to compute the Euler number of a 2-D binary image reported in literature in terms of time, we observe in Table 4 that our proposals are, of course, not the fastest but neither the slowest. Both MLP-based proposals are slower that the methods reported in [5, 14, 22, 17] but faster that the methods reported in [9, 11].

Compared with other ANN implementations, such the one reported in [34] where a Morphological Neural Network with Dendritic Processing (MNNDP) is trained to accomplish the same task, the corresponding average times over the set of 100 images used in this paper were of 402.0 s (MNNDP based implementation) and 75.8 s (MLP based implementation), respectively.

4   Conclusions and Future Trends

In this paper we have shown that an MLP can be used to correctly determine the Euler number of a 2-D binary image. Through an analysis of the local operations implied in the application of known formulations, (3) and (4), we have derived a specialized architecture.

Although our proposed MLP based implementation is slower that the sequential implementation of (3) and (4), it constitutes an original and interesting alternative for the automatic computation of the 2-D binary image Euler number by means of an MLP.

Although our proposed methodology can be adapted to any ANN model, we have presented results with the MLP in both cases of 4 and 8 connectivity. A parallel work in this same direction with morphological neural networks is reported in [34].

Nowadays, we also are working toward the natural extension of our proposal in the case of 3-D binary images where objects will now be represented by voxels and not by pixels as usual.

We are also working to obtain an efficient implementation of the MLP to be run a GPU platform under CUDA. For this we are first implementing or MLP in matrix form. Because CUDA allows to manage the execution threats in matrix form, each execution threat can be a value inside a matrix, this way it will not be necessary to iterate over an image as illustrated in Fig. 4. Each threat will execute over each image pixel the necessary operations of lecture, its processing until the MLP output. In theory, the execution time will be equivalent to only iteration of the ANN.

Acknowledgements. The authors would like to thank IPN-CIC under project SIP 20151187 and 20161126, and CONACYT under projects 155014 and 65 within the framework of call: Frontiers of Science 2015, for the economic support to carry out this research. The second authors thanks CONACYT for the economic support to carry out his Master studies.

References

1. Yang, H.S., Sengupta, S.: Intelligent shape recognition for complex industrial tasks. IEEE Control Syst. Mag. **8**(3), 23–29 (1988)
2. Snidaro, L., Foresti, G.L.: Real-time thresholding with Euler numbers. Pattern Recogn. Lett. **24**, 1533–1544 (2003)
3. Lin, X., Ji, J., Gu, G.: The Euler number study of image and its application. In: Proceedings of 2nd IEEE Conference on Industrial Electronics and Applications (ICIEA 2007), pp. 910–912 (2007)
4. Al Faqheri, W., Mashohor, S.: A real-time Malaysian automatic license plate recognition (M-ALPR) using hybrid fuzzy. Int. J. Comput. Sci. Netw. Secur. **9**(2), 333–340 (2009)
5. Gray, S.B.: Local properties of binary images in two dimensions. IEEE Trans. Comput. **20**(5), 551–561 (1971)
6. Dyer, C.R.: Computing the Euler number of an image from its quadtree. Comput. Vis. Graph. Image Process **13**, 270–276 (1980)
7. Beri, H., Nef, W.: Algorithms for the Euler characteristic and related additive functionals of digital objects. Comput. Vis. Graph. Image Process **28**, 166–175 (1984)
8. Beri, H.: Computing the Euler characteristic and related additive functionals of digital objects from their beentree representation. Comput. Vis. Graph. Image Process **40**, 115–126 (1987)
9. Chen, M.H., Yan, P.F.: A fast algorithm to calculate the Euler number for binary images. Pattern Recogn. Lett. **8**(12), 295–297 (1988)
10. Chiavetta, F., Di Gesú, V.: Parallel computation of the Euler number via connectivity graph. Pattern Recogn. Lett. **14**(11), 849–859 (1993)
11. Díaz de León S., J.L., Sossa, H.: On the computation of the Euler number of a binary object. Pattern Recogn. **29**(3), 471–476 (1996)
12. Bribiesca, E.: Computation of the Euler number using the contact perimeter. Comput. Math Appl. **60**, 1364–1373 (2010)
13. Sossa, H., Cuevas, E., Zaldivar, D.: Computation of the Euler number of a binary image composed of hexagonal cells. J. Appl. Res. Technol. **8**(3), 340–351 (2010)

14. Sossa, H., Cuevas, E., Zaldivar, D.: Alternative way to compute the Euler number of a binary image. J. Appl. Res. Technol. **9**(3), 335–341 (2011)

15. Imiya, A., Eckhardt, U.: The Euler characteristics of discrete objects and discrete quasi-objects. Comput. Vis. Image Underst. **75**(3), 307–318 (1999)

16. Kiderlen, M.: Estimating the Euler characteristic of a planar set from a digital image. J. Vis. Commun. Image Represent. **17**(6), 1237–1255 (2006)

17. Di Zenzo, S., Cinque, L., Levialdi, S.: Run-based algorithms for binary image analysis and processing. IEEE Trans. Pattern Anal. Mach. Intell. **18**(1), 83–89 (1996)

18. Sossa, H., Cuevas, E., Zaldivar, D.: Computation of the Euler number of a binary image composed of hexagonal cells. JART **8**(3), 340–351 (2010)

19. Sossa, H., Rubio, E., Peña, A., Cuevas, E., Santiago, R.: Alternative formulations to compute the binary shape Euler number. IET-Comput. Vis. **8**(3), 171–181 (2014)

20. Yao, B., Wu, H., Yang, Y., Chao, Y., He, L.: An Improvement on the Euler number computing algorithm used in MATLAB. In: IEEE Region 10 Conference on TECNON 2013–2013, Xi'an, China, 22–25 October 2013

21. He, L., Chao, Y., Suzuki, K.: A linear-time two-scan labelling algorithm. In: Proceedings of IEEE International Conference on Image Processing (ICIP 2007), pp. V-241–V-244, San Antonio, TX, USA, September 2007

22. He, L.F., Chao, Y.Y., Susuki, K.: An algorithm for connected-component labeling, hole labeling and euler number computing. J. Comput. Sci. Technol. **28**(3), 468–478 (2013)

23. He, L., Chao, Y.: A very fast algorithm for simultaneously performing connected-component labeling and Euler number computing. IEEE Trans. Image Process. **24**(9), 2725–2735 (2015)

24. Yao, B., He, L., Kang, S., Chao, Y., Zhao, X.: A novel bit-quad-based Euler number computing algorithm. SpringerPlus **4**(735), 1–16 (2015)

25. Yao, B., Kang, S., Zhao, X., Chao, Y., He, L.: A graph-theory-based Euler number computing algorithm. In: Proceeding of the 2015 IEEE International Conference on Information and Automation, pp. 1206–1209, Lijiang, China, August 2015

26. MuCulloch, W.S., Pitts, W.H.: A logical calculus of the ideas immanent in nervous activity. Bull. Math. Biophys. **5**, 115–133 (1943)

27. Rosenblatt, F.: The perceptron: a probabilistic model for information storage and organization in the brain. Psychol. Rev. **65**, 386–408 (1958)

28. Rosenblatt, F.: Principles of Neurodynamics: Perceptron and Theory of Brain Mechanisms. Spartan, Washington, DC (1962)

29. Ritter, G.X., Sussner, P.: An introduction to morphological neural networks. In: Proceedings of the 13th International Conference on Pattern Recognition, vol. 4, pp. 709–717 (1996)

30. Sussner, P.: Morphological perceptron learning. In: IEEE ISIC/CIRA/ISAS Joint Conference, pp. 477–482 (1998)

31. Ritter, G.X., Beaver, T.W.: Morphological perceptrons. Int. Joint Conf. Neural Netw. **1**, 605–610 (1999)

32. Ritter, G.X., Iancu, L., Urcid, G.: Morphological perceptrons with dendritic structure. In: 12th IEEE International Conference in Fuzzy Systems (FUZZ 2003), vol. 2, pp. 1296–1301 (2003)

33. Sossa, H., et al.: 2-D Binary Image Efficient Euler Number Computation, Paper under preparation

34. Sossa, H., Carreón, A., Guevara, E., Santiago, R.: Computing the 2-D image Euler number by an artificial neural network. In: Accepted to Be Presented at IJCNN 2006, Vancouver, Canada, 24–29 July 2016

Edge Detection in Time Variant Scenarios Based on a Novel Perceptual Method and a Gestalt Spiking Cortical Model

Juan Ramírez-Quintana[✉], Mario Chacon-Murguia,
and Alma Corral-Saenz

PVR Lab, Chihuahua Institute of Technology, Chihuahua, Mexico
{jaramirez,mchacon,adcorral}@itchihuahua.edu.mx

Abstract. Based on recently neurocomputational models inspired on neural synchronization for perceptual grouping, we propose in this paper the Gestalt Spiking Cortical Model (GSCM) and the Perceptual Grouping segmentation (PGSeg). The GSCM is a network based on the mechanisms of perceptual grouping models designed to detect scene attributes with excitatory and inhibitory inputs. PGSeg is a neuroinspired method designed to detect object edges presented in video sequences that involve time variant scenarios. Experimental results using videos from the perceptual computing and ChaDet2014 databases, show that PGSeg has better performance regarding edge detection and edge coherence through video sequences.

Keywords: Spiking neural networks · Edge detection · Background modeling

1 Introduction

Object perception is a complex human visual perception ability that involves interpretation of multiple features such as motion, depth, color and contours. Among these features, contour integration is a special case of perceptual grouping generated in the lower layers of the visual cortex that allows psychophysiological measures which contributes to integrate other features such as depth and movement [1]. There are many theories about how neurons develop visual perceptions skills, and some of them postulate that neural groups represent object features through synchronization of their pulse activity [2]. This synchronization plays an important role in perceptual grouping and Gestalt principles [3]. Based on these theories, different models have emerged such as oscillatory neural networks and Spiking Neural Networks (SNN). Among them, Pulse-Coupled Neural Networks (PCNN) have been used on contour integration, edge detection and other similar applications in image processing [4]. Several models based on PCNN have been reported in the literature to deal with these applications [5–7]. However, most of them do not consider the Gestalt principles given by neural synchronization and do not consider the most recently theories about contour perception in visual cortex. Therefore, we propose in this paper the Gestalt Spiking Cortical Model (GSCM), a PCNN simplified model based on the Gestalt rules generated from neural synchronization. Furthermore, we developed the method termed Perceptual Grouping

© Springer International Publishing Switzerland 2016
J.F. Martínez-Trinidad et al. (Eds.): MCPR 2016, LNCS 9703, pp. 54–63, 2016.
DOI: 10.1007/978-3-319-39393-3_6

Segmentation (PGSeg) that applies the GSCM model to perform edge objects detection in a similar way to the neurocomputational models that describes the process in the lower layers of visual cortex for contour integration and perceptual grouping. Edges caused by changes in object color, dynamic background conditions, lines inside objects, reflections and shadows are not considered contours of objects. PGSeg was designed for coherent edge object detection in video sequences that involve time variant scenarios, while classical edge detection methods were designed for still images with time invariant scenes. Furthermore, PGSeg consider background modeling of complex scenarios while other spatio-temporal edge detection methods as the presented in [8], consider only static backgrounds. In this paper, edge object detection refers to detect only contours of objects. This paper is organized as follows. Section 2 describes the GSCM. Section 3 describes the PGSeg method. Section 4 shows the results and Sect. 5 describes the conclusions.

2 Gestalt Spiking Cortical Model (GSCM)

Based on the perceptual models presented in [2, 3], and stimulus and inhibitor connections in the internal activity of Perceptual Grouping LISSOM presented in [1], the GSCM was defined as follows:

$$
\begin{aligned}
U(x,y,n) = U(x,y,n-1) \cdot \exp(-\alpha_F) \\
+ S(x,y,n)(1 - Y(x,y,n-1) * W_S) - I(x,y,n)(1 - Y(x,y,n-1) * W_I)
\end{aligned}
\tag{1}
$$

$$
Y(x,y,n) = \begin{cases} 1 & U(x,y,n) > E(x,y,n) \\ 0 & \textit{otherwise} \end{cases}
\tag{2}
$$

$$
E(x,y,n+1) = E(x,y,n) \cdot \exp(-\alpha_E) + Y(x,y,n)
\tag{3}
$$

where (x, y) is the pixel position in the frame, n is the iteration index, $U(x, y, n)$ is the internal activity of a neuron, $E(x, y, n)$ is the dynamic threshold and $Y(x, y, n)$ is the neuron response. α_F and α_E are the exponential decay factors of $U(x, y, n)$ and $E(x, y, n)$ respectively. $U(x, y, n)$ has two inputs: stimulus input $S(x, y, t)$ and inhibitor input $I(x, y, n)$. W_S, is the synaptic weights of the $S(x, y, n)$ and W_I is the synaptic weights of $I(x, y, n)$. The weights of GSCM are Gaussians defined by:

$$
W(\sigma_v) = w(x,y,\sigma_v) = \exp\left(-\frac{(x - x_\omega)^2 + (y - y_\omega)^2}{\sigma_v^2} \right) - \delta(x_\omega, y_\omega), \quad v = \{S, I\}
\tag{4}
$$

(x_ω, y_ω) is the center of weights, σ_v is the neighborhood radius, which depends of the scenario conditions. The Gaussian behavior is because of [2, 3], that indicates that Gestalt rules such as similarity can be implemented with Gaussians connections between oscillating neurons.

3   Perceptual Grouping Segmentation Method

PGSeg is a hierarchical method and is illustrated in Fig. 1. The first layer is the input that corresponds to a frame of a video sequence. The second layer is a module inspired on lateral geniculate nucleus (LGN) of the visual cortex that performs an edge soft detection. The third layer is based on the behavior of Orientation Receptive Fields (ORF) of the primary visual cortex. The aim of the ORF layer is to generate an orientation map and to improve the edge soft detection, which will be the input patterns for next layer. The following layer called perceptual grouping, finds the object edges using two GSCM networks: the first GSCM is used to model the background of the video sequence, and the second one detects the lines that are going to be the input to the edge detection layer. PGSeg includes background modeling of complex scenarios, therefore it may seem more complicated than classic edge detection methods. In the next subsections all layers will be explained.

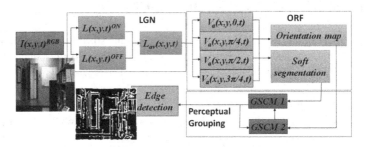

Fig. 1. PGSeg method.

The experiments were performed using different videos from the perceptual computing (http://perception.i2r.a-star.edu.sg/bk_model/bk_index.html) and ChaDet2014 (changedetection.net) databases, which are popular in literature. The selected videos have time-variant scenarios with different conditions, which are described in [9].

3.1   Input Layer

The input layer acquires a frame $I(x, y, t)^{RGB}$ of a video sequence, where t is the frame index. PGSeg does not require color information, therefore, the input layer extract the Value component (V) of HSV color space computed from the RGB information.

3.2   Lateral Geniculate Nucleus Layer

The LGN layer is inspired in the Receptive Fields (RF) located in the Retinal Ganglion Cells to Lateral Geniculate Nucleus on the visual cortex. Those RFs are modeled as simple cells whose response is given by:

$$RF(x,y)^c = \sum_a \sum_b I(x-a, y-b, t)^V G(a,b), \quad c = \{ON, OFF\} \quad (5)$$

where $G(x, y)$ is the response of the RFs, given by difference of gaussians (DoG) defined in [10]. In order to simplify the method, PGSeg uses for next layers arithmetic average of $L(x, y, t)^{ON}$ and $L(x, y, t)^{OFF}$, $L_{av}(x, y, t)$. Figure 2 shows the response of $L_{av}(x, y, t)$, which is a feature map that describes edges generated by the objects of the scenario.

 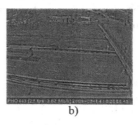

a)   b)

Fig. 2. $L_{av}(x, y, t)$ response. (a) Streetlight video, t = 100. (b) $L_{av}(x, y, t)$.

3.3 Orientation Receptive Fields Layer

After the LGN layer process, the visual information derived from RFs is projected to the primary visual cortex (V1). V1 has a cortical layer that consists on a set of receptive fields selective to orientation features (ORFs) [11]. Based on these ORFs, we design for PGSeg a layer with a set of orientation selective filters which were inspired in the model presented in [10], and defined by:

$$ORF(x, y, \theta) = \exp\left(-\frac{[(x - x_c)\cos\theta - (y - y_c)\sin\theta]^2}{\sigma_d^2} - \frac{[(x - x_c)\cos\theta + (y - y_c)\sin\theta]^2}{\sigma_f^2} \right)$$

(6)

where $ORF(x, y, \theta)$ is selective to lines with orientations similar to θ, (x_c, y_c) is the center of the filter, σ_d and σ_f define the size of width and length of the filter. The ORF are modeled as simple cells [11] and defined with:

$$V_a(x, y, \theta, t) = L_{av}(x, y, t) * RFO(x, y, \theta) \qquad (7)$$

In PGSeg, the orientation filters are selected with orientation $\theta = \{0, \pi/4, \pi/2, 3\pi/4\}$, $\sigma_d = 3$ and $\sigma_f = 1$. The values of θ were selected to simplify the calculus (as S_1 units of HMAX model in [12]), σ_d and σ_f were defined by experimentation. In models such as HMAX [12], the next layer of ORF is a set of nonlinear cells that select the highest magnitude. In the same way, PGSeg finds the ORF with the highest magnitude to have a better response of the edges, as follows:

$$V_{rfo}(x, y, t) = \max(V_a(x, y, \theta, t)) \qquad (8)$$

In addition, PGSeg defines an orientation map, as in the LISSOM models [10], which are generated by obtaining the orientation with the greatest magnitude as follows:

$$I_\theta(x, y, t) = \max\left(\left(\frac{4}{\pi}\right)\theta(x, y, t) + 1\right) \tag{9}$$

$\theta(x,\ y,\ t)$ is the orientation value in each pixel. Figure 3 shows the response of $V_{rfo}(x,\ y,\ t)$ and $I_\theta(x,\ y,\ t)$ (orientations are coded with colors). For pixels that belong to edges (edges could be noise or edge objects), $I_\theta(x,\ y,\ t)$ has the same value through the time. For the rest of the pixels, $I_\theta(x,\ y,\ t)$ have random values through the time. Therefore, it is possible to find the parts of the scenario where there are edges if we analyze differences of $I_\theta(x,\ y,\ t)$ and $I_\theta(x, y, t - 1)$. PGSeg analyzes those differences in order to find edges of the scenario that can be edges of objects. Then, next accumulator is computed:

$$AI_\theta(x, y, t+1) = AI_\theta(x, y, t) + |I_\theta(x, y, t) - I_\theta(x, y, t - 1)| \tag{10}$$

On initial conditions, pixels in $AI_\theta(x,\ y,\ t)$ are zero. After processing, $AI_\theta(x,\ y,\ t)$ has values close to zero in pixels that belongs to edges. This information is used by PGSeg to find possible edges. Figure 3(c) shows in black parts of the scenario where there are different edges. $V_{rfb}(x,\ y,\ t)$ and $AI_\theta(x,\ y,\ t)$ are the input patterns for the next layer, which will be discussed later.

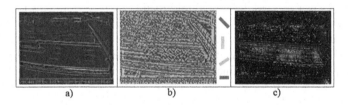

a)   b)   c)

Fig. 3. *StreetLight* video response at the frame $t = 100$. (a) $V_{rfo}(x, y, t)$ (b) $I_\theta(x, y, t)$. (c) $AI_\theta(x, y, t)$.

3.4   Perceptual Grouping Layer

The perceptual grouping layer has two GSCM as Fig. 4 shows. The first GSCM (GSCM1) is used for background modeling, and the second one (GSCM2) is used to classify the scenario in two classes: object-edges and no-edges. GSCM1 iterates once on each frame ($n = t$), but in the case of the GSCM2, iteration index is restarted each eight frames. The weights W_S and W_I of PGSeg were defined based on LEGION [12] and PG LISSOM model [1], in which, there are local excitatory and global inhibitory connections. Then, for PGSeg we must have $\sigma_I > \sigma_S$, and according to the experiments, $\sigma_I = 2$ and $\sigma_S = 0.5$.

The input stimulus $S_1(x,\ y)$ of the GSCM1 is $V_{rfo}(x,\ y,\ t)$ and for the inhibitory input $I_1(x,\ y)$ is $1 - Y_2(x,\ y,\ t)$, where $Y_2(x,\ y,\ t)$ is the output of the GSCM2. $V_{rfo}(x,\ y,\ t)$ contains a soft detection of lines and edges, and $Y_2(x,\ y,\ t)$ is used to reduce noise, which will be discussed later. The output of the GSCM1, $Y_1(x,\ y,\ t)$, is a set of pulses which are summed over time to generate the background modeling of $V_{rfo}(x,\ y,\ t)$. In consequence, the background model is obtained by

$$SY_1(x, y, t) = E_1(t)SY_1(x, y, t - 1) + Y_1(x, y, t) \tag{11}$$

Initially, $SY_1(x,\ y,\ t)$ is zero, $E_1(t)$ in the range of $0 < E_1(t) < 1$ is an entropy difference measure that depends on changes of the scenario composition and conditions.

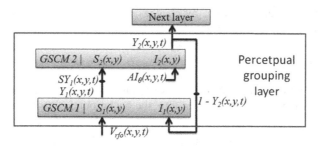

Fig. 4. Architecture of the GSCM module.

If there are changes in the scenario composition, $E_1(t)$ decreases, causing a faster background update since $SY_1(x, y, t)$ has less influence in the background modeling than $AY_1(x, y, t)$. If there are no changes in the scenario composition, $E_1(t) \approx 1$, then, $SY_1(x, y, t)$ has more influence than $Y_1(x, y, t)$ to model the background. $E_1(t)$ is given by:

$$E_1(t) = \tfrac{1}{2}(|e(t) - e(t-1)| - |(|e(t) - e(t-1)|) - 1| + 1) \tag{12}$$

$e(t)$ is the entropy of each frame, which, according to [13] is related to the composition change in the scenario. Figure 5 shows the results of background modeling.

a)   b)

Fig. 5. Background modeling. (a) StreetLight frame $t = 100$. (b) $SY_1(x, y, t)$.

In GSCM2, the input to $S_2(x, y)$ is $SY_1(x, y, t)$ normalized in range [0,1], and the input to $I_2(x, y)$ is $1-AI_\theta(x, y, t)$. $SY_1(x, y, t)$ is used to generate edges in $Y_2(x, y, t)$, while $AI_\theta(x, y, t)$ inhibits neurons in $Y_2(x, y, t)$ connected to regions that do not contain information of the edge objects candidates. The output of $Y_2(x, y, t)$ respect time is the following. In first iteration all neurons are activated in $Y_2(x, y, t)$. In second iteration few neurons are activated or in some cases none is activated. Neurons connected to pixels associated with dark-light changes are activated in the third iteration, and neurons connected to pixels associated with light-dark changes are activated in the fourth iteration. In the next three iterations, neurons associated to pixels with noise are activated. However, neurons in $Y_2(x, y, t)$ connected to regions that do not have edges, have no response after first iteration because of the inhibition of $AI_\theta(x, y, t)$. The next iterations have same behavior but with more noise. Therefore, every time that t is a multiple of eight, $Y_2(x, y, n)$ is restarted. $Y_2(x, y, n)$ is used as a feedback factor in the GSCM1 as shown in Fig. 4, because it helps to improve the response in $Y_1(x, y, t)$ since $Y_2(x, y, t)$ can help to inhibit the response in neurons connected to pixels with noise in

$V_{rfo}(x, y, t)$ related to edges caused by changes in object color, dynamic background conditions, lines inside objects, reflections and shadows.

3.5   Edge Detection Layer

The information obtained by the pulses of iterations of light-dark and dark-light edges is used for edge objects detection. Therefore, the background contours are obtained by

$$Y_{acum}(x, y, t) = \sum_{p=3,4} Y_2(x, y, p) \tag{13}$$

$$I_s(x, y, t) = \frac{1}{2}(|Y_{acum}(x, y, t)| - |(Y_{acum}(x, y, t)) - 1| + 1) \tag{14}$$

where p is the module of $t/8$; $p = \{3, 4\}$ are the iterations with contours information (edge objects). The iterations $p = \{1, 2, 5, 6, 7\}$ are related to the class no-edges. Figure 6 shows the result of PGSeg with an outdoor scenario.

 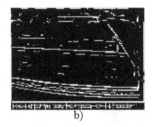
a)   b)

Fig. 6. Edge detection with PGSeg. (a) *StreetLight*, frame $t = 100$. (b) $I_s(x, y, t)$.

4   Results of PGSeg

The metric F measure (F1) [14] was used to compare the performance of PGSeg with the Canny, LoG, Roberts, Prewitt and Sobel methods which sometimes had been used for comparison purposes in literature [15–18]. This comparison consists on obtaining F1 between the ground truth and each $I_s(x, y, t)$ resulting from the PGseg processing. Then, the average of each F1 of $I_s(x, y, t)$ is obtained ($\mu F1$) for each video sequence. The videos used to measure the performance, were selected based on the situations that generate time-variant scenarios, and those videos are: Watersurface (WS), Subway Station (SS), Lobby (LB), Cubicle (CU) and Park (PK). The WS, SS and LB videos were obtained from the data base of Perceptual Computing and the CU and PK videos were obtained from ChaDet2014. The ground truths are images of edges that represent the contours of background objects. The edges caused by changes in object color, dynamic background, lines inside objects, reflections and shadows are considered false edges in this work. Figure 7 shows a frame of each video sequence and Fig. 8 shows the ground truths.

As Fig. 8 shows, all the videos have people crossing the scenario, as dynamic objects. Moreover, each video has different situations: in the case of the WS video, the sea generates a dynamic background on the scenario; the SS video has a dynamic background caused by the electric stairs and the light reflections in the floor could cause false edges;

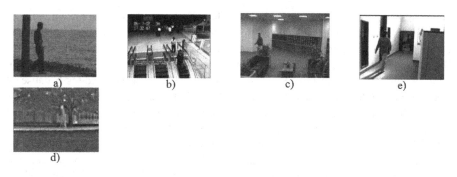

Fig. 7. Video sequence for validation. (a) 'WS' video, t = 500. (b) 'SS' video, t = 500. (c) 'LB' video, t = 370. (d) 'CU' video, t = 5000. (e) 'PK' video, t = 500.

Fig. 8. Ground truths. (a) 'WS' video. (b) 'SS' video. (c) 'LB' video. (d) 'CU' video. (e) 'PK' video.

the LB video has sudden illumination changes; the CU video has several shadows that can cause false edges. Finally, the PK video was recorded with a thermic camera with camouflage issues. Table 1 shows the results of $\mu F1$ for each video sequence, where PGseg has the better results. The parameters of Canny, LoG, Roberts, Prewitt and Sobel were selected based on the best $\mu F1$ results for each method in each video.

Table 1. Results of $\mu F1$.

Video/method	PGSeg	Canny	LoG	Roberts	Prewitt	Sobel
WS	0.5798	0.23093	0.28701	0.28149	0.33062	0.3309
LB	0.2677	0.2551	0.1996	0.1795	0.1825	0.1825
SS	0.2761	0.2525	0.195	0.1331	0.1632	0.1626
CU	0.5801	0.4892	0.3806	0.2717	0.2718	0.2755
PK	0.3095	0.2382	0.2687	0.2573	0.2072	0.2081

Figure 9 shows the results of each method with the WS video. PGSeg has adequate results, but with a few noise. However, Canny and LoG generates noise due dynamic background, also, the edge detection is different between one frame and another, although the scenario does not have changes its composition. The Roberts, Prewitt and Sobel methods fail detecting the edges and generate noise. PGSeg generates better results because $Y_2(x, y, t)$ inhibit the noise in the results in the background modeling, and $AI_\theta(x, y, t)$ inhibits the neurons in $Y_2(x, y, t)$ that are connected to object that generate false edges in the dynamic background. Furthermore, the feedback of $Y_2(x, y, t)$ in $Y_1(x, y, t)$ allows a stable edge detection through the time, allowing edge coherence through video WS. In Table 1, on the column of PGSeg, the lowest value of $\mu F1$ was obtained with LB video. In this video, all methods were affected by false edges caused by the plants, couches and reflections. However, among the methods, PGSeg has better performance because even with illumination changes, edge detection results remain constant. In the SS video, all methods generate false edges because of reflections on the floor and the time and date shown in the display, but PGSeg generates better results since it has a better performance detecting appropriately the edges of the electric stairs. In the case of the CU video, PGSeg has better performance because this method has coherence results in time and the rest of the methods were affected by shadows. In the video PK, all methods were affected by false edges of a wall, but PGseg has better results because generated less noise than the others in areas of the scenario that have a tree and a garden. Results are not showed for videos LB, CU, SS and PK for space reasons.

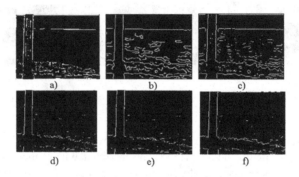

Fig. 9. Edge detection results with frame t = 460 of WS the methods: (a) PGSeg. (b) Canny. (c) LoG. (d) Roberts. (e) Prewitt. (f) Sobel.

5 Conclusions

In this paper we propose a Spiking Neural Network known as GSCM, which was applied in a novel method proposed also in this paper to detect edge called PGSeg. GSCM generates pulses from an internal activity that is based on an excitatory input and an inhibitory input. PGSeg is a method inspired in lower layers of the visual cortex and uses the GSCM to generate the edges of objects in the background model of a video sequence with time varying scenario. Results showed that PGSeg has better performance than other edge methods on detecting edges without noise caused by dynamic background, illumination changes, shadows, and reflections.

The parameters of the GSCM are used as constants in PGSeg. Hence, as future work, the GSCM model is going to be modified such that the parameters of the model can be adjusted based on scenario conditions to improve the performance of PGSeg in edge detection or scenario analysis with any method that uses GSCM.

Acknowledgment. This research was supported by Fomix CONACYT-Gobierno del Estado de Chihuahua under grant CHIH-2012-C03-193760 and PRODEP ITCHI-PTC-025.

References

1. Choe, Y., Miikkulainen, R.: Contour integration and segmentation with self-organized lateral connections. Biol. Cybern. **90**, 75–88 (2004). Springer
2. Ursino, M., Magosso, E., Cuppini, C.: Recognition of abstract objects via neural oscillators: interaction among topological organization, associative memory and gamma band synchronization. IEEE Trans. Neural Netw. **20**(2), 1871–1884 (2009)
3. Yu, G., Slotine, J.-J.: Visual grouping by neural oscillator networks. IEEE Trans. Neural Netw. **20**(12), 1871–1884 (2009)
4. Wang, Z., Ma, Y., Cheng, F., Yang, L.: Review of pulse-coupled neural networks. Image Vis. Comput. **28**(1), 5–13 (2010)
5. Gu, X.: A new approach to image authentication using local image icon of unit-linking PCNN. In: International Joint Conference on Neural Networks, pp 1036–1041 (2006)
6. Wang, Z., Ma, Y.: Medical image fusion using m-PCNN. Inf. Fusion **9**(2), 176–185 (2008)
7. Chen, Y., Ma, Y., Kim, H.D., Park, S.K.: Region-baed object recognition by color segmentation using a simplified PCNN. IEEE Trans. Neural Netw. Learn. Syst. **26**(8), 1682–1697 (2015)
8. Karamiani, A., Farajzadeh, N.: Detecting and tracking moving objects in video sequences using moving edge features. In: Scientific Cooperations International Workshops on Electrical and Computer Engineering Subfields, pp 88–92 (2914)
9. Brutzer, S., Höferlin B., Heidemann, G.: Evaluation of background subtraction techniques for video surveillance. In: Conference on Computer Vision and Pattern Recognition, pp. 1937–1944 (2011)
10. Miikkulainen, R., Bednar, J.A., Choe, Y., Sirosh, J.: Computational Maps in the Visual Cortex, 1st edn. Springer Science Media Inc., New York (2005)
11. Krüger, N., Jansen, P., Kalkan, S., Lappe, M., Leonardis, A., Piater, J., Rodriguez-Sanchez, A., Wiskott, L.: Deep hierarchies in the primary visual cortex: what can we learn for computer vision. IEEE Trans. Pattern Anal. Mach. Learn. **35**(8), 1847–1871 (2013)
12. Orozco-Rodriguez, H., Chacon-Murguia, M., Ramírez-Quintna, J.: A neural bioinspired scheme for head pose recognition. In: Signal Processing and Signal Processing Education Workshop, pp. 403–408 (2015)
13. Ramirez-Quintana, J., Chacon-Murguia, M.: Self-adaptive SOM-CNN neural system for dynamic object detection in normal and complex scenarios. Pattern Recogn. **48**(4), 1137–1149 (2015)
14. Maddalena, L., Petrosino, A.: A self-organizing approach to background subtraction for visual surveillance applications. IEEE Trans. Image Process. **17**(7), 168–1177 (2008)
15. Mofrada, M.H., Sadeghi, S., Rezvanian, A., Meybodi, M.R.: Cellular edge detection: combining cellular automata and cellular learning automata. Int. J. Electron. Commun. **69**(9), 1282–1290 (2015)
16. Uguz, S., Sahin, U., Sahin, F.: Edge detection with fuzzy cellular automata transition function optimized by PSO. Comput. Electr. Eng. **43**(1), 180–192 (2015)
17. Liu, X., Fang, S.: A convenient and robust edge detection method based on ant colony optimization. Opt. Commun. **353**(1), 147–157 (2015)
18. Jiang, W., Lam, K.-M., Shen, T.-Z.: Efficient edge detection using simplified Gabor wavelets. IEEE Trans. Syst. Man Cybern. **39**(4), 1036–1047 (2009)

An Effective Image De-noising Alternative Approach Based on Third Generation Neural Networks

Manuel Mejía-Lavalle[✉], Estela Ortiz, Dante Mújica, José Ruiz, and Gerardo Reyes

Departamento de Ciencias Computacionales, Centro Nacional de Investigación y Desarrollo Tecnológico (CENIDET), Cuernavaca, Mexico
{mlavalle,estela_or,dantemv,josera,greyes}@cenidet.edu.mx

Abstract. Searching to reduce the noise effect in gray scale digital images, an efficient and effective alternative approach that combines a kind of Pulse-Coupled Neural Network and the Median estimator is proposed to remove Salt and Pepper noise. The proposed approach is based on a simplified Third Generation Neural Network called Intersection Cortical Model (ICM). Using the ICM output images, we can detect which pixel position corresponds to Salt and Pepper noise. Then, a selective Median filter is used for suppressing the Salt and Pepper noise only over the previously detected noisy pixels. The performance of the proposed approach is evaluated by simulating different impulsive noise densities. Simulation results show that method's effectiveness is 32 % better and 225 % faster than conventional Median filter noise suppression. Results are measured by the Peak Signal to Noise Ratio, Mean Absolute Error and Normalized Mean Square Error metrics.

Keywords: Image de-noising · Third Generation Neural Networks · Pulse-Coupled Neural Network · Intersection Cortical Model · Salt and Pepper noise

1 Introduction

Noise in digital images causes problems in its analysis, recognition, classification and interpretation. Salt and Pepper noise is an impulse noise type that commonly affects digital images and is the result of defective sensors or poor transmission channels. The pixels that present this error are visually different from their neighbors, since their gray values tend to be extremely high or low (gray level 0 or 255).

The Median filtering can suppress Salt and Pepper noise successfully; in this well-known technique, the central pixel of the filtering window is replaced by the median value. Despite its simplicity, its main disadvantage is that, when remove the noise also removes image details [1]. Other better techniques are known, but they are more sophisticated and more computer time consuming, for example by using fuzzy techniques [2], or variants of the median filter [3–5]. A different or alternative way of addressing this problem has begun to be explored experimentally; such is the case of the Third Generation Neural Networks, also called Pulse-Coupled Neural Networks (PCNN).

PCNN is a simplified mathematical model proposed by Eckhorn [6], and it is based on the timing of the pulses released in the visual cortex of the mammals

© Springer International Publishing Switzerland 2016
J.F. Martínez-Trinidad et al. (Eds.): MCPR 2016, LNCS 9703, pp. 64–73, 2016.
DOI: 10.1007/978-3-319-39393-3_7

(with this research, John C. Eccles, Alan L. Hodgkin and Andrew F. Huxley won the 1963 Medicine Nobel Prize [6]). The pulse timing of the PCNN is used especially for the detection of noisy pixels because these are activated before or after the not noisy pixels [7].

Different simplified models of PCNN have been developed to work in Artificial Intelligence and Computer Vision context; two of the main variations are the Intersection Cortical Model (ICM) and the Spiking Cortical Model (SCM). In [8] they are used these methods combining them with local Median filter, morphological filter and Wiener filter to reduce Salt and Pepper and Gaussian noises.

The alternative approach proposed in this paper consists of two general phases: (a) the ICM model detects the pixels affected by Salt and Pepper noise, and (b) a Median filter selectively suppresses the previously detected noise pixels. The rest of the paper is organized as follows: in Sect. 2 they are described PCNN and ICM Third Generation Neural Networks; Sect. 3 is devoted to present our proposed alternative de-noising approach; Sect. 4 details experiments and discuss results; Sect. 5 concludes and address future research directions.

2 PCNN and ICM Paradigms

PCNN is a neural network paradigm that emulates biological neurons in the visual cortex of mammals and has been applied in a variety of domains of digital image processing such as noise removal, object detection, feature extraction, image fusion, optimization, image thinning, segmentation, shadows removing, among others.

There are relevant differences between traditional Artificial Neural Networks and PCNN, both in configuration and operation. The PCNN not requires training and its only function is to classify pixels by levels of intensity; in this model every pixel of the digital image corresponds to one neuron. The neurons firing threshold is dynamic and each neuron receives inputs from other neurons through synapses (linking process). These characteristics make neighboring neurons with similar intensity to fire at the same time in certain regions, phenomenon called "synchronous pulse firing" [9]. Each neuron corresponds to a pixel, for this reason the feeding, the linking and the threshold are of the same size of the processed image.

The original model of PCNN has some limitations in practice, when is used for image processing; for example, the great number of connections among neurons can result in a computer memory problem; another difficulty is related with the adequate operating parameters tuning.

For this reason in [10] it is proposed a simplified PCNN iterative model for image processing. In this paper Intersection Cortical Model (ICM) is used; its diagram is showed in Fig. 1. The ICM is a case of the PCNN when there are no linking neurons; the neuron's feeding inputs are composed of the feedback input F_{ij} and the last output Y_{ij}, where F_{ij} only accepts the external stimulus S_{ij} (gray level of each pixel normalized between 0 and 1). When F_{ij} is greater than dynamic threshold T_{ij}, the ICM neuron iteratively outputs sequential binary pulse series Y_{ij}. Each entry retains its previous state attenuated by a decay factor.

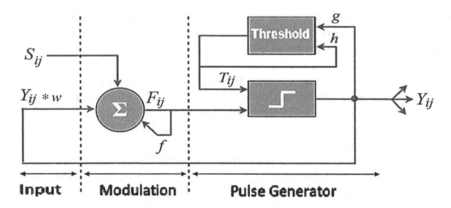

Fig. 1. ICM model diagram [10]

Based on the simplified PCNN model, the iterative computing formulations of the ICM can be simulated with the following functions [10]:

$$F_{ij}[n] = fF_{ij}[n-1] + \sum_{kj} w_{ijkl}Y_{ij}[n-1] + S_{ij} \tag{1}$$

$$Y_{ij}[n] = \begin{cases} 1 & \text{si } F_{ij}[n] > T_{ij}[n] \\ 0 & \text{in other case} \end{cases} \tag{2}$$

$$T_{ij}[n] = gT_{ij}[n-1] + hY_{ij}[n-1] \tag{3}$$

where $[n]$ denotes the current iteration, w is a synaptic weighting matrix that links a neuron with its neighbors, and finally f, g and h are adjust parameter coefficients, typically $g < 1.0$, $f < g$ and h is a large value (20 is a value normally used-recommended in specialized literature).

3 Alternative ICM De-noising Approach Proposed

In our work, all of the ICM neurons are linked mutually in the same mode and their outputs only have two states: firing or non-firing (0 or 1). This model is faster than PCNN model and the noisy pixel's gray values can be adjusted to reduce the noise effect without affect image borders. The main procedure of the proposed de-noising approach has four steps, as follows:

Step 1. Input the normalized noisy digital image to the ICM neural network, obtain the Y_{ij} output and find the high light areas according to firing synchronously. The light pixels in the output image are considered Salt noise.
Step 2. Find the location of correspondent salt noise pixels in this ICM output image and adjust them applying a Median filter.

Step 3. Duplicate the original image and invert its pixels gray value, then repeat step 1 and step 2 with this "negative" image, in order to detect Pepper noise.

Step 4. Find the neurons fired in advance in each ICM output iterative procedure and modify the gray values of their corresponding pixels applying Median filter.

Figure 2 depicts our alternative ICM de-noising approach.

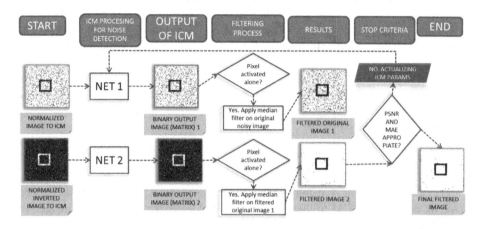

Fig. 2. Alternative ICM de-noising approach proposed

4   Experiments and Results

Experimental evaluation were carried out on the well-known gray scale digital images Lena, Peppers and Baboon, with sizes of 128 × 128 and 512 × 512 pixels, just as Fig. 3 shows. During the experimental process, the images were corrupted with impulsive noise densities of 5 %, 10 %, 15 %, 20 %, 25 %, 30 %, 40 %, 50 %, 60 % and 70 %. Then the normalized image from 0 to 1 enters to the ICM model as S_{ij}. The $F_{ij}[n-1]$, $Y_{ij}[n-1]$ and $T_{ij}[n-1]$ values were initially set to 0.

Fig. 3. Noisy test images: 10 % Salt and Pepper noise

The other parameters were empirically selected as follows in simulation.

- Internal-weighting matrix w formed with Gaussian weights as a function of the neuron neighbor distance:

$$w = \begin{bmatrix} 0.5 & 1 & 0.5 \\ 1 & 0 & 1 \\ 0.5 & 1 & 0.5 \end{bmatrix}$$

- Parameters: $f = 0.9$, $g = 0.8$ y and $h = 20$.
- In *Step 2*, the Median filter is applied. Let xij denote pixels with coordinates (i, j) in noisy image, and X_{ij} denote the set of pixels in $(2K + 1) \times (2K + 1)$ neighborhood window W centered at x_{ij}. In our case we use $K = 1$ and then W size is 3×3 pixels; this was chosen primarily to ensure better details preservation. Equation (4) resume these concepts.

$$X_{ij} = \left\{ x_{i-K,j-K}, \dots, x_{ij}, \dots, x_{i+K,J+K} \right\} \tag{4}$$

- Median operator is defined as:

$$m_{ij} = median\left(X_{ij}\right) \tag{5}$$

As can be seen in Fig. 4, using Lena of 128×128 pixels with 30 % Salt and Pepper noise, we can exploit ICM to distinguish noisy and noise-free pixels in an image for processing them.

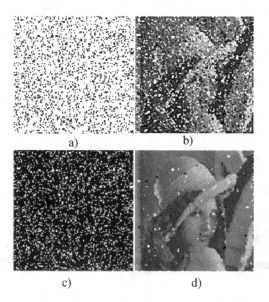

a) b)

c) d)

Fig. 4. Lena with 30 % noise (a) First iteration output pattern from ICM, (b) Resulted de-noised image, (c) Third iteration output pattern from ICM, (d) Resulted de-noised image

For the ICM based method it is difficult to wipe off the noise when its pixel value is similar to the background; for that reason we include-contribute the intuitive, useful and pragmatic idea of processing also the inverse gray level. When noisy pixels are detected for the ICM output, then the Mean filter was applied for de-noising.

Final results were evaluated by means of three well known metrics, which are formulated as follows [11, 12]:

(a) PSNR (Peak Signal to Noise Ratio in dB), which is used to measure the ability of impulse noise suppression (6): the bigger PSNR is, the effect of de-noising is.

$$PSNR = 10 \log_{10} \frac{f(m,n)^2}{\frac{1}{MN} \sum\limits_{m=1}^{M} \sum\limits_{n=1}^{N} [f(m,n) - f'(m,n)]^2} \tag{6}$$

(b) MAE (Mean Absolute Error), indicating filtering quality (7), that is, preserving fine details must be minimized.

$$MAE = \frac{1}{MN} \cdot \sum\limits_{m=1}^{M} \sum\limits_{n=1}^{N} |f(m,n) - f'(m,n)| \tag{7}$$

(c) NMSE (Normalized Mean Square Error), a better noise image filtering method can often result in the less NMSE (8).

$$SMSE = \frac{\sum\limits_{m=1}^{M} \sum\limits_{n=1}^{N} [f(m,n) - f'(m,n)]^2}{\sum\limits_{m=1}^{M} \sum\limits_{n=1}^{N} [f(m,n)]^2} \tag{8}$$

In (6) and (7) M, N denote the image's rows and columns; $f(m, n)$ is the non-processing image and $f'(m, n)$ is the de-noising resulting image.

Figure 5 shows a Lena zoom in order to observe that our approach is visualized better (detail preservation) than when only a Median filter is applied.

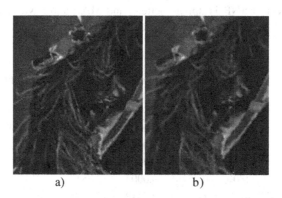

a)   b)

Fig. 5. Detail Lena 10 % noise (a) De-noised image applying our alternative ICM de-noising approach, (b) De-noised image obtained with only traditional median filter method

To determine the ICM iteration number required to perform the noise image filtering, the noise reduction process was measured iteration by iteration. The relation between the de-noising performance denoted by PSNR and MAE (Y axis) vs. the ICM iteration number (X axis) is showed in Fig. 6.

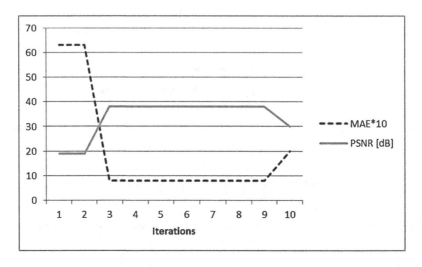

Fig. 6. The effect of ICM's iterations vs. noise suppression (Lena 10 % noise)

We can see that the PSNR and MAE metrics have no changes after three iterations; therefore, it is not necessary to conduct further ICM iterations and calculations; this observation allows us to reduce effectively computer processing time (225 % faster).

To show the detail preservation performance for each method, various contamination noise levels were applied to Lena, Peppers and Baboon digital images. According to Table 1, the first two rows indicate the images corrupted by Salt and Pepper noise with 5 % and 10 % respectively. Columns show the three metrics obtained results that are used to compare both methods: our alternative ICM approach vs. traditional Median filter; as can be seen, PSNR is upper (better), and MAE and NMSE are lower (again better) using our ICM approach than the traditional Median filter method.

Table 1. De-noising methods performances

Noisy image	Our alternative ICM approach filter			Traditional median filter		
	PSNR [dB]	MAE	NMSE	PSNR [dB]	MAE	NMSE
Lena 5 %	41.9741	0.3614	0.0003	33.2361	3.2268	0.0027
Lena 10 %	37.0746	0.8885	0.0011	32.4863	3.4124	0.0033
Peppers 5 %	41.8626	0.3435	0.0002	34.4246	2.8941	0.0013
Peppers 10 %	36.9480	0.8558	0.0007	33.2819	3.1012	0.0017
Baboon 5 %	32.5673	1.1571	0.0019	23.310	10.8228	0.0162
Baboon 10 %	28.5731	2.7620	0.0048	23.0225	11.1927	0.0173

So, from Table 1, it is observed that our proposed method shows better performance, for experiments when noise density is 5 and 10 %. Additionally, from Fig. 7, it is observed that Baboon and Peppers with 10 % noise density show the proposed approach restoration ability. It is visualized (human eye) that our proposed method preserves the edges and lines of the image while the Median filter has a smoothing effect on all the details. Also Fig. 7 shows that our approach based on ICM Neural Network virtually eliminates image noise.

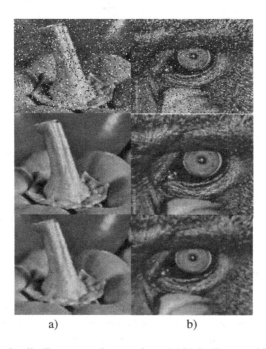

a)   b)

Fig. 7. Preservation details. From up to down, column (a) Noisy Peppers 10 %, de-noised image by our ICM and by median filter; column (b) Noisy Baboon 10 %, de-noised image by our ICM and by median filter.

In order to compare the effectiveness of the proposed ICM de-noising approach vs. traditional Median filter, another series of experiments were carried out varying noise degrees. Lena corrupted by Salt and Pepper noise with various noise densities from 5 % to 70 %, with 5 % and 10 % increments was used for experiments that are presented and resumed in Fig. 8.

The comparative PSNR graphical illustration for each method is showed. In particular, in the case of 5 % corrupted image, the proposed algorithm was approximately 8.74 dB better (32 %) than the Median filter. Additionally, MAE and NMSE scores were measured for each method. The proposed approach produced the lowest (better) MAE and NMSE values with various noise densities under 45 %.

Finally we can observe the behavior of our ICM de-noising approach vs. Median filter when increasing noisy degree beyond 45 % noise: the ICM performance is decreasing according to Fig. 8 (X axis shows noise degree expressed as probability,

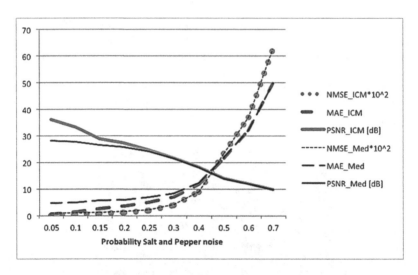

Fig. 8. Filtering performance curves on Lena with several Salt and Pepper noise degrees

and Y axis shows obtained values for the three evaluation metrics). As can be seen, when noise is upper than 70 %, the de-noised image lost important details for both methods, as PSNR, MAE and NMSE depicts: this result is predictable because very high noise density is presented.

5   Conclusion and Future Work

Noise reduction on digital images is essential for processing them. An efficient and effective alternative filtering approach is proposed to reduce the noise of Salt and Pepper; this alternative is based on the integration of the simplified model PCNN called ICM and is used to effectively detect noisy pixels and then combined with a selective Median filter to eliminate-reduce noise.

Experimental results show that the proposed method's effectiveness noise suppression and details preservation for gray scale images is 32 % better than traditional Median filter (as PSNR, MAE and NMSE metrics report) and 225 % faster because the PCNN can be implemented in a parallel processor and the median operator can be applied by an indexed method. Quantitative data show that only three iterations are enough to obtain a de-noised image using ICM for Salt and Pepper noise removing.

For future works, the proposed ICM based approach will also be evaluated on color digital images, with Gaussian and Speckle noise, and on a public collection of thousands of images. We should consider further adjustments and development of the proposed approach to increase its performance and compare it to some of the other selective de-noising methods to ensure is competing with more standard methods and other recent research.

References

1. Sreenivasulu, P., Chaitanya, N.K.: Removal of salt and pepper noise for various images using median filters: a comparative study. IUP J. Telecommun. **6**, 54–70 (2014)
2. Gao, G., Liu, Y.: An efficient three-stage approach for removing salt & pepper noise from digital images. Optik - International J. Light Electron Opt. **126**(4), 467–471 (2015)
3. Sharma, A., Chaurasia, V.: Removal of high density salt-and-pepper noise by recursive enhanced median filtering. In: IEEE 2nd International Conference on Emerging Technology Trends in Electronics, Communication and Networking (2014)
4. Zhang, C., Wang, K.: A switching median–mean filter for removal of high-density impulse noise from digital images. Optik - International J. Light Electron Opt. **126**(9–10), 956–961 (2015)
5. Chang, J.: Applying generalized weighted mean aggregation to impulsive noise removal of images. In: IEEE Proceedings of the 2014 International Conference on Machine Learning and Cybernetics, pp. 13–16 (2014)
6. Lindblad, T., Kinser, J.M.: Image Processing Using Pulse-Coupled Neural Networks, 2nd edn, pp. 11–23. Springer, Berlin (2005)
7. Johnson, J.L., Padgett, M.L.: PCNN model and applications. IEEE Trans. Neural Netw. **10**, 480–498 (1999)
8. Ma, Y., Zhan, K., Wang, Z.: Applications of Pulse-Coupled Neural Networks, pp. 6–23. Springer, Berlin (2010)
9. Zhang, J.Y., Li, B.Y.: Feature extraction on image smoothness based on PCNNs. Comput. Simul. Mag.-Room 9, 103–105 (2003)
10. Ekblad, U., et al.: The intersecting cortical model in image processing. Nucl. Instrum. Methods Phys. Res. **525**(1), 392–396 (2004)
11. Lui, C., Zhang, Z.: Sonar images de-noising based on pulse coupled neural networks. In: Congress on Image and Signal Processing, pp. 403–406 (2008)
12. Wang, Z., Ma, Y., Cheng, F., Yang, L.: Review of Pulse-Coupled Neural Networks. Image Vision Comput. J. **28**, 5–13 (2010). Elsevier

Toward the Labeled Segmentation of Natural Images Using Rough-Set Rules

Fernando J. Navarro-Avila, Jonathan Cepeda-Negrete,
and Raul E. Sanchez-Yanez[✉]

Universidad de Guanajuato DICIS, Salamanca, Guanajuato, Mexico
{jf.navarroavila,j.cepedanegrete,sanchezy}@ugto.mx

Abstract. This article introduces an approach that integrates color and texture features for the segmentation of natural images. In order to deal with the vague or imprecise information that is typically shown in this kind of scenes, our method consists in a supervised classifier based on rules obtained using the rough-set theory. Such rough classifier yields a label per pixel using as inputs only three color and three textural features computed separately. These labels are used to carry out the image segmentation. When comparing quantitatively the results from this work with state-of-the-art algorithms, it has shown to be a competitive approach to the image segmentation task. Moreover, the labeling of each pixel offers advantages over other segmentation algorithms because the outcome is intuitive to humans in two senses. On one hand, the use of simple rules and few features facilitate the understanding of the segmentation process. On the other hand, the labels in the segmented outcomes provide insight into the image content.

Keywords: Rough-set rule · Supervised classifier · Labeled segmentation · Natural image

1 Introduction

Image segmentation is defined as the process of partitioning an image into its constituent components, called segments or clusters of pixels [1]. The resulting segments are collections of pixels that share a similarity with respect to some feature or property. Over the last years, image segmentation has resulted to be of the most difficult tasks in computer vision and image processing, but also a promising approach towards object recognition and image understanding, mainly because each region within an image usually corresponds to an object. According to Ilea and Whelan [2], the image segmentation task has been developed mostly using color and texture properties in a large number of approaches, including region growing, edge detection, clustering and histogram-based, among others. In fact, the current research is leaning toward the integration of features and methods.

The use of color and texture information collectively, has strong links with the human perception and, in many scenarios, the use of color or texture solely

© Springer International Publishing Switzerland 2016
J.F. Martínez-Trinidad et al. (Eds.): MCPR 2016, LNCS 9703, pp. 74–83, 2016.
DOI: 10.1007/978-3-319-39393-3_8

is not sufficient to describe the image content. The segmentation of natural images exemplifies this problem, mainly because these images present significant inhomogeneities in color and texture [2,3]. In addition, scenes are often complex, with a degree of randomness and irregularity. Likewise, the strength of texture and color attributes can vary considerably from image to image when adding distortions due to uneven illumination, scale changes, and other sources.

Rough set theory, proposed by Pawlak in [4], is a promissory approach to deal with inconsistency of data. The theory has been proposed to solve a wide number of problems, particularly those related to artificial intelligence and cognitive science as machine learning, data mining and pattern recognition, but applications in image processing are relatively few. The rough set theory offers advantages when compared to other approaches. (I) There is no need of additional or preliminary information regarding the data. (II) Data quantification is carried out together with the rule induction. (III) Redundant information is reduced in two ways: when identical objects are represented several times or when some attributes do not contribute to the classification (superfluous attributes). (IV) The minimum set of attributes that preserves the knowledge in the original data is found. (V) Decision rules are induced automatically from data. (VI) The rough set theory can ease the interpretation of results.

Although several methods of color-texture image segmentation have been proposed, rule-based classifiers for image segmentation have not been widely explored. Our approach takes into account the fact that natural images are taken under different light conditions and other irregularities present by the nature of image acquisition. At first, we use color constancy in an image preprocessing step, and afterward a rough-set-rule classifier is utilized as an alternative to image segmentation approaches. This classifier integrates features allowing the processing of imprecise, incomplete or uncertain information of the real world. Furthermore, a rule-based classifier induces intuitive decision rules, thus enabling the understanding of the image segmentation process. Besides, the labeling of pixels, and consequently, of regions, provides a higher level of knowledge regarding the image content.

The rest of this paper is organized as follows. Section 2 describes the methodology used throughout our work. It discusses the selection of color and texture features, and the rough set rules obtained by the classifier. Section 3 includes the experimental results in the test series, and the observations from the data obtained. Finally, the concluding remarks are given in Sect. 4.

2   Methodology

In this section, we describe the methodology implemented for the natural image segmentation task. An explanation of the process is provided, as well as the criteria to choose the methods and the parameters used for the system configuration.

The block diagram shown in Fig. 1 depicts the system used for image segmentation. Regarding this diagram, two general processes are distinguished: training and testing stages.

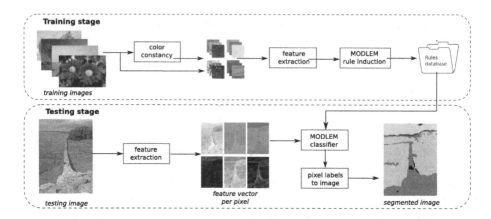

Fig. 1. Block diagram of the rough-based system for image segmentation.

The purpose of the training stage is to induce decision rules in order to generalize and label future unknown pixels using color and texture features. The set of training images is constituted by instances to be described. Each input image is a representative sample of a desired class. In our study, the number of classes is limited to four ("Foliage", "Soil", "Sky" and "Water") and, additionally, an "Indeterminate" class is included for denoting outliers. At the training stage, the first step consists in the application of a color constancy algorithm to the input image. In this study, the Gray-World assumption [5] and the White-Patch [6] algorithms are used to process the training images and increase the number of samples.

2.1   Feature Extraction

The second step consists of the feature extraction. The extraction of color features in this work requires a color space that shows minimum correlation between components, as a consequence providing more information of the scene and thus improving the segmentation. A color space with perceptual characteristics in its components is needed. In other words, little changes in each component are translated into changes perceived by humans. The CIELAB color space has the proper characteristics for this work. This color space considers one channel for Luminance L and two color channels a and b. We use the well-known equations in literature and the D65 white reference [7]. The image is transformed from the RGB color space to CIELAB giving as a result three images, each containing one of the color features used in the system (L, a^*, b^*). These three images are smoothed using a simple mean filter.

In addition to color features, the textural content exhibited in all images plays an important role in image analysis in a wide range of applications as medical imaging, industrial processes and robotics. Despite texture is an intuitive concept, there is no formal definition for a visual texture. Tamura *et al.* [8] mention that "A texture can be considered as a macroscopic region. Its structure is attributed to repetitive or primitive patterns whose elements are arranged according to a rule of position."

The processed image after the color constancy algorithm is transformed to gray scale, and the texture features are extracted on a neighborhood centered over each pixel using a $w \times w$ window. Homogeneity and entropy are extracted using sum and difference histograms [9] and the standard deviation map is extracted as well, obtaining three texture maps. These images are also smoothed using a low-pass filter. This procedure is repeated for each image in the training set obtaining an $m \times n$ matrix, where m is the number of samples and n is the number of features. In our study, n equals to six, combining only three color features and three texture features.

For computing the Sum and Difference Histograms (SDH) [9], the relative displacement vector ($\mathbf{V}$) between two picture elements is an SDH parameter, and in our study, it is defined as the composition of the Cartesian product $R \times \theta$, where $R = \{1, 2\}$ and $\theta = \{0, \pi/4, \pi/2, 3\pi/4\}$. Homogeneity and entropy are calculated using

$$\text{Entropy} = -\sum_i P_s(i) \cdot \log(P_s(i)) - \sum_j P_d(j) \cdot \log(P_d(j)) \tag{1}$$

$$\text{Homogeneity} = \sum_j (1 + j^2)^{-1} \cdot P_d(j) \tag{2}$$

where P_s and P_d are the normalized SDH.

The last texture feature used throughout the system is a standard deviation map. Texture regions within an image show more intensity variations than those pixels in homogeneous regions. A measure of those variations is used to determine the boundaries of such textured regions. Moreover, different textures have distinct variations in intensity allowing to discriminate those textures.

The computation of the standard deviation image T is obtained for each pixel in the image in a desired neighborhood consisting of a square region containing $k = (2d + 1)^2$ pixels and centered at the current pixel position. The parameter d is the number of pixels from the central pixel to a side of the window. Equations (3) and (4) are used to compute σ. In these equations I_i represents the intensity of the i_{th} pixel of the neighborhood, μ and σ are the first and second statistical moments about zero, respectively.

$$\mu = \frac{1}{k} \sum_{i=1}^{k} I_i, \tag{3}$$

$$\sigma = \sqrt{\sum_{i=1}^{k} (I_i - \mu)^2 / (N - 1)}. \tag{4}$$

Now that the features have been defined, the system requires a pixel wise classifier in order to assign labels to pixels according to the input features. In the training stage, the last step consists of the induction of rules using MODLEM [10], which is an algorithm based on rough set theory. The rules induced in this step are stored on a database of rules.

The testing stage is similar to the training one. For each testing image, the feature extraction is performed using the same window size, but this time the resulting matrix is of the size $P \times N$, where P is equal to the number of pixels within the test image and N still being the number of features. Each pixel generates a feature vector. The matching of this vector with the stored rules yields a label. Therefore, the process of segmentation involves the labeling of pixels at each position on the testing image, obtaining a matrix of labels with the same size that the image. The last step in the testing stage is the conformation of a map of labels as an image.

2.2   The MODLEM Classifier

After the feature extraction process, a decision table is obtained. This decision table can be inconsistent due to unbalanced data or confusing samples. Therefore, decision rules can be generated from rough approximations [11]. For this matter, rule-based classifiers are suitable algorithms. These algorithms iteratively create a set of rules for each class or concept. In addition, one of the main advantages of using rule-based classifiers is that a comprehensive description for the class or the concept is generated.

The rules allow the control of the complexity of the classifier describing the class, in order to simplify the understanding of the classification process, making this procedure intuitive to humans. Usually, decision rules are of the form IF (*antecedent*) THEN (*consequence*). The left-hand side of the rule is called antecedent; this expression refers to an attribute and its value. The right-hand side of the rule defines the consequence, in our case the label given for the class.

In this work, the MODLEM classifier is used. It is a sequential covering algorithm introduced by Stefanowski in [12] as a modification of the induction rule algorithm developed by Grzymala-Busse [13]. This rule induction algorithm has been chosen because there is no need for data discretization as it is computed simultaneously with the rule induction.

This algorithm generates a minimal set of decision rules for every decision concept (decision class or its rough approximation in case of inconsistent examples). Such a minimal set of rules attempts to cover all positive examples of the given decision concept denoted as B, and not to cover any negative examples $(U \setminus B)$.

The main rule induction scheme is described in the following steps:

1. Create a first rule by choosing sequentially the "best" elementary conditions according to the chosen *"Find best condition function"*.
2. When the rule is stored, all learning positive examples that match this rule are removed from consideration.
3. The process is repeated while some positive examples of the decision concept remain still uncovered.
4. Then, the procedure is sequentially repeated for each set of examples from a concept or category.

After the rules are induced in the training stage, then the testing stage consists of using those rules to predict a class assignment for pixels in an unseen image. This is done evaluating the matching of the new feature description with the condition parts of decision rules. This may result in unique matching to rules from one single class. However, two other ambiguous cases are possible: matching to more rules indicating different classes or the feature description does not match any of the rules. In these cases, it is necessary to apply proper strategies to solve these conflict cases [14].

The rule induction process is illustrated in Fig. 2, completing the whole system. This figure describes the classifier training stage, where the input to the classifier is a joint color-texture vector feature per sample to be classified. In the figure, the rule induction for the "Foliage" class is shown.

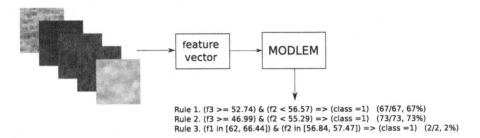

Rule 1. (f3 >= 52.74) & (f2 < 56.57) => (class =1) (67/67, 67%)
Rule 2. (f3 >= 46.99) & (f2 < 55.29) => (class =1) (73/73, 73%)
Rule 3. (f1 in [62, 66.44]) & (f2 in [56.84, 57.47]) => (class =1) (2/2, 2%)

Fig. 2. Sample of rule induction using MODLEM (Color figure online)

To measure our system performance, an evaluation strategy is needed. This evaluation consists of a standard dataset, an evaluation measure and a specific strategy for training and testing stages. This strategy is discussed in the following section.

3   Experimental Results

In this section, we present the results obtained by our method called Labeled Rough-Set-Based Segmentation (LRS). The goal of test series has been to evaluate the performance of our approach in comparison with other state-of-the art algorithms. In these tests, we have applied our approach to natural images using four categories: "Foliage", "Soil", "Sky" and "Water".

For the experiments, the Berkeley segmentation dataset (BSDS300) [15] is used. This dataset consists of 300 natural images, either color or grayscale. As our method utilizes color and texture features, only the color images have been used. Moreover, for each image in the dataset, a set of ground-truth images made by humans is available, and it is used to quantify the reliability of a given method. The dataset also presents a diversity of content, including landscapes and portraits, resulting in a challenge for any segmentation algorithm.

In order to assess the segmentation performance of our method, three widely used metrics are adopted: Probabilistic Rand Index (PRI) [1], Variation of information (VoI) [16] and Global Consistency Error (GCE) [15]. Each metric provides a measure obtained from the comparison of a segmented image and the corresponding ground-truth. We have used the source code provided by Yang *et al.* [17] to evaluate our system performance using PRI, VoI and GCE.

As a general explanation, the PRI counts the number of pixel pairs whose pixels are consistent, for both, the ground-truth and the segmentation result. If the segmented image and the ground-truth images have no matches, then the PRI is zero, giving the minimum value. On the other hand, the maximum value of 1 is achieved if both, the outcome and all the ground-truth images are identical. According to Yang *et al.* [17], the PRI describes a good correlation with the human perception through the hand-labeled segmentations. The VoI metric calculates conditional entropies between distributions of class labels. The GCE evaluates the extent to which a segmentation map can be considered as a refinement of another segmentation. For both, the VoI and the GCE metrics, the segmentation is better if the values are closer to zero.

3.1 Qualitative Evaluation

Three parameters are of special interest in our study: (i) the inspection window to compute the texture features (homogeneity and entropy); (ii) the inspection window for the standard deviation map and (iii) the number of samples used to train the pixel classifier. We have exhaustively searched for these parameters, finding to be the best a 23×23 window for the first parameter and an 11×11 window for the second one. Regarding the number of samples, it has been found that using more than 90 representative samples per class does not improve the segmentation results. It is important to mention that a post- processing stage is performed after the segmentation results are obtained. It consists of a simple median filter of 3×3 size applied to the segmented image. The final goal of this procedure is to eliminate extremely small regions and improve segmentation results.

The segmentation approach proposed in this work has been carried out with the 300 images from the BSDS300. Three examples are given in Fig. 3 where the images in the row (a) correspond to original images taken from the Berkeley dataset. The row (b) shows human-made segmentations taken from the ground-truth images. Notice that the reference images were colored randomly. The row (c) shows the images obtained by our system. Here, each pixel is associated to a class. Green, brown, light blue and dark blue correspond to "Foliage", "Soil", "Sky" and "Water" classes, respectively. Those pixels that the system is unable to classify based on the given features are colored in black.

It is important to point out some details of our results. The outcome obtained in row (c) of Fig. 3 shows a segmentation similar to the ground-truth. Our system is able to classify the pixels within the image based on color and texture features in a correct manner. However, our system cannot distinguish grass from trees based on the features used. In the same manner, the ability of abstraction of

Fig. 3. Example showing qualitative results: (a) Input image, (b) Human-made reference segmentation (not labeled) and (c) Segmented image using our approach. Last row: Color labels for each class in our approach (Color figure online).

a human to say if a house is in front of another as shown in the ground-truth image is a task that is beyond the scope of this work.

The results show that the final goal of this work has been achieved. Those images containing the trained classes are well segmented. The integration of color and texture using rough-set-rules was a successful approach for the image segmentation task.

3.2   Performance Evaluation

In our evaluation, we have compared our approach (LRS) with three state-of-the-art algorithms that use both, color and texture features: the J-image Segmentation (JSEG) method proposed by Deng and Manjunath [18], the Compression-based Texture Merging (CTM) approach introduced by Yang *et al.* [17], and the Clustering-based image Segmentation by Contourlet transform (CSC) presented by An and Pun [19]. The average performance of each method using the three quantitative measures, PRI, VoI and GCE is presented in Table 1.

Under specific circumstances, the results obtained by our method are comparable to those obtained using state-of-the-art methods, as can be seen in Table 1. Considering that these results have been obtained using a reduced number of classes to partition the universe of features, and that the qualitative results shown in Fig. 3 are good, our method is well placed among other approaches.

Table 1. Average performance of the segmentation algorithms.

Method	PRI	VoI	GCE
JSEG (300)	0.774	**2.134**	0.196
$\text{CTM}_{\gamma=0.1}$ (300)	0.756	2.464	**0.176**
CSC (100)	**0.796**	2.732	0.225
LRS (300)	0.602	2.529	0.254

Besides, for many real-world applications, it can be more important to have labeled clusters of pixels than achieving a higher accuracy in segmentation results.

4   Conclusions

The segmentation based on rough-set-rules has resulted to be a promissory tool for the integration of color and texture features. Our method has proven to be an efficient and robust approach for feature integration. Even in those cases that our quantitative results are below the state-of-the-art approaches, our system provides an important feature over other methodologies: for each pixel within the image, there is a related label that corresponds to the pixel categorization. This means that our results not only provide the segmented image, but each segment of the image represents a class, which is intuitive to humans. Having a label for each segment provides basic information to make further assertions about the image, for example, if the segment with label X is besides, above, below or under another segment with label Y. We believe that our approach is a first step toward image understanding.

Acknowledgments. Jonathan Cepeda-Negrete thanks the Mexican National Council on Science and Technology (CONACyT), scholarship 290747 (Grant No. 388681/254884) and to the University of Guanajuato (PIFI-2015 program) for the financial support provided.

References

1. Unnikrishnan, R., Pantofaru, C., Hebert, M.: Toward objective evaluation of image segmentation algorithms. IEEE Trans. Pattern Anal. Mach. Intell. **29**(6), 929–944 (2007)
2. Ilea, D.E., Whelan, P.F.: Image segmentation based on the integration of colour-texture descriptors - a review. Pattern Recognit. **44**(10), 2479–2501 (2011)
3. Hossein, M., Shankar, R.R., Allen, Y.Y., Shankar, S.S., Yi, M.: Segmentation of natural images by texture and boundary compression. Int. J. Comput. Vis. **95**, 86–98 (2011)
4. Pawlak, Z.: Rough sets. Int. J. Comput. Inf. Sci. **11**(5), 341–356 (1982)

5. Buchsbaum, G.: A spatial processor model for object colour perception. J. Frankl. Inst. **310**(1), 1–26 (1980)
6. Land, E.H., McCann, J.: Lightness and retinex theory. J. Opt. Soc. Am. **61**(1), 1–11 (1971)
7. Schanda, J.: Colorimetry: Understanding the CIE System. Wiley, Hoboken (2007)
8. Tamura, H., Mori, S., Yamawaki, T.: Textural features corresponding to visual perception. IEEE Trans. Syst. Man Cybern. **8**(6), 460–473 (1978)
9. Unser, M.: Sum and difference histograms for texture classification. IEEE Trans. Pattern Anal. Mach. Intell. **8**(1), 118–125 (1986). (PAMI)
10. Stefanowski, J.: The rough set based rule induction technique for classification problems. In: Proceedings of 6th European Conference on Intelligent Techniques and Soft Computing EUFIT, vol. 98 (1998)
11. Stefanowski, J.: The bagging and n^2-classifiers based on rules induced by MOD-LEM. In: 4th International Conference of Rough Sets and Current Trends in Computing, vol. 3066, pp. 488–497 (2004)
12. Stefanowski, J.: On rough set based approaches to induction of decision rules. Rough Sets Knowl. Disc. **1**(1), 500–529 (1998)
13. Grzymala-Busse, J.W.: LERS - a system for learning from examples based on rough sets. Intelligent Decision Support: Handbook of Applications and Advances of the Rough Sets Theory, vol. 11, pp. 3–18. Springer, The Netherlands (1992)
14. Stefanowski, J.: On combined classifiers, rule induction and rough sets. In: Peters, J.F., Skowron, A., Düntsch, I., Grzymała-Busse, J.W., Orłowska, E., Polkowski, L. (eds.) Transactions on Rough Sets VI. LNCS, vol. 4374, pp. 329–350. Springer, Heidelberg (2007)
15. Martin, D., Fowlkes, C., Tal, D., Malik, J.: A database of human segmented natural images and its application to evaluating segmentation algorithms and measuring ecological statistics. In: Proceedings of 8th IEEE International Conference on Computer Vision, vol. 2, pp. 416–423 (2001)
16. Meilă, M.: Comparing clusterings-an information based distance. J. Multivar. Anal. **98**(5), 873–895 (2007)
17. Yang, A.Y., Wright, J., Ma, Y., Sastry, S.S.: Unsupervised segmentation of natural images via lossy data compression. Comput. Vis. Image Underst. **110**(2), 212–225 (2008)
18. Deng, Y., Manjunath, B.S.: Unsupervised segmentation of color-texture regions in images and video. IEEE Trans. Pattern Anal. Mach. Intell. **23**(8), 800–810 (2001)
19. An, N.Y., Pun, C.M.: Color image segmentation using adaptive color quantization and multiresolution texture characterization. SIViP **8**(5), 943–954 (2014)

Dynamic Object Detection and Representation for Mobile Robot Application

Jose J. Lopez-Perez, Victor Ayala-Ramirez$^{(\boxtimes)}$,
and Uriel H. Hernandez-Belmonte

División de Ingenierías, Campus Irapuato-Salamanca DICIS,
Universidad de Guanajuato DICIS, Carr. Salamanca-Valle Km. 3.5+1.8,
Palo Blanco, 36700 Salamanca, Mexico
{jjlp,hailehb}@laviria.org, ayalav@ugto.mx

Abstract. This paper presents how to classify the sensorial information
to set the label of the cells in the occupation grid map of an unknown
environment where several robots are wandering and how to use these
labels to complete a path execution task. Reactive navigation of a mobile
robot needs to identify the objects that encounters in its way towards a
specified goal. The robot does not know the environment but it knows the
position of its goal. The robot needs to build a map during its navigation
to the goal. A pattern classification task is needed to update the state
of the map grid cells as more information is acquired by the sensors.
We propose to use a finite state machine approach to set the correct
label according to the objects in the environment. The occupancy grid
map is crucial to complete the navigation task safely. We have tested our
approach in several scenarios with a varying number of mobile obstacles.
The results show that our method works efficiently in a set of benchmark
scenarios of varying degree of complexity.

1 Introduction

Autonomous navigation in unknown and dynamic environments is classified as
one of the more challenging tasks in the mobile robotic community [4]. This
capability is needed for the robot to navigate safely in environments populated
either by other robots or people. In such an environment, the other robots or
people behave in ways difficult to model and the autonomous robot has to react
in order to guarantee the safety of all the entities sharing the same environment. Another scenario where safe navigation is required is the development of
autonomous vehicles. Safety also needs to be guaranteed to let the autonomous
vehicles to complete the navigation tasks without colliding among them.

We propose to use an occupancy grid map representation for the unknown
environment. The occupancy grid is a well known approach in mobile robotics
that has shown to be robust even if it shows also several disadvantages [1]. Some
recent works use occupancy grids in the safe navigation context. For example,
Li and Ruichek use it to map urban environments [5]; Wang *et al.* use it to track

© Springer International Publishing Switzerland 2016
J.F. Martínez-Trinidad et al. (Eds.): MCPR 2016, LNCS 9703, pp. 84–93, 2016.
DOI: 10.1007/978-3-319-39393-3_9

dynamical objects [9] and Xin *et al.* use it for representing the dynamic environment of an autonomous vehicle [10]. In this paper, we also use an occupancy grid in a reactive safe navigation task.

We use the information provided by a laser range finder (LRF) to classify the cells in the occupancy grid and a finite state machine (FSM) to determine if obstacles are moving or steady.

Rest of this paper is organized as follows: in Sect. 2, we present the main details of our approach. The implementation details, the tests and the results are presented in Sect. 3. Finally, our conclusions are presented in Sect. 4.

2   Methodology

2.1   Path Execution Task

The robot considered in this paper has to execute a safe navigation task. The current position $S = (x_s, y_s)$ and the goal position coordinates $G = (x_g, y_g)$ are known. The environment dimensions $W \times H$ are also known.

The mobile robot is provided with proprioceptive and exteroceptive sensors. The proprioceptive sensors are used to compute the current position of the robot. The exteroceptive sensors let the robot to acquire information about its surrounding area. In this navigation task, the robot uses a laser range finder (LRF) to get the measures from the obstacles that are near to it.

A LRF sensor operates by sending a laser beam in a given number L of orientations from the current position of the mobile robot. The time of flight lectures that the LRF records are used to compute how far the obstacles are. However, these measurements give no information about the nature of the obstacle.

The robot uses an occupancy grid to represent the information acquired from the environment during the execution of the path towards its goal. The occupancy grids were originally proposed by Elfes in [1] and they are widely used for mobile robot path planning tasks [10]. This approach discretizes the environment into a grid of cells of some dimensions. Figure 1(a), shows how an environment is discretized into an occupancy grid and the implicit inaccuracies in the scenario representation. Essentially, each cell can be labelled as a free cell or as an obstacle cell (see Fig. 1(b)). If the dimensions are too small, the representation of a small map can take a lot of memory space and can not be useful for the planning task. In our case, we discretize the environment as a grid of cells of dimensions $0.5\,m \times 0.5\,m$. In our work, we use the occupancy grid for two objectives: (i) to have a partial representation of the explored part of the scenario and (ii) to ease the planning algorithm when a re-planning is needed.

Using the occupancy grid approach, the mobile robot computes a path from the start position S to the goal position G by using a $A^\star$ planning algorithm [7,8]. Initially, the robot does not know any information about the environment, so it considers that all the map is free space.

2.2   Occupancy Grid Update

As the robot advances in the execution of its planned path, it needs to update the current map representation M according to the planned path P and to the obstacle information obtained from the LRF sensor L. After processing the current lectures of the LRF sensors, an updated map M' and a re-planning flag f_{rp} are computed. This enables the robot to handle the mobile and steady obstacles present in the environment.

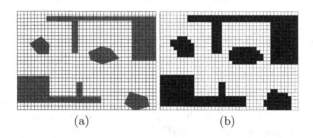

| (a) | (b) |

Fig. 1. (a) Scenario representation using occupancy grids. (b) Cells are labeled as *obstacle cells* (in black color) and as *free cells* (in white color) (Color figure online)

The cells in the maps M and M' can be labeled using four possible values: *non explored cell, explored cell, steady obstacle cell,* and *mobile obstacle cell.* The labeling procedure depends on the computation of two functions:

f1 that serves to determine which cells are occupied by an obstacle, either steady or mobile obstacle. This function uses the collision points of each of the laser measures $L = \{l_1, l_2, \ldots, l_L\}$ with L being the maximum number of lectures taken by the LRF.

f2 that serves to update the label of all the cells inside the polygon formed by the laser measures to the *explored cell* status.

For the function f_1, we obtain the collision points $l_j(x_j, y_j)$ for each of the measurements of the LRF. The distance from these points $l_j(x_j, y_j)$ to each of the center positions $c_i(x_i, y_i)$ of the map grid are computed. If the distance $d(lj, c_i)$ is smaller than the safety radius of the mobile robot r, the function return 1, otherwise it returns 0 (See Fig. 2).

$$d(l_j, c_i) = \sqrt{(x_j - x_i)^2 + (y_j - y_i)^2} = \|l_j - c_i\| \tag{1}$$

$$f_1 = \begin{cases} 1 & \text{si } d(l_j, c_i) \leq r \\ 0 & \text{si } d(l_j, c_i) > r \end{cases} \tag{2}$$

For the function f_2, we generate a polygon using the laser collision points $l_j(x_j, y_j)$ as vertices. If a cell is inside the polygon, f_2 takes the value of 1,

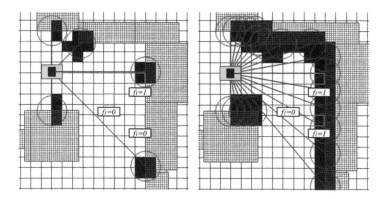

Fig. 2. Computation of f_1 function. *Steady obstacle cells* in black, *non explored cells* in white, measurement of the LRF in blue (Color figure online).

otherwise it returns a value of 0. The computation to know if the point $c_i(x_i, y_i)$ is in inside the polygon is made by tracing a vertical line and counting the number of intersections with the laser measurement polygon. If there is an odd number of intersections, the function f_2 returns 1, otherwise it returns 0 (See Fig. 3).

The transition between the states that a cell can have is described via the FSM diagram in Fig. 4. Note that the combination $(f_1, f_2) = (1, 1)$ is not represented there. That is because it can not occur in any state. When f_1 is 1, f_2 takes automatically the value of zero. All the cells are initially labeled as *non explored cells*.

If a cell changes from the state *non explored* to the state steady obstacle or from the state *explored* to the state *mobile obstacle*; and if this cell is also part of the planned path P, the f_{rp} flag is enabled. If this flag is enabled, the robot must do a re-planning using the partially known environment.

3 Results

3.1 Implementation Details

The proposed system was developed in a modular form to be able to implement it in frameworks like ROS (*Robot Operating System*) or ARIA (*Advanced Robot Interface for Applications*). These are the frameworks used by the robots available at the Laboratory for Vision, Robotics and Artificial Intelligence (LaViRIA) of the University of Guanajuato: a Husky A200 mobile robot named CompaBot (in Fig. 5(a)) and a Pioneer 3-AT named XidooBot (in Fig. 5(b)).

3.2 Test Protocol

The tests were performed using a benchmark set of 34 scenarios shown in Fig. 6. They are available from the *motion planning repository* [2]. These scenarios are

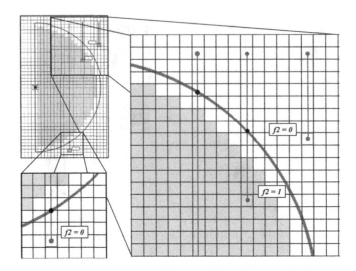

Fig. 3. Computation of the f_2 function. *Explored cells* in yellow, *non explored cells* in white, LRF measurement polygon in blue (Color figure online).

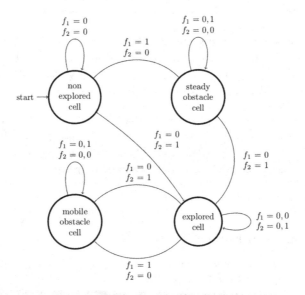

Fig. 4. State transition diagram for each cell of the occupancy grid.

very similar to those used for real world applications. These set include some difficult problem like robot traps, narrow corridors, zones with a large density of obstacles, labyrinths, rooms, etc.

For each of the test scenarios, 3 different navigation tasks were generated, i.e., a different start and goal position of the robot were chosen. For each navigation

Fig. 5. (a) CompaBot, a Husky A200 mobile robot from Clearpath Robotics Inc. (b) XidooBot, a Pioneer 3-AT mobile robot designed by Adept MobileRobot.

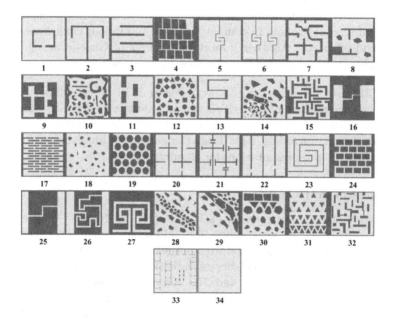

Fig. 6. Benchmark set of test scenarios from the repository of the Technical University of Prague [2].

task, both the start and goal position are in the free space of the environment under use. This procedure was repeated 30 times for each configuration and navigation task with a number of existing mobile obstacles on stage that was increased from 0 to 10. For the mobile obstacles, we have used several instances of the Pioneer 3-AT mobile robot that wander around the free space, simulating the human-robot interaction in a scenario.

3.3   Gazebo and RViz Simulation Using ROS

In order to implement the proposed system for safe navigation of mobile robots in dynamic unknown environments, we implement it on the ROS testbed. This enable us to implement it in a Husky A200 platform given that this platform is supported by ROS. The main simulation tools for ROS are Gazebo and RViz. In the following paragraphs, we describe the main features of these tools:

ROS is a development platform for writing robot software. It is composed of a collection of tools, libraries and standard messaging protocols that simplify the robot software development task for different robotic hardware [6].

Gazebo is a robot simulator that is useful for rapid prototyping and test of robot algorithms. It can be used to design robots and to perform the analysis of the simulation data in realistic environments. It also offers the capability of simulation of teams of robots in complex outdoor and indoor environments with accuracy and efficiency. Another important features of Gazebo are the robust physics engine and its high quality graphics rendering [3].

RViz is a 3D data visualization tool available in ROS. It can show the sensor data and the current state of ROS. RViz can be used to visually show the current state of the robot in a virtual model. It can also be used to display real-time sensor information, including data from the cameras, the sonar, the infrared devices and the map representation, for example [6].

3.4   Quantitative Results

Figure 7 shows the success rate for each test scenario. In most of the scenarios under test the success rate is over 90 %. However, we can observe that the scenarios with narrow corridors and with only one route to access the goal position, the system exhibits a low success rate (Fig. 6 scenario 15 - *maze*, 23 - *square_spiral*, 26 - *tunnel_twisted* and 27 - *tunnel*). The factors that cause the mobile robot failing to reach the goal position is the continuous obstruction of the mobile obstacles in its route or the presence of an error in the detection and classification of the mobile obstacles.

3.5   Qualitative Results

Here we present the qualitative results of applying the proposed method in a Husky A200 mobile robot in the Gazebo simulator. The robot model uses a LRF with 640 measurements and a vision field of 180° in front of the robot. We show the results in the *back_and_forth* test scenario (Fig. 6 scenario 3). The aforementioned scenario was built in Gazebo using the same size specifications than in the benchmark set. Figure 8 depicts the component modules of the ROS implementation. The proposed navigator, labeled as *dpn* (dynamic planner navigator), publishes the map that is building (*/compabot/map*) and the robot velocities (v y ω) to the simulator (*/compabot/husky/cmd_vel*). It also publishes the

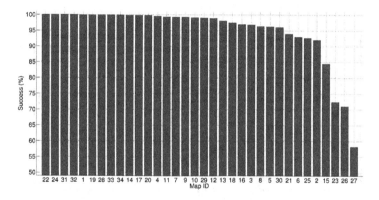

Fig. 7. Result of the success rate (%) vs map. In most scenarios test the success rate is over 90 %.

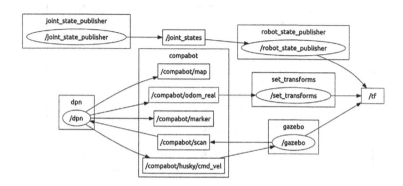

Fig. 8. Connection diagram for the ROS implementation of the proposed system.

current position of the mobile robot (*/compabot/odom_real*) and the information about its planned path (*/compabot/marker*), and it gets the laser scan data (*/compabot/scan*) from the simulator.

Some aspects of a mobile robot simulation are shown in Fig. 9. In the left column, we present the Gazebo display output and in the right column we present the RViz display output. The Gazebo output serves to render the current state of the virtual world. In our case, it renders the Husky A200 model and the models of the P3-AT robots that are wandering in the environment. In the RViz output, we can observe the model of the Husky A200 mobile robot and the executed path (in yellow), the planned path (in red),the map created by the mobile robot where the white cells are explored cells. The steady obstacles are represented using black cells and the moving obstacles are colored using a gray color, We can observe how the map is evolving to reflect the current knowledge about the environment.

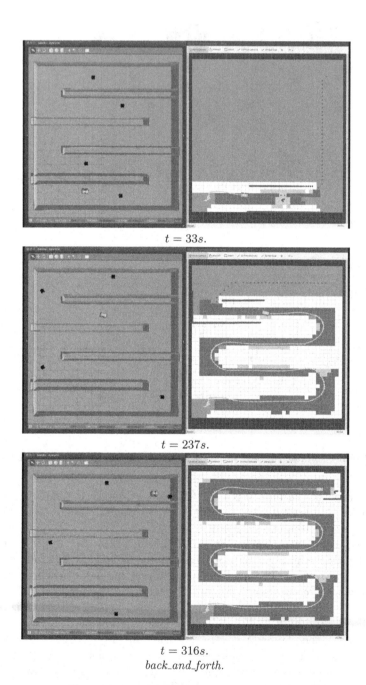

$t = 33s.$

$t = 237s.$

$t = 316s.$
back_and_forth.

Fig. 9. Simulation results for the proposed navigator in Gazebo (in the left) and RViz (in the right). The timestamp t of each frame is shown below the snapshot.

4   Conclusions

In this work, we have proposed a cell labeling method for building the map of a dynamic unknown scenario. The method is based in a Finite State Machine approach. We have presented the rules that govern the transition for the cell labels of the occupancy grid map as the robot acquires information from its sensors. The method was tested using scenarios of different degrees of complexity and it has shown to be efficient and accurate. For the tests, we have used the ROS testbed. Future work will include to extend the tests to real robots. The use of this approach as a part of a simultaneous localization and mapping (SLAM) method will also be performed.

Acknowledgments. Jose J. Lopez-Perez and Uriel H. Hernandez-Belmonte would like to acknowledge CONACYT for the financial support through the educational scholarships with numbers 291047/736591 and 229784/329356 respectively.

References

1. Elfes, A.: Using occupancy grids for mobile robot perception and navigation. Computer **22**(6), 46–57 (1989)
2. Intelligent and Mobile Robotics Group: Motion planning maps. http://imr.ciirc.cvut.cz/planning/maps.xml. Accessed 29 Sep 2015
3. Koenig, N., Howard, A.: Design and use paradigms for gazebo, an open-source multi-robot simulator. In: Proceedings of the 2004 IEEE/RSJ International Conference on Intelligent Robots and Systems, vol. 3, pp. 2149–2154. IEEE (2004)
4. Laugier, C., Chatila, R.: Autonomous Navigation in Dynamic Environments. Springer Tracts in Advanced Robotics. Springer, Heidelberg (2007). Chap. Preface
5. Li, Y., Ruichek, Y.: Occupancy grid mapping in urban environments from a moving on-board stereo-vision system. Sensors **14**, 10454–10478 (2014)
6. Martinez, A., Fernández, E.: Learning ROS for Robotics Programming. Packt Publishing Ltd, Birmingham (2013)
7. Murphy, R.: Introduction to AI Robotics. MIT Press, Cambridge (2000)
8. Siegwart, R., Nourbakhsh, I.R., Scaramuzza, D.: Introduction to Autonomous Mobile Robots. MIT Press, Cambridge (2011)
9. Wang, D.Z., Posner, I., Newman, P.: Model-free detection and tracking of dynamic objects with 2D lidar. Int. J. Robot. Res. **34**(7), 1039–1063 (2015)
10. Xin, Y., Liang, H., Mei, T., Huang, R., Du, M., Sun, C., Wang, Z., Jiang, R.: A new occupancy grid of the dynamic environment for autonomous vehicles. In: Proceedings of the 2014 Intelligent Vehicles Symposium, pp. 787–792. IEEE (2014)

Saliency Detection Based on Heuristic Rules

Diana E. Martinez-Rodriguez, Victor Ayala-Ramirez$^{(\boxtimes)}$,
and Uriel H. Hernandez-Belmonte

División de Ingenierías, Campus Irapuato- Salamanca DICIS,
Universidad de Guanajuato, Carr. Salamanca-Valle Km. 3.5+1.8, Palo Blanco,
36700 Salamanca, Mexico
{demr,hailehb}@laviria.org, ayalav@ugto.mx

Abstract. Detecting salient regions in images aims at finding regions which contains relevant information, where a more detailed process can be applied. Saliency detection is useful in many computer vision tasks such as object segmentation, object detection, image retrieval, place recognition, among others. In this paper, we propose a method based on heuristic rules that uses color and spatial features which allows to get a good approximation to the salient region in a very low time compared with other methods in the state of the art. The tests were performed over the images of a benchmark dataset.

1 Introduction

Humans can easily identify the relevant objects in a scene because their inherent mechanism of visual attention. Human visual system models suggest that humans selectively process perceived information instead of taking all in mind [18]. The visual attention model attributed to Neisser states that there are two stages in the visual saliency task: a pre-attentive stage and an attentive stage. In the first stage, the features are detected and in the second stage, the visual system finds relationships between them.

Visual saliency computation methods try to emulate the human visual attention methods. The first computational approach for saliency detection was proposed by Itti *et al.* [8]. The main contribution of this work is the proposal of the saliency map. This map is an image-like representation where the intensity is proportional to the relevance of the corresponding pixels.

Saliency detection has been widely used in computer applications. It has been used in object segmentation [11], object recognition [17] adaptive image compression [4] and place recognition [19]. In particular, saliency detection is useful because it reduces the computational cost of these tasks by focusing the process in reduced regions instead of processing the entire image.

Nothdurft [16] and Ma and Zhang [15], coincide in the statement that there is not a feature *per se* that captures human attention. For example, the saliency of an object does not depends on a particular color but on its contrast with respect to its neighborhood.

© Springer International Publishing Switzerland 2016
J.F. Martínez-Trinidad et al. (Eds.): MCPR 2016, LNCS 9703, pp. 94–103, 2016.
DOI: 10.1007/978-3-319-39393-3_10

In recent works, the color contrast has been received too much attention. Luo *et al.* [14] consider the global color contrast as the key property for saliency and they also assign more importance to the center of the image. Ma and Zhang [15] consider local contrast in the CIELuv space, it computes the pixel saliency by using a local window. Liu and Gleicher [12] create a Gaussian pyramid of the image in order to be invariant to scale, the distance is computed by using a L_2 norm in CIELuv color space.

Saliency detection methods can be classified as biologically based [8], purely computational based [5,14], or a combination of both [3]. We propose a pure computational method that uses color and spatial features. The proposed approach reduces the computational cost by using a grid representation of the image. Each cell is composed of a rectangular region of a given size. This operation does not significantly reduce the quality of the saliency detection because the humans are attracted by objects and not by individual pixels. In a second step, cells that are connected and that exhibit similar color properties are grouped into non regular regions. Heuristic rules are applied to determine which regions are the seeds of the background and foreground regions. Another set of heuristic rules is used to assign the rest of the regions either into the salient foreground or the background. The final foreground is a meaningful region, in the sense proposed by [17]. The main advantage of our method is that obtains comparable results to the methods in the state of the art but in considerably less time.

We consider to find salient regions as a binary classification labeling: salient or not salient. This is to avoid the problem of thresholding the image, something that is necessary for most of the applications, as in the object segmentation problem. We have found that the thresholding method affects significantly the results.

The content in this paper is organized as follows: In Sect. 2 the proposed approach is described. In Sect. 3 we present the tests and the results. A comparison with previous approaches is also presented. Finally conclusions and perspectives are presented in Sect. 4.

2 Heuristic Rules Based Saliency Detection

In this section a Heuristic Rules based Saliency Detection method is presented (HRSD for short). The proposed approach looks for a meaningful region that satisfies a set of constraints both in color contrast and in spatial arrangement to consider it as a salient region. The objective of the method is to partition the image into two regions: the foreground region (R_F) and the background region (R_B). Each of these regions are constructed through a number of steps described below.

1. An input image is partitioned into a grid of rectangular regions, each of them containing $m \times n$ pixels. Each of these regions will be assigned either to R_F or to R_B. Let us name the cells of the grid as C_i, $i \in \{1, 2, \ldots, u \times v\}$ with u and v being respectively the number of rows and columns resulting from the grid partition of the input image.

2. Each cell is characterized by the mean value $\bar{C}_i$ of the color coordinates of all the $m \times n$ pixels belonging to it. We use the YUV color space to represent pixel color in this work. That is:

$$\bar{C}_i = [\bar{C}_{iY} \ \bar{C}_{iU} \ \bar{C}_{iV}] \tag{1}$$

Each of the components of $\bar{C}_i$ is computed as follows:

$$\bar{C}_{iY} = \frac{1}{m \times n} \sum_{C_i} c_{iY}(i, j) \tag{2}$$

where $C_{iY}(i, j)$ is the color coordinate of the pixels in the region R_i. Similarly for U and V coordinates, we have:

$$\bar{C}_{iU} = \frac{1}{m \times n} \sum_{C_i} c_{iU}(i, j) \tag{3}$$

$$\bar{C}_{iV} = \frac{1}{m \times n} \sum_{C_i} c_{iV}(i, j) \tag{4}$$

3. Cells with similar color features are then grouped using a connected component labeling-like procedure explained later in this paper. As a result, we obtain a list of component regions R_i, $i \in \{1, 2, \ldots, r\}$, with r the total number of connected components found in the image. Each of these regions groups cells that are similar among them with respect to its mean color $\bar{C}_i$ and that are spatially connected.

4. The cells of each region R_i that are located in the boundary cells of the image are counted and the sum is recorded as B_i.

5. The initial selection of R_B and R_F is done by choosing the pair of more contrastive regions in $R = \{R_1, R_2, \ldots, R_r\}$. For doing this, we compute a table of distances D where the element d_{ij} is the distance between the color mean values of the regions R_i for each pair of regions R_i and R_j, $i \neq j$. The color distance is computed using an Euclidean distance in the YUV color space, weighted by a factor depending on the size of both regions as shown in Eq. 5. Where the variables $size_i$ and $size_j$ are the sizes of the regions i and j respectively. The variable $size$ is the number of pixels of the whole image.

$$dist_{ij} = d_{ij} \left(\frac{size_i + size_j}{size} \right) \tag{5}$$

$$dcolor = \sqrt{(Y_1 - Y_2)^2 + (U_1 - U_2)^2 + (V_1 - V_2)^2}$$

For the more contrastive pair, R_B is chosen as the region covering the larger number of cells.

We have chosen to use the YUV color space because it has shown better performance when compared to CIELab, CIELuv and HSI color spaces for image representation in our experiments. The parameters were tuned experimentally to optimize the performance evaluation measure of the system.

There exist some heuristic rules that R_F and R_B should satisfy. If that is not the case, the next distance in the rank of region color distances is used to choose the R_F and R_B and the verification of the heuristic rules is repeated.

The heuristic rules are as follows:
(a) The salient object is in the center of the image. The R_F (foreground) is limited to have at most 5 cells in the boundary of the image. This is to avoid selecting a region of the background as the foreground.
(b) The size of the representative objects must be above a 3 cell area threshold. The R_F and R_B selected have to be initially composed by at least 3 cells. This is to try to avoid choosing an artifact of the image as a salient region.
6. In the following step, the rest of the regions are grouped either to R_F or to R_B. This procedure is guided by another set of heuristic rules that includes spatial relationships. The rules to determine if a region R_i is salient are the following:

(a) $dist_{R_i R_F} < dist_{R_i R_B}$,
(b) R_i does not contain cells in the contour of the image.

In the Fig. 1, a graphical block diagram of our method is presented.

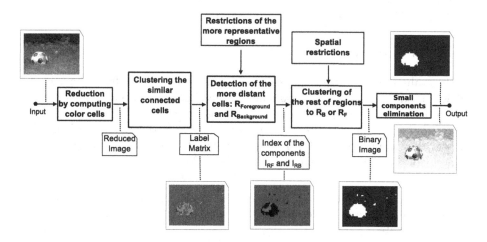

Fig. 1. Heuristic Rules based Saliency Detection Method

2.1 Color Connected Component Labeling (CCL) Procedure

As it was mentioned before, we generate regions from cells of a grid. The CCL task is performed by extending the work by Hernadez-Belmonte *et al.*

Fig. 2. Reduced Connectivity Mask [7].

[7]. The key concept of this work is the use of a Reduced Connectivity Mask (RCM) and the use of a lookup table that determines if the regions need to be connected as a component or not. We consider a neighborhood of these cells in the grid, as it is shown in the Fig. 2. Let as assume that d is the cell under analysis. The scanning of the image using the RCM, is performed in a left to right, top to bottom sequence.

- If the cells (b, c) are similar, join their labels.
- If cell (d) is similar to one of the neighbor labels (a, b, c). In the Table 1 the operations in each case are presented.
- If d is not similar to the other cells, create a new label.

The criteria for defining the similarity of two cells is the Euclidean distance in YUV coordinates. YUV coordinates include one luminance (brightness) and two chrominance (color) components [10]. If the distance between two colors is lower than a threshold we consider that colors are similar. We selected color because is a very important feature, but another features (e.g. texture-related features) could also be used by using an appropriate threshold.

3   Tests and Results

For the evaluation of our method we used the standard images from the MRSA, a widely used image dataset for saliency evaluation. We compare our approach with other state of the art methods using the standard F metrics. We also compare our system in execution time with the best approach in or knowledge [5]. Our system was implemented in Language C and the tests were executed using an Intel Core i7-4700MQ machine with 8 GB RAM.

Table 1. Operations which have to be computed in each case for the color CCL procedure. (1) The two grids are similar, (0) The two grids are not similar (-) It is not necessary to verify.

d, b	d, a	d, c	Operation
1	-	-	Join d to b.
0	1	-	Join d to a.
0	0	1	Join d to c.
0	0	0	Create a new label

3.1   Test Protocol

In order to evaluate our method, we use the MSRA image dataset images. The images in this dataset present a variety of situations: there are images for indoor scenes and outdoor scenes; there are also images including natural and artificial objects. The salient regions in these images represent humans, animals, plants, and objects. Achanta *et al.* provide the ground truth for a subset of 1000 images of the original dataset MSRA [1]. We use same subset of 1000 images to compare the proposed method to the other methods in the same conditions. The metrics used to evaluate the results are the well known *precision* and *recall* metrics combined into the *F-measure* presented in the Eq. 6. The *precision* and *recall* metrics are computed using the Eqs. 7 and 8. Where B are the salient pixels detected by the HRSD method method and G are the salient pixels in the ground truth; x and y are the coordinates of the pixel under analysis. We use $\beta = 0.3$ to weight precision more than recall. That is more convenient in objects segmentation, and that is used by most of the automatic saliency methods

$$F = \frac{(1+\beta)(P \cdot R)}{\beta(P) + R} \tag{6}$$

$$P = \frac{\sum_{(x,y)} B(x,y)G(x,y)}{\sum_{(x,y)} B(x,y)} \tag{7}$$

$$R = \frac{\sum_{(x,y)} B(x,y)G(x,y)}{\sum_{(x,y)} G(x,y)} \tag{8}$$

3.2   Results

The best results of the Heuristic Rules based Saliency Detection (HRSD) method were obtained by using the YUV color space for image representation, and a cell size of 8×8 pixels.

In the Fig. 3, we present some qualitative results obtained with the proposed method. In the Fig. 4 a histogram of the F-measure results for the entire dataset is presented, we can see that more of 400 images obtain a very high F-measure value between 0.9 and 1.0.

3.3   Comparisons

At first, we present the results of the F-measure of some methods in the state of the art. This is in order to establish how well perform our system against the other methods. The results presented are the reported results in the works [9,14] with the codes provided by original authors. Finally, we show the comparison with [5]. We use the last version of the implementation provided by the authors. This is in order to compare the executing time with our system. The both methods were tested in the same machine.

Most of the saliency methods obtain a continuous saliency values. For comparison purpose we need a binary image. To binarize the image we need to choose

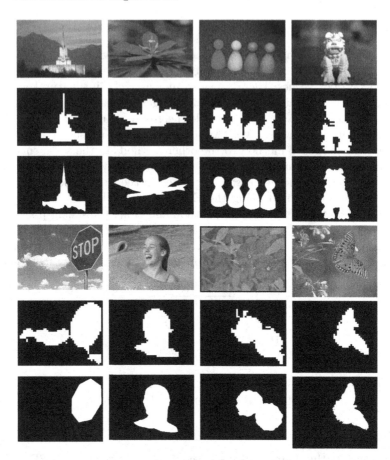

Fig. 3. Qualitative examples of the output of the HRSD method compared to the ground truth. In the first and fourth rows the original images are shown; in the second and fifth rows the HRSD output is presented; in the third and sixth rows the available ground truth is shown.

a threshold. For this reason, the results reported in works by Luo et al. [14] and Kannan et al. [9] may vary from the results reported by the original authors when a non optimal threshold is used.

In the Fig. 5(a), we present the results reported by Luo [14] including the methods labeled CA [6], KD [13], RC [5], OS [21], using Otsu as the method to choose the threshold. In the Fig. 5(b), we present the results reported by Kannan [9] for the methods labeled CA [6], FT [1], RC [5], UL [20], HC [5] using Eq. 9 to compute the threshold. At the end of both graphs, we have added a column with our results. In the comparisons presented, they report different results for RC and CA, this is caused by the use of a different method to define the threshold.

$$Th = 2 \sum S_{(x,y)} \tag{9}$$

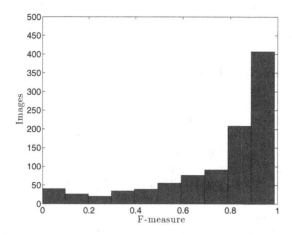

Fig. 4. Histogram of the F-measure of the 1000 images resulting from the application of the HRSD method.

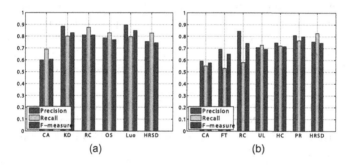

Fig. 5. (a) Results comparison by Luo [14]. (b) Results comparisons by Kannan [9]. We add to both graphics the average results of HRSD method (Color figure online)

In Table 2 we present the results obtained by our approach when compared with the RC method, using the code available from https://github.com/MingMingCheng/CmCode.git. We compare our results with the RC method [5] because it is, in our best knowledge, the faster method in saliency detection with relatively good results. The time of most of the other methods which performs in several seconds. As we can see, the F-measure value is practically the same,

Table 2. Comparison results in F-measure and Time.

Method	F-measure	Time
RC	0.77	41.2 ms
HRSD	0.74	20.8 ms

but the time needed by the proposed approach to perform the computation is about the half of the time spent by the RC method.

4   Conclusions and Perspectives

In this paper, we presented a method to find the salient regions in images. This method considers in addition to the use of color contrast among regions, the use of spatial information to determine the salient regions.

We process the image by using a grid of regularly spaced cells in both horizontal and vertical directions and we group similar image cells by using a color connected component labeling algorithm in the YUV space. The regions formed in such way are then classified into foreground or background regions according to a heuristic set of rules.

The results obtained by our approach are comparable to the results obtained by the state of the art method proposed by Cheng. However, the method computes the saliency output in half the time in the average than that method.

Future work will be directed towards implementing a computational intelligence algorithm for the automatic setting of the parameters of the approach. This could improve the efficiency of heuristic rules for the correct foreground association of the grouped regions. Another line of research will be to make the heuristic rules adaptive to take advantage of specific features of each image.

Acknowledgments. Diana E. Martinez-Rodriguez and Uriel H. Hernandez-Belmonte would like to acknowledge CONACYT for the financial support through the educational scholarships with numbers 291047/736576 and 229784/329356 respectively.

References

1. Achanta, R., Hemami, S., Estrada, F., Susstrunk, S.: Frequency-tuned salient region detection. In: IEEE Conference on Computer Vision andPattern Recognition (CVPR), pp. 1597–1604 (2009)
2. Borji, A., Sihite, D.N., Itti, L.: Salient object detection: a benchmark. In: Fitzgibbon, A., Lazebnik, S., Perona, P., Sato, Y., Schmid, C. (eds.) ECCV 2012, Part II. LNCS, vol. 7573, pp. 414–429. Springer, Heidelberg (2012)
3. Bruce, N., Tsotsos, J.: Attention based on information maximization. J. Vis. **7**(9), 950 (2007)
4. Chen, T., Cheng, M.M., Tan, P., Shamir, A., Hu, S.M.: Sketch2photo: internet image montage. ACM Trans. Graph. **28**(5), 124 (2009)
5. Cheng, M., Mitra, N.J., Huang, X., Torr, P.H., Hu, S.: Global contrast based salient region detection. IEEE Trans. Pattern Anal. Mach. Intell. (PAMI) **37**(3), 569–582 (2015)
6. Goferman, S., Zelnik-Manor, L., Tal, A.: Context-aware saliency detection. IEEE Trans. Pattern Anal. Mach. Intell. (PAMI) **34**(10), 1915–1926 (2012)
7. Hernandez-Belmonte, U.H., Ayala-Ramirez, V., Sanchez-Yanez, R.E.: Enhancing CCL algorithms by using a reduced connectivity mask. In: Carrasco-Ochoa, J.A., Martínez-Trinidad, J.F., Rodríguez, J.S., di Baja, G.S. (eds.) MCPR 2012. LNCS, vol. 7914, pp. 195–203. Springer, Heidelberg (2013)

8. Itti, L., Koch, C., Niebur, E.: A model of saliency-based visual attention for rapid scene analysis. IEEE Trans. Pattern Anal. Mach. Intell. **11**, 1254–1259 (1998)

9. Kannan, R., Ghinea, G., Swaminathan, S.: Salient region detection using patch level and region level image abstractions. IEEE Signal Process. Lett. **22**(6), 686–690 (2015)

10. Kekre, H.B., Thepade, S.D., Athawale, A., Parkar, A.: Using assorted color spaces and pixel window sizes for colorization of grayscale images. In: International Conference and Workshop on Emerging Trends in Technology, pp. 481–486 (2010)

11. Ko, B.C., Nam, J.Y.: Object-of-interest image segmentation based on human attention and semantic region clustering. J. Opt. Soc. Am. A **23**(10), 2462–2470 (2006)

12. Liu, F., Gleicher, M.: Region enhanced scale-invariant saliency detection. In: IEEE International Conference on Multimedia and Expo (ICME), pp. 1477–1480 (2006)

13. Liu, Z., Xue, Y., Shen, L., Zhang, Z.: Nonparametric saliency detection using kernel density estimation. In: IEEE International Conference on Image Processing (ICIP), pp. 253–256 (2010)

14. Luo, S., Liu, Z., Li, L., Zou, X., Le Meur, O.: Efficient saliency detection using regional color and spatial information. In: European Workshop on Visual Information Processing (EUVIP), pp. 184–189 (2013)

15. Ma, Y.F., Zhang, H.J.: Contrast-based image attention analysis by using fuzzy growing. In: Proceedings of the Eleventh ACM International Conference on Multimedia, pp. 374–381 (2003)

16. Nothdurft, H.C.: Salience from feature contrast: additivity across dimensions. Vis. Res. **40**(10), 1183–1201 (2000)

17. Ren, Z., Gao, S., Chia, L.T., Tsang, I.W.H.: Region-based saliency detection and its application in object recognition. IEEE Trans. Circ. Syst. Video Technol. **24**(5), 769–779 (2014)

18. Sha, C., Li, X., Shao, Q., Wu, J., Bian, S.: Saliency detection via boundary and center priors. In: 6th International Congress on Image and Signal Processing (CISP), vol. 2, pp. 1066–1071 (2013)

19. Siagian, C., Itti, L.: Rapid biologically-inspired scene classification using features shared with visual attention. IEEE Trans. Pattern Anal. Mach. Intell. (PAMI) **29**(2), 300–312 (2007)

20. Siva, P., Russell, C., Xiang, T., Agapito, L.: Looking beyond the image: unsupervised learning for object saliency and detection. In: IEEE Conference on Computer Vision and Pattern Recognition (CVPR), pp. 3238–3245 (2013)

21. Zhang, X., Ren, Z., Rajan, D., Hu, Y.: Salient object detection through over-segmentation. In: IEEE International Conference on Multimedia and Expo (ICME), pp. 1033–1038 (2012)

Order Tracking by Square-Root Cubature Kalman Filter with Constraints

Oscar Cardona-Morales[1][(✉)] and German Castellanos-Dominguez[2]

[1] GIDTA Research Group, Universidad Católica de Manizales, Manizales, Colombia
ocardonam@ucm.edu.co
[2] Signal Processing and Recognition Group,
Universidad Nacional de Colombia, Manizales, Colombia
cgcastellanosd@unal.edu.co

Abstract. Condition monitoring of mechanical systems is an important topic for the industry because it helps to improve the machine maintenance and reduce the total operational cost associated. In that sense, the vibration analysis is an useful tool for failure prevention in rotating machines, and its main challenge is estimating on-line the dynamic behavior due to non-stationary operating conditions. Nevertheless, approaches for estimating time-varying parameters require the shaft speed reference signal, which is not always provided, or are oriented to off-line processing, being not useful on industrial applications. In this paper, a novel Order Tracking (OT) is employed to estimating both the instantaneous frequency (IF) and the spectral component amplitudes, which does not require the shaft speed reference signal and may be computed on-line. In particular, a nonlinear filter (Square-Root Cubature Kalman Filter) is used to estimate the spectral components from the vibration signal that provide the necessary information to detect damage on a machine under time-varying regimes. An optimization problem is proposed, which is based on the frequency constraints to improve the algorithm convergency. To validate the proposed constrained OT scheme, both synthetic and real-world application are considered. The results show that the proposed approach is robust and it successfully estimates the order components and the instantaneous frequency under different operating conditions, capturing the dynamic behavior of the machine.

Keywords: Order tracking · Non-stationary signals · Kalman filtering

1 Introduction

Vibration analysis of rotating machines is one of the most used techniques for fault diagnosis and condition monitoring due to its high performance and low implementation cost. Nowadays, the main challenge in vibration analysis is to

O. Cardona-Morales–This research is supported by research project "Estudio de la Compatibilidad electromagnética de los sistemas de comunicación. Fase I" developed by the Universidad Católica de Manizales.

J.F. Martínez-Trinidad et al. (Eds.): MCPR 2016, LNCS 9703, pp. 104–114, 2016.
DOI: 10.1007/978-3-319-39393-3_11

track and reduce the influence of changes during time-varying operation conditions and loads. In this regard, the order tracking (OT) techniques had been proposed, oriented to obtain the fundamental component features of the shaft reference speed (called basic order) and capture the dynamics of the measured vibration signals. The OT have shown to be useful within the analysis of non-stationary vibration signals, condition monitoring and fault diagnosis [5]. This technique allows to identify the rotation speed and the spectral/order components, which are fundamental to describe the state of both, the machine and its conforming mechanisms, during changing loads and speed regimes [11].

Particularly, a suitable estimation strategy is carried out using Kalman filtering in [5]; its improved version with increased precision, termed Vold-Kalman filtering (VKF_OT), was proposed in [14]. However, it requires measuring the shaft speed, which makes the order analysis still complex. The measurement of the shaft speed implies installing additional equipment near to the machine, which in certain situations is inconvenient. In [4], another approach is discussed consisting of a nonlinear least minimum squares algorithm, which estimates the amplitude, frequency, and phase of a non-stationary sinusoid, but the principal shortcomings come from its lack of robustness in the estimation procedure. In [6,15], a frequency tracker based on an oscillatory model, is introduced, where its parameters are calculated by Extended Kalman Filtering (EKF), obtaining the amplitude, phase, and mainly the frequency of a harmonic signal for de-noising in non-stationary signals. However, the tuning of model parameters is complex and requires an expertise degree.

In [12] an extended version of EKF frequency tracker for non-stationary harmonic signals is presented, where the time-varying amplitude is another state variable attached to the oscillatory model, outperforming the conventional methods aforementioned. Nonetheless, the increment in the amount of state variables implies more computational cost affecting the on-line tracking task. In contrast, in [8] the time-varying amplitude is estimated assuming the state variables as the in-phase and quadrature components of the signal, and computing the quadratic mean between those components. Besides, the EKF is based on the linearization by using Taylor's series expansion, however, the computation of Jacobian matrix induces high running complexity, limiting the application capability. To overcome the drawbacks of EKF on estimation accuracy, stability, convergence, among others., a novel nonlinear filtering approach is proposed in [1], termed Cubature Kalman Filter (CKF). In order to improve the CKF performance, a extended version was proposed in [3], so-called square-root cubature Kalman filter (SRCKF). The SRCKF propagates the probability distribution function in a simple and efficient way and it is accurate up to second order in estimating mean and covariance [3]. Based on explained above approaches, this paper discusses an OT approach with improved IF estimation by means of SRCKF, which introduces a frequency tracker that allows to capture the signal intrinsic dynamics, and thus, the OT deals with non-stationarity associated to several parts of the machine when determining the fundamental frequency of a vibration signal. Additionally, a simple method to incorporate state constraints is presented to improve the precision and the tolerance to the parameters initialization.

2   Materials and Methods

From the input signal $y(n)$, Order Tracking provides the estimation of modes and amplitudes present in each oscillation. The machine shaft speed is the *basic order*, while superior orders are related as the shaft speed harmonics. Thus, the shaft speed, $\eta = 60f$, is equivalent to shaft fundamental frequency of the machine, where η is given in revolutions per minute (*rpm*) and f in *Hz*.

2.1   Oscillatory Model and Instantaneous Frequency Estimation

The vibration signal, $y(n) \in \mathbb{R}$, acquired from a rotating machine can be represented as a superposition of K sinusoidal functions (termed *order components*), as follows:

$$y(n) = \sum_{k=1}^{K} a_k(n) \cos(k\omega(n)n + \varphi_k(n)) \tag{1}$$

where $a_k(n)$ and $\varphi_k(n)$ denote the amplitude and the phase of the k-th order component, respectively; $\omega(n) = 2\pi f_0(n)$ is the angular frequency of a rotational frequency $f_0(n)$. The variables $a_k(n)$, $\varphi_k(n)$ and $\omega(n)$ are time-varying.

Accordingly to [9], it is possible to extract both the instantaneous frequency (IF) and the order component amplitudes by the following state-space model:

$$
\begin{bmatrix} x_1(n+1) \\ \vdots \\ x_K(n+1) \\ x_{K+1}(n+1) \end{bmatrix} =
\begin{bmatrix} M(x_{K+1}(n)) & & \\ & \ddots & \\ & & M(Kx_{K+1}(n)) \\ & & 1 \end{bmatrix}
\begin{bmatrix} x_1(n) \\ \vdots \\ x_K(n) \\ x_{K+1}(n) \end{bmatrix} +
\begin{bmatrix} \xi_1(n) \\ \vdots \\ \xi_K(n) \\ \xi_{K+1}(n) \end{bmatrix}
\tag{2}
$$

$$
y(n) = \begin{bmatrix} h & \cdots & h & 0 \end{bmatrix}
\begin{bmatrix} x_1(n) \\ \vdots \\ x_K(n) \\ x_{K+1}(n) \end{bmatrix} +
\begin{bmatrix} v_1(n) \\ \vdots \\ v_K(n) \\ v_{K+1}(n) \end{bmatrix}
\tag{3}
$$

where the remaining terms of the state transition matrix are zero filled, and $x_k(n) = [x_c(n)\, x_d(n)]^T \in \mathbb{R}^{2\times 1}$ is the state variable vector, being $x_c(n) = a(n)\cos(\omega(n)n + \varphi(n))$ and $x_d(n) = a(n)\sin(\omega(n)n + \varphi(n))$, which are the in-phase and quadrature components, respectively. In consequence, the order component amplitude and IF estimation are computed by $a(n) = \sqrt{x_c(n)^2 + x_d(n)^2}$ and $\omega(n) = x_{K+1}(n)$. The matrix $M(x_{K+1}(n)) \in \mathbb{R}^{2\times 2}$ is a rotation matrix that is defined as follows:

$$
M(\omega(kn)) = \begin{bmatrix} \cos\omega(kn) & -\sin\omega(kn) \\ \sin\omega(kn) & \cos\omega(kn) \end{bmatrix} \tag{4}
$$

As regards the model parameters, $\xi(n) \sim \eta(0, Q(n)) \in \mathbb{R}^{2K\times 1}$ is the process noise, where $Q(n) \in \mathbb{R}^{2K\times 2K}$ is the covariance matrix of process noise; $h = \begin{bmatrix} 1 & 0 \end{bmatrix}$

forms the measurement matrix $H \in \mathbb{R}^{1 \times 2K+1}$, $v(n) \sim \mathcal{n}(0, r(n)) \in \mathbb{R}$ is the measurement noise, and $r(n) \in \mathbb{R}$ is the measurement variance.

It is worth noting that the model described in Eqs. (2) and (3) could be applied under the assumption that the speed does not present strong changes neither discontinue behaviors [7], it means, the following approximations hold: $a(n+1) \cong a(n)$, $\varphi(n+1) \cong \varphi(n)$ and $\omega(n+1) \cong \omega(n)$.

For the sake of simplicity, the process equation (Eq. (2)) could be rewritten in short form as

$$X(n+1) = \vartheta(n, X(n)) + w(n) \tag{5}$$

where $\vartheta(n, X(n))$ is the state transition nonlinear function. In this case, the estimation of state variable vector implies a set of nonlinear equations. Therefore, a recursive solution can be computed by means of nonlinear Kalman filtering.

2.2   Estimation of Model Parameters

As seen in Eqs. (2) and (3), parameter computation implies a recursive nonlinear analysis allowing to get an approximated solution when Gaussian noise is assumed but avoiding calculation of corresponding Jacobians of state variables. To this end, the Square-Root Cubature Kalman Filter (SRCKF), which is based on the recursive propagation of variable state moments (mean and variance), is suggested in [2], under the assumption that implicated nonlinear function, ϑ, should be reasonably smooth. In this case, a quadratic function near the prior mean is used assuming that it could properly approximate the given nonlinear function. To this end, the error covariance matrix should be symmetric and positive definiteness to preserve the filter properties on each update cycle. So hence, SRCKF uses a forced symmetry on the solution of the Ricatti equation improving the numerical stability of the Kalman filter [10], whereas the underlying meaning of the covariance is embedded in the positive definiteness [2].

The SRCKF algorithm that is described in Table 1 carries out the QR decomposition (termed triangularization procedure, $S = \text{tria}\{\cdot\}$), where the S is a lower triangular matrix and denotes a square-root factor [2]. Besides, aiming to parameterize the SRCKF is mandatory taking into account the process and measurement covariance parameters, where the main parameter is q because it comprises the information related with variances of the state estimates as follow:

$$diag(Q) = \begin{bmatrix} q_1^a \; q_1^a \; \cdots \; q_K^a \; q_K^a \; q_{K+1}^f \end{bmatrix} \tag{6}$$

where $q_i^a (i = 1, ..., K)$ denotes the amplitude variance of the order components and q_{K+1}^f denotes the frequency variance, which describes the dynamic behavior of the system.

2.3   State Estimation with Constraints

Constraints on states $x(n)$ to be estimated are important model information that is often not used in state estimation. Typically, such constraints are due to

Table 1. SRCKF algorithm (Part 1), [9]

Initialization:

1. Define the input values

$$y_{1:N}, \ \widehat{x}_0, \ P_{0|0} = S_{0|0}S_{0|0}^{\top}, \ Q_0, \ R_0$$

2. Define the cubature points

$$\phi_i = \sqrt{m/2}\left\{\begin{bmatrix}I_{m \times m} & -I_{m \times m}\end{bmatrix}\right\}$$

Tracking:

3. **for** $n = 1$ to N **do**

Time update

4. Evaluate the cubature points $(i = 1, 2, \ldots, m)$, where $m = 2K + 1$,

$$\chi_{i,n-1|n-1} = S_{n-1|n-1}\phi_i + \widehat{x}_{n-1|n-1}$$

5. Evaluate the propagated cubature points $(i = 1, 2, \ldots, m)$

$$\chi_{i,n|n-1}^{*} = \vartheta\left(\chi_{i,n-1|n-1}\right)$$

6. Estimate the predicted state

$$\widehat{x}_{n|n-1} = \frac{1}{m}\sum_{i=1}^{m}\chi_{i,n|n-1}^{*}$$

7. Estimate the square-root factor of prediction error covariance

$$S_{n|n-1} = \text{tria}\left\{\begin{bmatrix}\widehat{\chi}_{n|n-1} & S_{Q|n}\end{bmatrix}\right\}$$

where $\widehat{\chi}_{n|n-1} = \frac{1}{\sqrt{m}}\begin{bmatrix}\chi_{1,n|n-1}^{*} - \widehat{x}_{n|n-1} & \chi_{2,n|n-1}^{*} - \widehat{x}_{n|n-1} \cdots \chi_{m,n|n-1}^{*} - \widehat{x}_{n|n-1}\end{bmatrix}$

and $S_{Q|n}$ denotes a square-root factor of Q_{n-1}

Measurement Update

8. Evaluate the cubature points $(i = 1, 2, \ldots, m)$

$$\chi_{i,n|n-1} = S_{n|n-1}\phi_i + \widehat{x}_{n|n-1}$$

9. Evaluate the propagated cubature points $(i = 1, 2, \ldots, m)$

$$\psi_{i,n|n-1} = h\left(\chi_{i,n|n-1}\right)$$

10. Estimate the predicted state

$$\widehat{y}_{n|n-1} = \frac{1}{m}\sum_{i=1}^{m}\psi_{i,n|n-1}$$

11. Estimate the square-root of the innovation covariance matrix

$$S_{yy,n|n-1} = \text{tria}\left\{\begin{bmatrix}\mathcal{Y}_{n|n-1} & S_{R|n}\end{bmatrix}\right\}$$

where $\mathcal{Y}_{n|n-1} = \frac{1}{\sqrt{m}}\begin{bmatrix}\psi_{1,n|n-1} - \widehat{y}_{n|n-1} & \psi_{2,n|n-1} - \widehat{y}_{n|n-1} \cdots \psi_{m,n|n-1} - \widehat{y}_{n|n-1}\end{bmatrix}$

and $S_{R|n}$ denotes a square-root factor of R_n

12. Estimate the cross-covariance matrix

$$P_{xy,n|n-1} = \mathcal{X}_{n|n-1}\mathcal{Y}_{n|n-1}^{\top}$$

where $\mathcal{X}_{n|n-1} = \frac{1}{\sqrt{m}}\begin{bmatrix}\chi_{1,n|n-1} - \widehat{x}_{n|n-1} & \chi_{2,n|n-1} - \widehat{x}_{n|n-1} \cdots \chi_{m,n|n-1} - \widehat{x}_{n|n-1}\end{bmatrix}$

13. Estimate the Kalman gain

$$W_n = \left(P_{xy,n|n-1}/S_{yy,n|n-1}^{\top}\right)/S_{yy,n|n-1}$$

14. Estimate the updated state

$$\widehat{x}_{n|n} = \widehat{x}_{n|n-1} + W_n\left(y_n - \widehat{y}_{n|n-1}\right)$$

15. Estimate the square-root factor of the corresponding error covariance

$$S_{n|n} = \text{tria}\left\{\begin{bmatrix}\mathcal{X}_{n|n-1} - W_n\mathcal{Y}_{n|n-1} & W_n S_{R|n}\end{bmatrix}\right\}$$

16. **end**

---

physical limitations on the states. In Kalman filter theory, there is no general way of incorporating these constraints into estimation problem. However, the constraints can be incorporated in the filter by projecting the unconstrained Kalman filter estimates onto the boundary of the feasible region at each time step [16,17]. The numerical optimization at each time step may be a challenge in time-critical applications. In this section, a simple method introduced in [13] is applied to handle state constraints in the SRCKF.

Assume that the constraints of state variables are represented by box constraints as follow:

$$\boldsymbol{x}_L(n) \leq \boldsymbol{x}(n) \leq \boldsymbol{x}_H(n) \tag{7}$$

where subindexes L and H denote the lower and upper boundaries, respectively. The method is illustrated for $\boldsymbol{x}(n) \in \mathbb{R}^2$. In the case of a second order system, the feasible region by the box constraints can be represented by a rectangle as in Fig. 1. It is showed the illustration of the steps of constraint handling of the SRCKF algorithm from one time step to the next. At $t = n - 1$, the actual state $\boldsymbol{x}_n$, its estimate $\widehat{\boldsymbol{x}}_{n-1}$ and state covariance are selected. The constraints information can be incorporated in the SRCKF algorithm in a simple way during the time-update step. After the propagation of the sigma points (step 5.), the (unconstrained) transformed sigma points which are outside the feasible region can be projected onto the boundary of the feasible region and continue the further steps. In Fig. 1, at $t = n$ two sigma points which are outside the feasible region are projected onto the boundary (right plot in the figure). The mean and covariance with the constrained sigma points now represent the a priori state variable (x_n^{-SRCKF}) and covariance, and they are further updated in the measurement update step. The advantage here is that the new a priori covariance includes information on the constraints, which should make the SRCKF estimate more efficient (accurate) compared to the SRCKF estimate without constraints. Extension of the proposed method to a higher dimension, d, is straightforward. Alternative linear constraints, e.g., $C\boldsymbol{x} \leq d$ are easily included by projecting the sigma point violating the inequality normally onto the boundary of a feasible region. It is observed that the new (constrained) covariance obtained at a time step is lower in size compared to the unconstrained covariance. If in case, the estimate after the measurement update is outside the feasible region, the same projection technique can be extended. In a practical point of view, the boundaries are fixed according to maximum and minimum values that could take the state variables. In consequence, in case of the order components $L^a = \min y(n)$ and $H^a = \max y(m)$, whereas the IF constraints depend on the approximated knowledge of the machine speed range, where L^f and H^f are usually fixed as zero or idle speed and maximum speed, respectively.

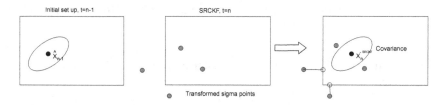

Fig. 1. Illustration of sigma constrained points.

3 Simulation Study

A synthetic signal is desing to validate the performance of considered OT schemes for the closed-order component identification, as recommended in [14]. The synthetic signal comprises three order components including 1, 4 and 4.2. The assumed reference shaft speed linearly increases from 0 to 1800 rpm for 5 s. A sampling frequency of 1 kHz is used through this simulation. The Table 2 illustrates the amplitudes assigned to these order components. Order amplitude level is set as time-varying, since it is assumed that, most of the machine mechanisms have different vibration levels.

Table 2. Spectral components composing the in synthetic signal

Order numbers	1	4	4.2
Amplitude	Linearly increasing from 0 to 10	Linearly increasing from 0 to 3	Linearly increasing from 0 to 2.5

In Fig. 2, the time-frequency representation of the synthetic signal is shown, as well as its generative time series. It can be observed the difficult to distinguish closed-order components using methods based on Fourier transforms. Also, it is worth noting that the amplitude differences between the first order and its harmonics make almost insignificant the low-frequency information. As a result, it generates a wrong representation of required components. Instead, when using OT techniques based on parametric models, it is possible to capture properly the information about the behavior of each order component. Computation parameters that influence tracking performance, such as the correlation matrix of process noise and the error covariance propagation, are investigated here. The algorithms are tested under different parameter values, i.e., two values for variance of process noise and two values for error covariance propagation, in order to shows the improvement achieved with constraints. The Fig. 3 shows the waveform reconstruction (WR), amplitude (A) and frequency estimation errors. It is possible to see a high performance (accuracy) in the WR, specially when SRCKF with constraints is used. It is important notice that the frequency estimation, by SRCKF without constraints (segmented line), presents a inverse behavior, which is a wrong interpretation of the algorithm and it may produce errors during the signal analysis.

3.1 Case Study: Wind Turbine if Estimation - CMMNO2014 Contest

This experiment consisted of estimating the instantaneous speed in *rpm*, or instantaneous frequency in Hz, from a wind turbine operating under nonstationary conditions. The information given hereafter, as well as the signal,

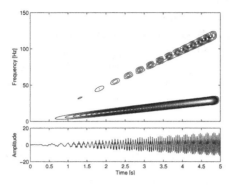

Fig. 2. Time-Frequency representation from synthetic signal obtained by STFT (hamming window, 512 frequency samples, and 50 % overlap)

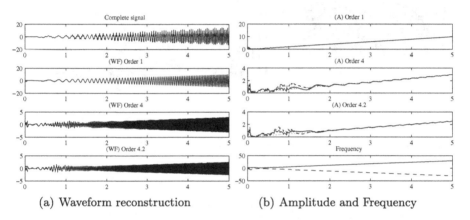

(a) Waveform reconstruction   (b) Amplitude and Frequency

Fig. 3. Estimated order components of the synthetic signal using the parameters $p = 1e^{-1}$, $q^a = 1e^{-10}$, $q^f = 1e^{-12}$ and $r = 1e^{-9}$. (··) Original signal, (- -) unconstrained and (-) constrained estimation

have been kindly provided by **Maïa Eolis** to solve the contest in the framework of the International Conference on Condition Monitoring of Machinery in Non-stationary Operations (CMMNO), December 15–16, 2014 Lyon-France[1]. The provided signal comes from an accelerometer located on the rotor side of the gearbox (high speed shaft) casing in the radial direction, and the speed of the main shaft (also called low speed shaft) is between 13 and 15 *rpm* during the recording. The sampling frequency is 20 KHz and the acquisition time is 547 s approx. As regards to high-speed shaft estimation, from the kinematics of the machine the boundaries $[L^f, H^f]$ from the desired IF are defined between 25.99 Hz and 29.98 Hz. However, after to carried out the testing, it was found that the minimum boundary must be fixed at 15 Hz. In Fig. 4 is shown the

---

[1] Contest rules link: http://cmmno2014.sciencesconf.org/conference/cmmno2014/pages/cmmno2014_contest_V2.pdf.

provided signal, where it is possible to observe the signal in time and frequency ((a) and (b) parts, respectively). In frequency domain are marked the harmonics obtained using an harmonic algorithm discussed in [9], by the Fourier transform computation from 20 s signal segment. Here, the harmonic 26.2 Hz is used as the first order, obtaining in total a set of 26 orders. In addition, the SRCKF parameters associated to process and measurement covariances are fixed as $q_i^a = 10^{-3}$, $q^f = 10^{-10}$ and $r = 10^{-11}$.

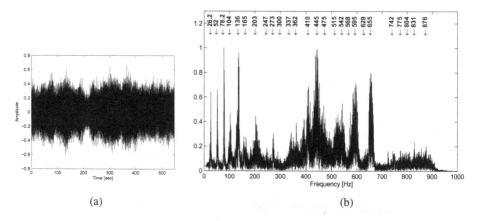

(a)   (b)

Fig. 4. Provided signal by CMMNO2014 contest in time and frequency domain.

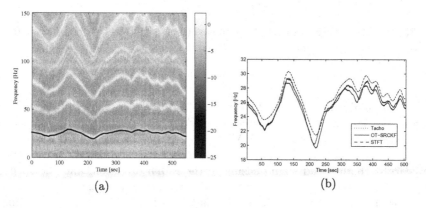

(a)   (b)

Fig. 5. IF estimated by IAS-OT model from CMMNO2014 contest wind turbine signal: (a) time-frequency representation highlighting the estimation with black line, and (b) a comparison with the tacho reference.

As a result, Fig. 5 displays the IF estimated using the proposed constrained OT model. Obtained IF is highlighted with a black line on the time-frequency representation Fig. 5(a), where it is possible to see that the estimation match with the high speed shaft, ranging from 20Hz to 30Hz, which confirms the boundaries fixed into the model. A comparison with the tacho reference is shown in Fig. 5(b), and besides, the IF estimation (red line) using a traditional method based on time-frequency representation (noted as STFT), which consists of tracking the maxima values in the STFT [18]. It is worth noting that using the aforementioned method was achieve the fifth place in the contest. In that sense, the proposed IAS-OT model allows to improve the result obtained using the based-STFT method, reaching a relative error under $\pm 3\%$ despite the fact that the intervals $[220 - 250]$ and $[320 - 350]$s there are a delay between the reference and the estimated IF.

4   Conclusions

The study proposes, derives and implements a novel constrained SRCKF approach based on a nonlinear state-space model, where it is possible simultaneously extract multiple order/spectral components together with IF estimation and decouple close orders. An improvement of the tracking algorithm is rendered incorporating constraints to the state variables, allowing to have a better performance of the SRCKF. The SRCKF captures the dynamic behavior of the system in terms of the IF estimation, which is an advantage when it is necessary to analyze machines where the reference shaft speed cannot be measured. Therefore, the proposed approach is an useful tool to compensate the non-stationary operating conditions of the machine and it contributes with the diagnostic analysis. The future work will be centered into optimization of the initialization of the model parameters and validation the proposed scheme in other kind of applications.

References

1. Arasaratnam, I., Haykin, S.: Cubature kalman filters. IEEE Trans. Autom. Control **54**(6), 1254–1269 (2009)
2. Arasaratnam, I., Haykin, S., Hurd, T.: Cubature kalman filtering for continuous-discrete systems. IEEE Trans. Theory Simul. **58**(10), 4977–4993 (2010)
3. Arasaratnam, I., Haykin, S.: Cubature kalman smoothers. Automatica **47**(10), 2245–2250 (2011)
4. Avendano-Valencia, L., Avendano, L., Ferrero, J., Castellanos-Dominguez, G.: Improvement of an extended kalman filter power line interference suppressor for ECG signals. Comput. Cardiol. **34**, 553–556 (2007)
5. Bai, M., Huang, J., Hong, M., Su, F.: Fault diagnosis of rotating machinery using an intelligent order tracking system. J. Sound Vib. **280**, 699–718 (2005)
6. Bittanti, S., Saravesi, S.: On the parameterization and design of an extended kalman filter frequency tracker. IEEE Trans. Autom. Control **45**, 1718–1724 (2000)
7. Borghesani, P., Pennacchi, P., Randall, R., Ricci, R.: Order tracking for discrete-random separation in variable speed conditions. Mech. Syst. Signal Process. **30**(1–2), 1–22 (2012)

8. Cardona-Morales, O., Avendano-Valencia, L., Castellanos-Dominguez, G.: Instantaneous frequency estimation and order tracking based on kalman filters. In: IOMAC2011 International Operational Modal Analysis Conference (2011)
9. Cardona-Morales, O., Avendaño, L.D., Castellanos-Domínguez, G.: Nonlinear model for condition monitoring of non-stationary vibration signals in shipdriveline application. Mechan. Syst. Signal Process. **44**, 134–148 (2014). special Issue on Instantaneous Angular Speed (IAS) Processing and Angular Applications
10. Grewal, M., Andrews, A.: Kalman Filtering: Theory and Practice Using Matlab. Wiley, New York (2001)
11. Guo, Y., Chi, Y., Huang, Y., Qin, S.: Robust ife based order analysis of rotating machinery in virtual instrument. J. Phys. Conf. Ser. **48**, 647–652 (2006)
12. Hajimolahoseini, H., Taban, M., Abutalebi, H.: Improvement of extended kalman filter frequency tracker for nonstationary harmonic signals. In: International Symposium on Telecommunications (IST 2008). pp. 592–597 (2008)
13. Kandepu, R., Foss, B., Imsland, L.: Applying the unscented kalman filter for nonlinear state estimation. J. Process Control **18**, 753–768 (2008)
14. Pan, M., Wu, C.: Adaptive vold-kalman filtering order tracking. Mechan. Syst. Signal Process. **21**, 2957–2969 (2007)
15. Scala, B.L., Bitmead, R.: Design of an extended kalman filter frequency tracker. IEEE Trans. Signal Process. **44**, 525–527 (1994)
16. Simon, D., Chia, T.: Kalman filtering with state equality constraints. IEEE Trans. Aerosp. Electron. Syst. **38**, 128–136 (2002)
17. Ungarala, S., Dolence, E., Li, K.: Constrained extended kalman filter for nonlinear state estimation. In: 8th International IFAC Symposium on Dynamics and Control Process Systems, Cancun - Mexico. vol. 2, pp. 63–68 (2007)
18. Urbanek, J., Barszcz, T., Antoni, J.: A two-step procedure for estimation of instantaneous rotational speed with large fluctuations. Mechan. Syst. Signal Process. **38**(1), 96–102 (2013). condition monitoring of machines in non-stationary operations

Contour Detection at Range Images Using Sparse Normal Detector

Alejandra Cruz-Bernal$^{(\boxtimes)}$, Dora-Luz Alamanza-Ojeda,
and Mario-Alberto Ibarra-Manzano

Departamento de Ingeniería Electrónica, DICIS, Universidad de Guanajuato,
Salamanca, Guanajuato, Mexico
{a.cruzbernal,dora.almanza,ibarram}@ugto.mx
http://www.dicis.ugto.mx

Abstract. The object surfaces on the Range images can be easily treated as elevations, at each point of these surfaces. The Sparse Normal Detector technique focuses in the extraction of keypoints, from homogeneous surface of Range images. Additionally, the contour of the objects in the scene can be represented through these points. First, the homogeneity feature is computed by means of the Sum and Difference Histogram technique, producing the Homogeneity image. Then, the corresponding dense normal vectors of the surface formed by this image are computed. A normal probability density function is used to select the most outstanding dense vectors, yielding the Sparse Normal descriptor. These vectors form new flat directional surfaces. The final detection of interesting point is performed using the Sparse Keypoints Detector technique. The experimental test involves a qualitative analysis, using the Middlebury and DSPLab dataset, and a quantitative evaluation of repeatability.

Keywords: Range image · Sparse Normal Detector · Keypoints detection · Contour detection

1 Introduction

A Range image provides the geometrical information (depth), not only the 2D information of a RGB image. Moreover, the feature extraction task in Range images is generally invariant to scale, rotation and illumination [1]. RGB-D images are acquired by means of a low cost 3D acquisition system, such as the Microsoft Kinect sensor [14].

On the other hand, the interesting point detection is an essential phase to develop a local feature extractor [2]. In this phase, the data, (for instance, texture in intensity images) is obtained to characterize the keypoints. Recently, Steder *et al.* [12] have introduced the Normal Aligned Radial Feature (NARF) for 3D object recognition from a Range image. Some other local detectors and descriptors for 2D or their version for 3D images are: Harris corner detector [5,6], SURF (Speed Up Robust Feature) [7], and FAST (Features from Accelerated Segment Test).

© Springer International Publishing Switzerland 2016
J.F. Martínez-Trinidad et al. (Eds.): MCPR 2016, LNCS 9703, pp. 115–124, 2016.
DOI: 10.1007/978-3-319-39393-3_12

The proposal presented here accomplishes a robust and balanced transformation from dense to a sparse process. First, the surface of the Homogeneity image H_m is constructed from the texture features of the Range image. Moreover, the dense normal vectors referred here as N_D of the Homogeneity image are computed. After in the *sparse process*, a Gaussian distribution is used to select a set with the more representative normal vectors (N_D) at each x-, y-, z-direction, forming a *sparse normal vectors* referred here as $\mathbf{N}_S$. Additionally, is carried out an analysis on the neighborhood of each component of $\mathbf{N}_S$ through the *pdf* (Probability Density Function), in accordance to describe this particular region. Afterwards, the interesting points are obtained from each directional surface highlighting the contour of the objects in the scene. Experimental tests have been performed with two different datasets: (1) the benchmark Middlebury proposed by *Pal et al.* in [3] and *Hirsmüller et al.* in [8], and (2) our DSPLab dataset [9]. Finally, a comparative analysis among our proposal and different proposals for key points detection considered in the-state-of-art demonstrates a high performance at least in almost all the test.

The main contributions of this study are: first, the use of the homogeneity texture feature as a local surface descriptor applied to Range images. In particular, it is proposed to highlight the homogeneous regions because they represent the smooth curvatures of the Range image. Second, this proposal in the sparse process allows a transformation from $\mathbb{R}^3$ to $\mathbb{R}$ space, implying a significant reduction in the cardinality of the descriptor vector ($\mathbf{N}_S$ vector). These descriptors allow the representation of the scene through the separation of the forefront and the background planes. Finally, objects are defined by their keypoints highlighting their contours, with a low computational cost. In this paper, firstly, the proposed technique is described in Sects. 2 and 3. Section 4 discusses the results with a deep qualitative and quantitative analysis. Finally, brief conclusions are discussed in Sect. 5.

2   Methodology

This section describes each phase of our proposal based on the Sparse Normal Detector (SND) technique. Figure 1 illustrates a global block diagram of the proposed strategy. The Range images used are the Middlebury dataset and the Depth images acquired by a Kinect sensor, in which the income image could contain either controlled conditions or real indoor scenes. In the *dense process*, the homogeneity texture feature constructs a dense surface from the Range image, to highlight the uniform regions. Then, the normal vectors (called dense normal vectors) of the Homogeneity image are computed. Later, in the *sparse process*, $\mathbf{M}_x, \mathbf{M}_y$ and $\mathbf{M}_z$ surfaces are built up by selecting the dense normal vectors in accordance to a Gaussian distribution. This process is called Sparse Normal descriptor (SN-descriptor), which carries out a transformation from $\mathbb{R}^3$ to $\mathbb{R}$. The vectors contained in the $\mathbf{M}_x, \mathbf{M}_y$ and $\mathbf{M}_z$ surfaces are the sparse normal vectors, with which are represented the most distinctive values of these surfaces. Finally, the interesting points corresponding to each object in the scene are computed from the sparse vectors, as well as of the pdf information computed.

SPARSE NORMAL DETECTOR TECHNIQUE
(SND)

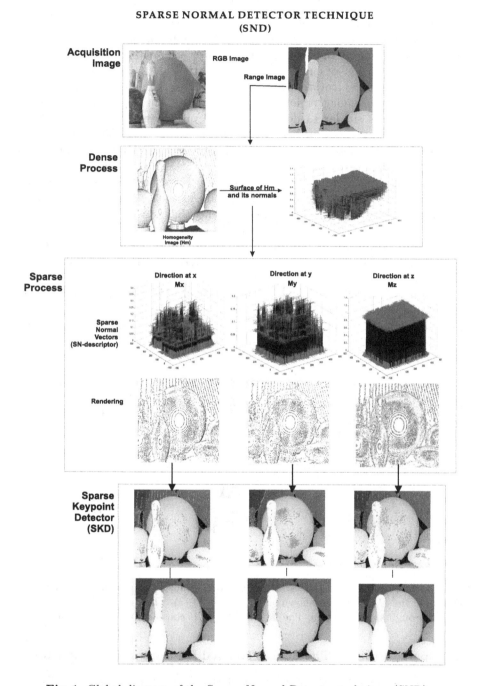

Fig. 1. Global diagram of the Sparse Normal Detector technique (SND).

2.1   Dense Porcess: The Homogeneity Image (H_m)

The Range image is a gray-level image with areas of smooth variation of intensity representing the depth of different objects. Therefore, in this paper we suggest that, the extraction of the homogeneity feature is similar to obtain a surface with a minimal change of energy. Furthermore, the change in the flow obtained from the homogeneity feature forms geodesic curves that highlight the contour and the curvature among regions [13]. In this case, the representation of the geodesic curves is similar to depict the iso-elevation lines (lines that, join of the points of equal value in height), usually used on topographic maps. On another hand, the information obtained from the Homogeneity image allows us represent objects as minimal surfaces (see Fig. 2(c)), with a Gaussian curvature established as a positive hemisphere [4,11], see Fig. 2 top row. Therefore, the extraction of normal vectors for all the points in such a hemisphere allows a dense representation of the Range image surface. These vectors are orthogonal to all tangent vectors in the H_m surface yielding the dense descriptor referred as $\mathbf{N}_D$. The H_m image contains the homogeneity texture feature of the Range image, and it is obtained using the Sum and Difference of Histograms technique (SDH) presented in [10]. A window size of 3×3 pixels is applied to the SDH, yielding an range image of $K \times L$ size and it is defined in the range 0 to 255 grey levels. This process is illustrated in the first block of the Fig. 1. Note that, the equal values of homogeneity are depicted as iso-elevation lines (see first block, Homogeneity image Fig. 1).

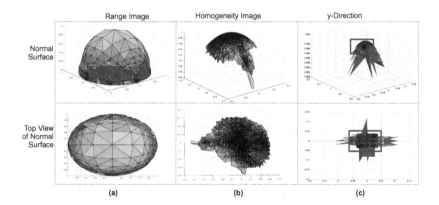

(a)   (b)   (c)

Fig. 2. Column (a) depicts the surface of the Range image as a positive hemisphere, column (b) the H_m surface is shown as a highlight of (a). The surface in **y**-direction is formed through *sparse* vectors. The labelled region is considered a smooth flat.

2.2   Sparse Process: The Sparse Normal Descriptor (SN-Descriptor)

The Sparse Normal descriptor process (referred here as SN-descriptor), divides the dense normal vectors ($\mathbf{N}_D$) into clusters with the most representative values per surface for each $\boldsymbol{x}$-, $\boldsymbol{y}$- and $\boldsymbol{z}$- direction. This process is illustrated in the first step of the Sparse process in Fig. 1. When the surface of H_m is decomposed in its normal vectors, the homogeneous surface is *"broken up"*. That is, new flat surfaces called $\mathbf{M}_x$, $\mathbf{M}_y$ and $\mathbf{M}_z$ are built up at $\mathbb{R}^2$ space. Hence, components of each directional surface are referred as *sparse* normal vectors ($\mathbf{N}_S$). Here, for every $\mathbf{N}_S$ vector on $\mathbf{M}_x$, $\mathbf{M}_y$ or $\mathbf{M}_z$ surfaces are mapping of the H_m surface to $\mathbb{R}^3$. Additionally, the neighborhood corresponding to each $\mathbf{N}_S$ vector is associated with a similar six neighbor connectivity. Then, it is carried out by an analysis of pdf in this neighborhood, to obtain the mean values with respect to intensity level (depth or disparity) and homogeneity feature. Finally, this allows to establish a descriptors set, for every one of the directional surfaces, where each component of this set is a vector of $1 \times m$ size. Therefore, descriptors are into $\mathbb{R}$ space, but, both are inmerse into $\mathbb{R}^2$ space.

2.3 Keypoint Detection: SKD Technique

The Sparse Keypoint Detector technique (referred as SKD) is shown in last block of the Fig. 1. From every one of the directional surfaces, the interesting points are extracted through of a pdf analysis. This is carried out for each neighborhood corresponding to one of the $\mathbf{N}_S$ vector, contained in the $\mathbf{M}_x$, $\mathbf{M}_y$ or $\mathbf{M}_z$ surfaces. Additionally, it is considered a 95 % confidence interval. Thereby, all sparse normal vectors form clusters in accordance to similarity values of their normals corresponding to the same neighborhood. Subsequently, these highlight vectors belong to the contour of objects in the scene.

Algorithm 1. Pseudo code of Sparse Normal Detector Technique.

Begin

1. Depth Image $\rightarrow$ Homogeneity Image H_m.
2. Compute Dense Normal Vectors $\mathbf{N}_D = [\mathbf{N}_{Dx}, \mathbf{N}_{Dy}, \mathbf{N}_{Dz}]$ from H_m.
3. Normal Vector ($\mathbf{N}_D$) $\rightarrow$ Sparse-Normal vector:
   a. $[\mathbf{N}_{Dx}, \mathbf{N}_{Dy}, \mathbf{N}_{Dz}] > 0$
   b. Select level of threshold $\mathbf{t}_x$ and $\mathbf{t}_y$ from $\mu(\mathbf{N}_D) \pm \sigma(\mathbf{N}_D)$, for $\mathbf{x}$- and $\mathbf{y}$-direction, respectively.
   c. Select level of threshold $\mathbf{t}_z$, from $\mu(\mathbf{N}_{Dz})$ or min ($\mathbf{N}_{Dz}$).
   d. Generate Sparse Normal vectors $[\mathbf{N}_{Dx}, \mathbf{N}_{Dy}, \mathbf{N}_{Dz}]$ using $\mathbf{t}_x$, $\mathbf{t}_y$ and $\mathbf{t}_z$.
4. Build up $\mathbf{M}_x$, $\mathbf{M}_y$ and $\mathbf{M}_z$ surfaces from $[\mathbf{N}_{Sx}, \mathbf{N}_{Sy}, \mathbf{N}_{Sz}]$.
5. Detect keypoints into $\mathbf{M}_x$, $\mathbf{M}_y$ and $\mathbf{M}_z$
   a. Select level of threshold to $\mathbf{t}_k$
   b. Sparse-Keypoints = $\mathbf{N}_S > \mathbf{t}_k$

End

---

3 Algorithm of the SND Technique

Algorithm 1 explains the computation of the SND technique. The normal vector $\mathbf{N}_D$ is computed from the H_m surface (first and second steps of the algorithm). Vector $\mathbf{N}_D = [\mathbf{N}_{Dx}, \mathbf{N}_{Dy}, \mathbf{N}_{Dz}]$ contains matrices of $K \times L$ size, which contain dense normal vectors in each one of directions. Moreover, $\mathbf{N}_D$ vector is not normalized and its values are in the range $[-1, 1]$. Here, the surface in H_m is defined as a positive hemisphere surface (see Subsect. 2.1) thus, only the outward-pointing normal vectors of $\mathbf{N}_D$ are considered (step 3(a)). This first selection is used to start the sparse process (steps 3(b-c)).

The sparse normal vectors are formed by selecting the components of $\mathbf{N}_D$ in accordance to a Gaussian distribution represented by the $\mathbf{t}$ vector (see step 3(d) of the algorithm). Here, the $\mathbf{N}_S$ vector is defined as $\mathbf{N}_S = [\mathbf{N}_{Sx}, \mathbf{N}_{Sy}, \mathbf{N}_{Sz}]$. It is built up from the $\mathbf{N}_D$ elements that are into the range $(0, 1)$. Thus, the elements of $\mathbf{N}_S$ are defined as all the normal vectors of $\mathbf{N}_D$ selected by the condition $\mathbf{N}_{Di} < \mathbf{t}$, where i is the i-th component of $\mathbf{N}_D$. Additionally, $\mathbf{t}$ is the vector consisting of three threshold levels in the range of $0 < \mathbf{t} < 1$, called $\mathbf{t}_x$, $\mathbf{t}_y$ and $\mathbf{t}_z$. As indicated in step 3(b) and 3(c) of the algorithm, the values of $\mathbf{t}$ are computed from the statistical information, mean and standard deviation, by analyzing each direction of vector $\mathbf{N}_D$. That is, $\mathbf{t}$ is given by

$$t_x = \begin{cases} \mu(N_{Dx}) + \sigma(N_{Dx}) \text{ if } & 0 < N_{Dx} \le t_u \\ \mu(N_{Dx}) - \sigma(N_{Dx}) \text{ if } & t_u < N_{Dx} < 1 \end{cases} \tag{1}$$

$$t_y = \begin{cases} \mu(N_{Dy}) + \sigma(N_{Dy}) \text{ if } & 0 < N_{Dy} \le t_u \\ \mu(N_{Dy}) - \sigma(N_{Dy}) \text{ if } & t_u < N_{Dy} < 1 \end{cases} \tag{2}$$

$$t_z = \begin{cases} \mu(N_{Dz}) & \text{if } \quad max(N_{Dz}) < 1 \\ max(N_{Dz}) - \sigma(N_{Dz}) & \text{if } \quad max(N_{Dz}) = 1 \end{cases} \tag{3}$$

where $\mathbf{t}_u$ is a level of threshold defined within 95 % confidence interval corresponding to mean of all components vectors of $\mathbf{N}_D$. Therefore, the computation of the thresholds in Eqs. (1) and (2), is determined in accordance to value of $\mathbf{t}_u$. The $\mathbf{N}_S$ vector has enough information to represent the most significant normal components of the H_m surface, at each $\boldsymbol{x}$-, $\boldsymbol{y}$- and $\boldsymbol{z}$- direction, respectively (see fourth step). Furthermore, one of the main contributions of this study is the number of elements contained in the $\mathbf{N}_S$ vector, which is established as lower than those of vector $\mathbf{N}_D$ by at least 80 % on average, in any of its three directions. Thus,

$$\begin{aligned} If \quad C_{Sx} &= \quad card(N_{Sx}) \quad \text{and} \quad C_{Dx} = card(N_{Dx}) \quad \text{then} \quad C_{Sx} \ll C_{Dx} \\ If \quad C_{Sy} &= \quad card(N_{Sy}) \quad \text{and} \quad C_{Dy} = card(N_{Dy}) \quad \text{then} \quad C_{Sy} \ll C_{Dy} \quad (4) \\ If \quad C_{Sz} &= \quad card(N_{Sz}) \quad \text{and} \quad C_{Dz} = card(N_{Dz}) \quad \text{then} \quad C_{Sz} \ll C_{Dz} \end{aligned}$$

Eq. (4) presents how the cardinality of $\mathbf{N}_{Sx}$ and $\mathbf{N}_{Sy}$ is closer to 10 % with respect to the cardinality of $\mathbf{N}_{Dx}$ and $\mathbf{N}_{Dy}$, respectively, and this percentage appears in virtually all the experimental results. In z-direction, the cardinality of $\mathbf{N}_{Sz}$ is closed to 30 % with respect to the cardinality of $\mathbf{N}_{Dz}$. The difference between the percentages of the sparse normal vectors in x- and y- directions with respect to z- direction is explained through of the homogeneity level presented at the H_m image. Therefore, at each direction is possible to form a new surface that represents the original H_m surface, with a minimal number of descriptors. Finally in the fifth step, the strategy used to detect the most significant key points is the Sparse Keypoints Detector (SKD). This strategy uses the data contained on the $\mathbf{M}_x$, $\mathbf{M}_y$ and $\mathbf{M}_z$ surfaces to rank the sparse key points based on the statistically defined thresholds $\mathbf{t}_{kx}$, $\mathbf{t}_{ky}$, $\mathbf{t}_{kz}$, corresponding to each $\mathbf{x}-$, $\mathbf{y}-$, and $\mathbf{z}-$ direction, respectively. These thresholds are defined by:

$$\mathbf{t}_{kx} = \mu(N_{Sx}) + \sigma(N_{Sx}); \quad \mathbf{t}_{ky} = \mu(N_{Sy}) + \sigma(N_{Sy}); \quad \mathbf{t}_{kz} = \mu(N_{Sz}) \quad (5)$$

The last block of the Fig. 1 shows a first example of the interesting points detected at each directional surface, using the SKD technique.

4   Experimental Results

The experimental tests have been performed using two datasets: the benchmark Middlebury provided in [3,8], as well as our DSPLab dataset [9]. The importance of testing two different datasets is to establish the robustness and repeatability of our algorithm through offline images obtained under several non-controlled conditions (changes of intensity, exposure and noise in indoor environments). The most representative results respect to DSPLab dataset are depicted in Fig. 3. Also, a qualitative analysis to compare the SKD technique with respect to other common key point detectors is carried out. Finally, the effectiveness of the method is tested using a quantitative analysis of the keypoint repeatability.

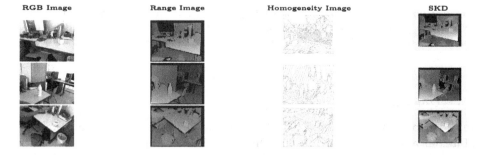

Fig. 3. Results of the homogeneity and extraction of the contour using the DSPLab database. First row, cube-rugby. Second row, white-bottle and last row, flower-cup. Notice that H_m images (third column) show a slightly loss of information. Last column shows the key points with which is defined the contour of the objects in the scene.

Figure 4 shows the most meaningful keypoints detected on the H_m surface using the approach presented in the Sect. 3. Thus, the results using the SKD technique can be compared with those of other techniques proposed in the state-of-the-art applied to range images (third to sixth column of Fig. 4). In particular, the applied techniques are NARF, Harris corner, SURF and FAST. The performance of the SKD shows a better index of repeatability than the other techniques (see Fig. 5). Qualitatively, it can be seen that the keypoints retrieve the contour of all the objects at the scene using e.g., sparse descriptors of the $\mathbf{M}_y$ surface.

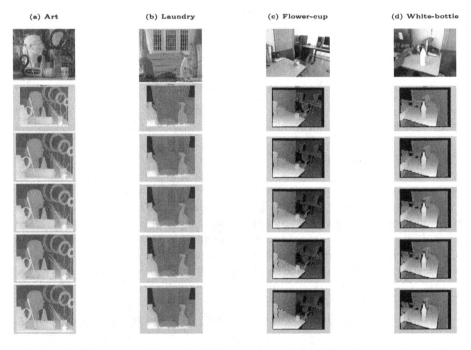

Fig. 4. Comparison between the proposal Sparse Keypoint Detector (SKD) and other techniques. First row color image. Second row SKD technique. Third row NARF detector. Fourth row Harris detector. Fifth row SURF detector and sixth row FAST detector.

4.1   Quantitative Analysis

In order to evaluate the quality of the proposed key point detector, an analysis of its repeatability was carried out. This analysis was applied to the Middlebury and DSPLab dataset. A Gaussian Filter and salt and pepper noise (**S&P**) were used to generate synthetic and diffused images. To this purpose, the Gaussian filter was generated with a typical mask of 3×3 pixels; in addition, the average level of noise used by the **S&P** was 5 %. The test images contain four types of noises: salt and pepper noise (**S&P**), diffuse image (**D**), salt and pepper combined with diffuse image (**S&P+D**) and diffuse image combined with salt and pepper (**D+S&P**). The repeatability rate is established as the number of interesting points found in the different synthetic images under the same process with

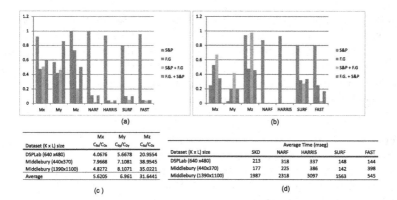

Fig. 5. Comparative analysis of the keypoints detectors depicted in Fig. 4. Graph (a) illustrates the results for images Middlebury and (b) DSPLab dataset; (c) presents the relation between sparse vectors and dense vectors (see Eq. (4)), and (d) show the performance with respect to time.

respect to the total points initially detected. Thereby, each interesting point of SKD is associated once to an interesting point in the other image. These points are first validated by means of the Sum of Squared Differences technique (SSD) to avoid false positive point detection. Figure 5(a, b) show the results for the Middlebury and the DSPLab dataset, respectively. In general, the best performance of the detectors is given for the synthetic images with **S&P** noise. The worst results are obtained for the diffuse image and their different combinations. On average, the repeatability rate is near to 0.5 for the SKD technique (in particular, M_x and M_y surfaces), 0.4 for SURF and FAST techniques, and 0.2 with other techniques. In the case of crowded scenes or similar textures under controlled indoor conditions, such an images Fig. 4(a, b), the SURF and FAST techniques present a low performance (close to Zero). It is important to note that under uncontrolled environmental conditions (see images Fig. 4(c, d) NARF and Harris techniques have shown a low performance (under to 0.2 in Fig. 5(b)). Figure 5(c) presents the relation between N_S with respect to N_D. For each technique in both datasets, it was obtained the average computation time of 100 executions, see Fig. 5(d). All of our data were obtained by using non-optimized Matlab implementations on an ordinary Intel(R) Core(TM)2 Duo 3.16 GHz CPU with 4 GB of RAM.

5 Conclusions

This study presented a novel approach for the contour detection by means of the interesting points from Range images. Although range images contain crowded or similar texture, the objects in the scene are separated from the background and the object contours are depicted using interesting points, with a minimal loss in the details of the scene. It is important to points out that the quantity of the descriptors is less to 6 % and 8 % of the total dense descriptors at **x**- and **y**-directions, respectively. This represents a reduction of near to 90 % in almost all tested cases. Additionally, the comparison with similar key-point detectors and the proposal presented here demonstrates that the SKD technique produces the best results in the detection of interesting points. Finally, the SND technique could be used in disciplines that imply processing of grey-level images.

References

1. Pears, N.C., Austin, J.: A machine-learning approach to keypoint detection and landmarking on 3d meshes. Int. J. Comp. Vis. **102**, 146–179 (2013)
2. Tombari, F., Salti, S., Di Stefano, L.: Performance evaluation of 3d keypoint detectors. Int. J. Comput. Vis. **102**, 128–220 (2013)
3. Pal, J.C., Weinman, J.J., Tran, L.C., Scharstein, D.: On learning conditional random fields for stereo. Int. J. Comput. Vis. **99**(3), 319–337 (2012)
4. Rodolà, E., Albarelli, A., Cremersa, D., Torsello, A.: A simple and effective relevance-based point sampling for 3d shapes. Pattern Recogn. Lett. **59**, 41–47 (2015)
5. Harris, C., Stephens, M.: A combined corner and edge detector. In: Alvey Vision Conference, pp. 147–151 (1988)
6. Rosten, E., Drummond, T.W.: Machine learning for high-speed corner detection. In: Leonardis, A., Bischof, H., Pinz, A. (eds.) ECCV 2006, Part I. LNCS, vol. 3951, pp. 430–443. Springer, Heidelberg (2006)
7. Bay, H., Tuytelaars, T., Van Gool, L.: SURF: speeded up robust features. In: Leonardis, A., Bischof, H., Pinz, A. (eds.) ECCV 2006, Part I. LNCS, vol. 3951, pp. 404–417. Springer, Heidelberg (2006)
8. Hirschmuller, H., Scharstein, D.: Evaluation of cost functions for stereo matching. In: IEEE Conference on Computer Vision and Pattern Recognition. CVPR 2007, pp. 17–22. IEEE Press, Minneapolis, MN, USA (2007)
9. Hernndez-Lpez, J.J., et al.: Detecting objects using color and depth segmentation with Kinect sensor. In: Procedia Technology, editor. The 2012 Iberoamerican Conference on Electronics Engineering and Computer Science, pp. 196–204. Elsevier Science, Mexico (2012)
10. Ibarra-Manzano, M.-A., Almanza-Ojeda, D.-L.: An FPGA implementation for texture analysis considering the real-time requirements of vision-based systems. In: Koch, A., Krishnamurthy, R., McAllister, J., Woods, R., El-Ghazawi, T. (eds.) ARC 2011. LNCS, vol. 6578, pp. 110–117. Springer, Heidelberg (2011)
11. Hedrich, J., Dietrich, P., Francois, G., Marcin, M.: Enhanced surface normal computation by exploiting RGBD sensory information. In: 2015 14th IAPR International Conference on Machine Vision Applications (MVA), Tokyo, USA, pp. 26–29 (2015)
12. Steder, B., Rusu, R.B., Konolige, K., Burgard, W.: NARF: 3D range image features for object recognition. In: International Conference on Intelligent Robots and Systems (IROS), vol. 44 (2010)
13. Sapiro, G.: Geometric Partial Differential Equations and Image Analysis. Cambridge University Press, Cambridge (2006)
14. MICROSOFT: The Kinect effect - how the world is using Kinect [Internet]. http://www.xbox.com/en-GB/kinect/kinect-effect

An Optimization Approach to the TWPVD Method for Digital Image Steganography

Ismael R. Grajeda-Marín, Héctor A. Montes-Venegas$^{(\boxtimes)}$,
J. Raymundo Marcial-Romero, J.A. Hernández-Servín, and Guillermo De Ita

Facultad de Ingeniería, Universidad Autónoma del Estado de México,
Cerro de Coatepec s/n, Toluca, Edo. de Mexico, Mexico
tione_210@hotmail.com, {hamontesv,jrmarcialr,xoseahernandez}@uaemex.mx,
deita@cs.buap.mx

Abstract. In Digital Image Steganography, Pixel-Value Differencing methods commonly use the difference between two consecutive pixel values to determine the amount of data bits that can be inserted in every pixel pair. The advantage of these methods is the overall amount of data that an image can carry. However, these algorithms frequently either overflow or underflow the pixel values resulting in an incorrect output image. To circumvent this issue, either a number of extra steps are added to adjust those values, or simply the pixels are deemed unusable and they are ignored. In this paper, we adopt the Tri-way Pixel-Value Differencing method and find an optimal pixel value for each computed pixel block such that their difference holds the maximum input data and neither underflow or overflow pixels exist.

Keywords: Optimisation · Steganography · Tri-way Pixel-Value Differencing · TWPVD

1 Introduction

Digital steganography is the set of techniques designed to conceal digital data (the payload) within a digital medium or carrier. Unlike related areas, such as cryptography or watermarking, steganography techniques aim to keep the existence of a message undetected and to continuously increase the amount of input data to be embedded [3].

In digital image steganography, the pixel intensities are used to hide the payload data. A common approach, and perhaps the simplest, is to use some form of Least Significant Bit (LSB) insertion method [1]. LSB methods replace b least significant bits of the carrier pixels with the same number of payload data bits. The less bits being replaced, the less altered the carrier image will be, but also the payload will be smaller. Some LSB substitution techniques have implemented an optimal pixel adjustment for data embedding to reduce the disruption of the carrier image [10]. Other steganographic methods include an assortment of transformation as well as masking and filtering techniques. Surveys and reviews of current methods are readily available in the literature [3,7].

© Springer International Publishing Switzerland 2016
J.F. Martínez-Trinidad et al. (Eds.): MCPR 2016, LNCS 9703, pp. 125–134, 2016.
DOI: 10.1007/978-3-319-39393-3_13

With the objective of increasing the amount of data that an image can carry, a set of techniques have been proposed that use the difference between two neighbour pixels to hide input data. This difference can be computed in any neighbouring direction. Wu and Tsai [11] proposed a Pixel Value Differencing (PVD) method that produces a stego-image with considerable payload data and a substancial image quality. Thereafter various approaches based on PVD have been produced [2,4,12].

Ideally, the payload must be recovered using only the resulting pixel values, and all pixels of the original image should be used to embed data in order to achieve a higher payload. However, many PVD methods yield overflow or underflow pixels (i.e., out of the valid range interval) and decide either to ignore or to somehow adjust the resulting pixel values. This, however, may lead to a lower payload or to include additional strategies to retrieve the embedded data [11] that may reveal the existence of a hidden message.

In this paper, we adopt the Tri-way Pixel-Value Differencing method and find an optimal pixel value for each computed pixel block such that their difference holds the maximum input data. Our method reduces the size of the search space and computes a much more smaller set of feasible solutions. In addition, two more strategies are discussed to further increase the size of the embedded payload. The method is designed in such a way that the resulting pixel intensities are never out of the valid interval and it uses all pixel blocks to carry payload data. A series of experimental results show the feasibility of the method.

We begin in Sect. 2 by covering the basics of the Pixel-Value Differencing method. Section 3 presents a detailed description of our two optimisation algorithm approaches. Section 4 presents several experimental results, and Sect. 5 concludes the paper.

2   Pixel-Value Differencing

The PVD method [11] assumes that the payload is a continuos stream of input bits that represent any type of digital data. The PVD embeds data using the intensity difference of two contiguous pixels. The idea is to modify these pixels by adding a decimal conversion of some input data bits in such a way that their value difference is kept to preserve the image quality. Regions in the image with larger differences in pixel intensities can carry more pieces of payload than others. This usually happens in the areas with evident edges and less frequently in smoother regions. The method provides a good embedding capacity but is prone to be detected using statistical based stego-analysis methods [5].

Chang *et al.* [2] proposed a modified version of the PVD named *Tri-way Pixel-Value Differencing* (TWPVD). Whereas the PVD inserts data in only one pixel pair, the TWPVD uses horizontal, vertical and diagonal diferences (hence its name) in 2×2 pixel blocks to hide input data, thus achieving a higher payload than the PVD in the carrier image. One problem arising in PVD based methods is that they frequently yield overflow/underflow pixel values. These pixels are either adjusted or ignored by the method, thus reducing the number of pixels available to carry data payload [2,6].

2.1   Tri-Way Pixel-Value Differencing

The Tri-way Pixel-Value Differencing (TWPVD) method was designed to get more pixels involved in the data embedding process [2]. The TWPVD divides the carrier image into non-overlapping blocks of 2×2 consecutive pixels. Three difference values are computed in each block from the values of two neighbour pixels in three distinctive directions. The first difference is computed between the pixel in the upper left corner, namely the pivot, and the pixel on its right. The second difference is between the pivot and the pixel in the opposite corner, and the third one is also between the pivot and the pixel below it. Each difference belongs to one of a predefined set of range intervals which, in turn, determines the number of bits to be inserted in every pixel pair. Each range interval R_k has a lower l_k and an upper u_k value listed in the form of a range table. The range table has been designed simply by computing each interval width using a power of two, either to provide large capacity or to provide high imperceptibility [11,12]. Other approaches have designed the range table based on the *perfect square number* [9], or have opted for entirely replacing the range table with a well crafted function based on the floor and ceiling functions [4].

Regardless of how these range intervals are produced, the TWPVD algorithm follows these steps:

1. Compute the differences $d_i = p_i - p_1$ within the pixel block $i \in \{1, 2, 3, 4\}$,
   where

p_1	p_2
p_3	p_4

2. Locate for each d_i the range k such that $l_k \leq |d_i| \leq u_k$
3. Compute the amount of input data bits t_i to be inserted in the difference i of the block p_i as follows:

$$
t_i = \begin{cases} 0 & \text{if } i = 1 \\ \lfloor \log_2(u_k - l_k + 1) \rfloor & \text{otherwise} \end{cases} \tag{1}
$$

4. Compute the decimal representation b_i of the t_i bits
5. A new d'_i is computed for each d_i

$$
d'_i = l_{k_i} + b_i \tag{2}
$$

6. Later the TWPVD uses each d'_i to compute the values of the resulting pixels p'_i using a well crafted set of rules [2]. We have adopted the TWPVD by replacing these rules with an *optimisation* strategy to determine the best pixel values that hold the maximum payload.

A closer look to this algorithm, reveals that it also produces over-flow/underflow pixel values that are simply skipped as data payload carriers. Worse still, TWPVD authors [2] do not seem to discuss how the extraction algo-rithm knows which pixels are being ignored [4]. This is fundamental to guarantee the integrity of the secret message.

3 An Optimisation Approach to Modify the TWPVD

Any PVD method can be seen as an optimisation problem as follows: Given d'_i and p_i, search for a solution p'_i subject to the following set of conditions:

1. Overflow/underflow must be prevented subject to $0 \leq p'_i \leq 255$
2. Retrieving the payload data is subject to $d'_i = |p'_i - p'_1|$, where p'_i and p'_1 are now variables to be searched as an optimization problem which will define the stego-image.
3. Distortion of the resulting image must be subject to minimize the objective function

$$f(p_i, p'_i) = \sum_{i=1}^{4} (p_i - p'_i)^2 \qquad (3)$$

We know that $p'_i = |d'_i|$ is a solution, *i.e.* $p_1 = 0$, that fulfills conditions 1 y 2, but does not fulfill condition 3 because it causes a major distortion to the resulting stego-image. Nonetheless, the solution shows that there exist at least one solution for any given input.

Since there are 4 pixels per block in the range [0..255], we can easily estimate the size of the search space to be 2^{32} possible pixel value combinations times the carrier image dimensions divided by 4. These solutions take far too long to be explored efficiently, as shown in Table 1.

Table 1. Comparison between Optimal-TWPVD and a simply Brute Force strategy added to the TWPVD. The time performance advantage is clear.

	bpp		PSNR		Time	
	BFTWPVD	OTWPVD	BFTWPVD	OTWPVD	BFTWPVD	OTWPVD
Barbara	2.54	2.54	36.50	36.50	471.60	9.56
Airplane	2.37	2.37	38.90	38.90	458.35	10.02
Boat	2.41	2.41	38.19	38.19	394.81	8.35
Goldhill	2.38	2.38	38.73	38.73	323.85	17.72
Lena	2.35	2.35	39.34	39.34	366.10	11.80
Average	2.41	2.41	38.33	38.33	402.94	11.49

One alternative is to reduce the size of the search space so that it can be explored in useful times.

Using equation from condition 2 it follows that

$$p'_i = \pm d'_i + p'_1 \qquad (4)$$

This evidently means that we can compute p'_i using the two following variables:

1. $\pm d'_i$ takes the different sign combinations for d'_i. These combinations are 8 because d'_1 is always 0 and d'_2, d'_3, d'_4 only can take 2 different values: one positive and one negative of equal magnitude.
2. p'_1 must be subject to $0 \le p'_1 \le 255$. This means that p'_1 only can take 256 distinct values.

This further reduces the size of the search space to 2^{11}. A search space of this size can be readily explored in its entirety. That is, all possible values for p'_1 must be combined with all possible values for $\pm d'_i$.

3.1   An Additional Optimisation Strategy

We now describe an additional optimisation strategy to further increase the payload inserted by the method from Sect. 3. Such strategy is based on the first derivative of the objective function with respect of p'_1 and discards the overflow/underflow solutions.

Using Eqs. 3 and 4, a quadratic function can be produced in terms of p'_1, namely:

$$f(p'_1) = \sum_{i=1}^{4} (\pm d_i + p'_1 - p_i)^2 \tag{5}$$

Eight different quadratic curves can be plotted from the eight different combinations of signs in $\pm d_i$. When computing the first derivative of these functions, a point for each curve can be found for which f is minimum:

$$p'_1 = \frac{1}{4} \sum_{i=1}^{4} p_i - \frac{1}{4} \sum_{i=1}^{4} \pm d_i \tag{6}$$

The resulting 8 candidate values for p'_1 can become 16 because Eq. (6) can yield real numbers that need to be converted into integers using both the *ceil* and *floor* functions.

In some cases the optimal point can be off the valid range or can even cause some of the other 3 pixels to be off. It is necessary then to move that point within the proper range as that value is potentially a solution.

Figure 1 shows 2 curves plotted using the objective function. These curves are bounded between a pair of dotted lines representing the upper and lower bounds valid for p'_1. It also shows that the points of minimum value are not always within the valid interval and is necessary to move that point to a valid area.

Equation (4) can yield valid intervals for each curve as $\max(\pm d_i) \le p'_1 \le \min(\pm d_i + 255)$. From this equation, we can define the adjustment function:

$$A(p, M, m) = \begin{cases} 0 & if\ M \le p \le m \\ M - p & if\ p < M \\ -(p - m)\ if\ p > m \end{cases} \tag{7}$$

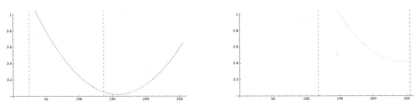

a) Optimal point *outside* the valid interval b) Optimal point *inside* the valid interval

Fig. 1. Two different objective function graphs

Therefore the optimal point in the valid range would be defined as:

$$p'_1 = p'_1 + A(p'_1, \max(\pm d_i), \min(\pm d_i) + 255) \tag{8}$$

As mentioned before, this point needs to be adjusted using the *ceil* or *floor* functions. Both functions yield an identical or extremely close value. Because there are 8 curves each with 2 solutions, we end up with a new search space of only 16 potential solutions.

The algorithm follows these steps:

1. Go through steps 1–5 of the algorithm from Sect. 2.1
2. Compute $s_i = \pm d_i + p'_1$ using the *ceil* or *floor* functions. Overflow/underflow solutions are discarded
3. The optimal solution is given by $p'_i = \min(f(p_i, s_i))$
4. Replace the original 2×2 pixel block with the optimal solution found
5. Repeat for each 2×2 pixel block of the carrier image

To recover the secret message, the inverse process is applied as follows:

1. Divide the carrier image into non-overlapping blocks of 2×2 consecutive pixels
2. Compute the differences $d_i = p_i - p_1$ within the pixel block $i \in \{1, 2, 3, 4\}$
3. For each d_i locate the table range $r_i = k$ such that $l_k \leq |d_i| \leq u_k$
4. Compute the number of inserted bits in each difference
$$t_i = \begin{cases} 0 & \text{if } i = 1 \\ \lfloor \log_2(u_{r_i} - l_{r_i} + 1) \rfloor & \text{otherwise} \end{cases}$$
5. The entire data payload is recovered by concatenating the binary representation of $b_i = d_i - l_{r_i}$

3.2 Inserting an Extra Bit

The method can insert an additional bit to further increase the secret message inserted in each 2×2 block with a minimal deterioration to the carrier image.

The *floor* and *ceil* functions yield two consecutive integer numbers that produce very close or even identical Objective Function results. This type of function curves constantly appear and are used as indication for inserting an additional bit of the secret message. This additional bit is called β. If $\beta = 0$, p'_1 must be even, if $\beta = 1$, p'_1 must be odd.

To find the optimal we say that $2c = p'_1 - \beta$ and modify Eq. 4 as follows:

$$p'_i = \pm d_i + 2c + \beta \tag{9}$$

$$f(c) = \sum_{i=1}^{4} (\pm d_i + 2c + \beta - p_i)^2 \tag{10}$$

Therefore, the valid interval for the optimisation problem is given by:

$$c = \frac{1}{8} \sum_{i=1}^{4} p_i - \frac{1}{8} \sum_{i=1}^{4} \pm d_i - \frac{1}{2} \beta \tag{11}$$

$$c = c + A(c, \max(-\frac{\beta}{2} - \frac{1}{2}(\pm d_i)), \min(-\frac{\beta}{2} - \frac{1}{2} \pm (d_i) + 255)) \tag{12}$$

The algorithm is also modified as follows:

1. Go through steps 1–5 of the algorithm from Sect. 2.1
2. Compute $s_i = \pm d_i + c$ using the *ceil* or *floor* functions. Overflow/underflow solutions are discarded
3. The optimal solution is given by $p'_i = \min(f(p_i, s_i))$
4. Replace the original 2×2 pixel block with the optimal solution found
5. Repeat for each pixel block of the carrier image

To recover the message payload, the same steps from Sect. 3.1 are used, and an extra bit 0 is added to the message if p_1 is even or a 1 otherwise.

4   Experimental Results

A set of images were used to test the performance of our algorithms and to compare our results to those previously published in the literature. All carrier images, shown in Fig. 2, are 8-bit grayscale images of size 512×512. These images belong to a larger set that have become a *de facto* standard in Image Processing and Computer Vision experiments for testing new developments. We also have chosen these images to compare our results with previous work by Peng et al. [8] and Hernandez-Servin et al. [4]. Both authors compared their own results with work previously published. In addition, we also compare the performance of our algorithm with the results of the TWPVD [2].

The peak signal-to-noise ratio (PSNR) is used to measure the difference between the original carrier image and the image with the message payload. The higher the PSNR, the better the quality of the stego image. The number of bits per pixel (bpp) for each test image, is computed simply by dividing the number of bits inserted by the number of pixels in the carrier image.

Airplane   Barbara   Boat   Goldhill   Lena

Fig. 2. Original images (first row). Resulting stego images using the Optimal-TWPVD (second row). Resulting stego images using the OTWPVD and Extra Bit Insertion (third row)

Table 2 shows a comparison between the Optimal-TWPVD and the Extra Bit Insertion algorithms. While the former shows a better performance than previous work (as shown in Table 4), the latter further increases the overall results in terms of both the amount of data payload (the *bpp*) inserted, and the image distortion measured with the PSNR in all images tested. This might seem expected as both the Optimal TWPVD and the Extra Bit Insertion strategies use every 2 × 2 block to carry data payload. No pixel block is ignored and no pixel overflow/underflow occurred.

Table 2. Comparison between the Optimal-TWPVD and the Optimal-TWPVD with Extra Bit Insertion

	bpp		PSNR		Time	
	OTWPVD	EOTWPVD	OTWPVD	EOTWPVD	OTWPVD	EOTWPVD
Barbara	2.54	2.79	36.50	36.43	9.56	16.21
Airplane	2.37	2.62	38.90	38.76	10.02	17.99
Boat	2.41	2.66	38.19	38.09	8.35	20.24
Goldhill	2.38	2.63	38.73	38.64	17.72	16.53
Lena	2.35	2.60	39.34	39.17	11.80	15.65
Average	2.41	2.66	38.33	38.22	11.49	17.32

We also compare our results with those from the TWPVD [2] in Table 3. Since our algorithms search for the optimal pixel values for each block, the results are superior in terms of both *bpp* and PSNR. The general notion is that less data embedded should result in less distortion of the carrier image, which is not observed by comparing the PSNR values of our experiments.

Table 3. Comparison between the TWPVD and a our Optimal-TWPVD

	bpp		PSNR		Time	
	TWPVD	OTWPVD	TWPVD	OTWPVD	TWPVD	OTWPVD
Barbara	2.54	2.54	36.38	36.50	1.55	9.56
Airplane	2.37	2.37	38.23	38.90	1.90	10.02
Boat	2.40	2.41	37.72	38.19	1.88	8.35
Goldhill	2.38	2.38	38.09	38.73	1.84	17.72
Lena	2.35	2.35	38.61	39.34	2.19	11.80
Average	2.41	2.41	37.81	38.33	1.87	11.49

A similar comparison with recent results by Peng *et al.* [8] and Hernandez-Servin *et al.* [4] is shown in Table 4. This table also shows favorable results in terms of both data payload carried and stego image quality.

Table 4. Comparison between our proposals Optimal-TWPVD & Extra Bit OTW-PVD, and Hernadez-Servin *et al.* [4] and Peng *et al.* [8]

bpp					PSNR			
	OTWPVD	EOTWPVD	[4]	[8]	OTWPVD	EOTWPVD	[4]	[8]
Barbara	2.54	2.79	1.38	1.20	36.50	36.43	36.04	30.75
Airplane	2.37	2.62	1.30	1.20	38.90	38.76	36.09	33.45
Boat	2.41	2.66	1.80	1.20	38.19	38.09	34.56	26.66
Goldhill	2.38	2.63	1.66	1.20	38.73	38.64	37.03	30.70
Lena	2.35	2.60	1.60	1.20	39.34	39.17	37.55	26.89
Average	2.41	2.66	1.55	1.20	38.33	38.22	36.25	29.69

5 Conclusions

This work designs an optimisation algorithm that modifies and improves the TWPVD [2] steganographic method. It is favourably compared against the TWPVD and also against recent results by Peng *et al.* [8] and Hernandez-Servin *et al.* [4].

Our results show improvements in several important aspects, namely, (1) Number of Bits per pixel inserted, (2) Better stego image quality measured with the PSNR, (3) No overflow/underflow pixels are produced, and (4) No blocks of pixels are skipped or ignored as data carriers.

The major merit of our algorithms is to reduce the feasible set of possible pixel values for each block such that the search for the best solution in terms of both data payload and stego image quality can be efficiently conducted.

There are a couple directions in which this work may proceed. The first logical next step is to use the method to hide data payload into color images. The obvious result would be to achieve a very large payload insertion, but the effects in color and general image distortion may either require to adapt or entirely change the algorithm.

References

1. Chan, C.K., Cheng, L.M.: Hiding data in images by simple LSB substitution. Pattern Recogn. **37**(3), 469–474 (2004)
2. Chang, K.C., Chang, C.P., Huang, P.S., Tu, T.M.: A novel image steganographic method using Tri-way Pixel-Value Differencing. J. Multimedia **3**(2), 37–44 (2008)
3. Cheddad, A., Condell, J., Curran, K., Kevitt, P.M.: Digital image steganography: survey and analysis of current methods. Sig. Process. **90**(3), 727–752 (2010). http://www.sciencedirect.com/science/article/pii/S0165168409003648
4. Hernández-Servin, J.A., Marcial-Romero, J.R., Jiménez, V.M., Montes-Venegas, H.A.: A modification of the TPVD algorithm for data embedding. In: Carrasco-Ochoa, J.A., Martí-nez-Trinidad, J.F., Sossa-Azuela, J.H., Olvera López, J.A., Famili, F. (eds.) MCPR 2015. LNCS, vol. 9116, pp. 74–83. Springer, Heidelberg (2015)
5. Mahajan, M., Kaur, N.: Adaptive steganography: a survey of recent statistical aware steganography techniques. Int. J. Comput. Netw. Inf. Secur. (IJCNIS) **4**(10), 76–92 (2012)
6. Mandal, J.K., Das, D.: Steganography using adaptive pixel value differencing (APVD) of gray images through exclusion of overflow/underflow (2012). CoRR arXiv:1205.6775
7. Mishra, M., Mishra, P., Adhikary, M.C.: Digital image data hiding techniques: a comparative study. ANSVESA **7**(2), 105–115 (2012)
8. Peng, F., Li, X., Yang, B.: Adaptive reversible data hiding scheme based on integer transform. Sig. Process. **92**(1), 54–62 (2012)
9. Tseng, H.W., Leng, H.S.: A steganographic method based on pixel-value differencing and the perfect square number. J. Appl. Math. **2013**(1), 1–8 (2013)
10. Wang, R.Z., Lin, C.F., Lin, J.C.: Image hiding by optimal LSB substitution and genetic algorithm. Pattern Recogn. **34**(3), 671–683 (2001)
11. Wu, D.C., Tsai, W.H.: A steganographic method for images by pixel-value differencing. Pattern Recogn. Lett. **24**(9), 1613–1626 (2003)
12. Wu, H.C., Wu, N.I., Tsai, C.S., Hwang, M.S.: Image steganographic scheme based on pixel-value differencing and lsb replacement methods. IEE Proc. Vis. Image Sig. Process. **152**(5), 611–615 (2005)

Real Time Gesture Recognition with Heuristic-Based Classification

Omar Lopez-Rincon and Oleg Starostenko[✉]

CENTIA, Department of Computing, Electronics, and Mechatronics,
Universidad de Las Americas Puebla, 72820 Cholula, Mexico
{omar.lopezrn,oleg.starostenko}@udlap.mx

Abstract. The recognition of human gestures in real time is still open problem due to low success rate of systems recently reported in scientific literature. This paper presents the proposed method and designed prototype for motion analysis and classification for human-computer interaction. The method is based on pattern recognition techniques of artificial vision without applying any markers or special sensors as well as utilizing low resolution cameras and simple hardware specifications. The proposed method provides interaction of user with computer via gestures in habitual and normal manner in order to activate system events (up, left and right) in real time. The proposed heuristic classifier recognizes specified gestures with an appropriate system context precision of 91.25 %. Comparing the obtained results with recent reports, the proposed approach provides satisfactory gesture recognition in real time with low resolution cameras.

Keywords: Pattern recognition · Gesture recognition · Artificial vision · Heuristic classifier

1 Introduction

Nowadays, the most common interfaces between human and computer systems still are the keyboard and the mouse, but the tendency in a short term has been focused on devices with touchscreen and gesture recognition with movements executed by the user [1]. The gestures are generated from body movements such as arms, hands, fingers, head, face or body [2].

Karam [3] reported that hands are the most used to execute gestures compared with any other body parts, as it is a natural part of human's communication, either for senti-ments as for intentions. Therefore they are the most adequate for the natural interaction with computers too. The research concerning pattern recognition is directed to systems that can identify human gestures as entries and process them to control devices, mapping those gestures as commands. The main technologies at the present moment are based on artificial vision and contact [4]. We will focus in the artificial vision (AV) recognition. The capacity to detect gestures using AV and pattern recognition, allows to explore a variety of interaction techniques to control different environments, for example: changing the music volume or manipulating the thermostat without approaching to it [5]. In devices based on touchscreens for gesture recognition, it is necessary to detect the

© Springer International Publishing Switzerland 2016
J.F. Martínez-Trinidad et al. (Eds.): MCPR 2016, LNCS 9703, pp. 135–144, 2016.
DOI: 10.1007/978-3-319-39393-3_14

beginning of the movement, called *gesture localization* [6], which is recognized at the moment of making contact with the surface or the sensible part of the device.

By keeping record of the executed movement over the surface or tactile sensor, the registered sequences are verified to evaluate if they match with the established classifications. If a system matches with any of the gestures to which the system responds, it is considered as an action by the user and then the system triggers an event in response. This kind of feedback doesn't exist in touchless systems. Several research groups are "on the run" in developing the standard scope for pattern recognition. There are several alternatives for pattern recognition, from complex devices, such as complete body suits, to non-invasive devices such as infrared depth cameras as Kinect early known as Prime-Sense of Israeli 3D sensing company [2]. There are also complex methods such as those which by using wireless network signal they detect the corporal movements and recognize the human gestures [5].

Through movement analysis using a web cam and user interface for a simple computing tasks system, new technologies are within reach of everyone. Using AI would be a practical way of solving gesture recognition [7]. Thus, the main purpose of this paper is gesture interpretation, using a static camera as well as to present a novel and fast method to classify gestures.

The rest of paper is organized as follows. Section 2 presents an analysis of well-known relevant methods of real time gestures recognition systems based in artificial vision. Section 3 describes the proposed method and Sect. 4 presents the heuristic and classification approaches. Section 5 shows experimental results and performance evaluation of the proposed method. Finally, Sect. 6 presents conclusions and future work.

2   Related Works for Real Time Gesture Recognition

According to Mitra [2], gesture recognition is the process where the user acts a gesture out and the receptor recognizes it as an input. According to this, we could interact with machines, sending them messages as a signal relating them with the environment and the system's syntaxes. In order to achieve this, the image processing and furthermore, feature extraction are required. Most of vision based systems, comprehend three stages: detection, following, and classification or recognition [8]. At the first stage the challenges are the hands recognition and segmentation of the desire region within the image. This process is imperative to eliminate irrelevant information from the background and then follow the movement as a sequence. Several characteristics had been taken into consideration in different methods to achieve this like color, shape, movement or templates [8]. Due to space-time variations, desired segmentation of the hand and correct movement tracking are still a higher challenge. Errors at this early stage of the process cause deviation of the real trajectory during movement tracking [9].

Coming up next, some of the most used methods from the past 5 years for gesture recognition with AI are summarized according to Athavale [4] (see Table 1). All of them work in real time, searching and detecting skin color, which makes them sensible to image color and lighting as well as users must have their hands uncovered.

Table 1. Summary of recent relevant methods for human gesture recognition.

Source:	Bhuyan 2014 [9]	Kshirsagar 2013 [10]	Palochkin 2014 [11]	Rios-Soria 2013 [12]	Lee 2013 [13]
Advantage	robust	easy	easy	easy	Fast/robust
Disadvantage	short gestures	weak	static gestures	static gestures	expensive
Features	TV[a]	TV	C-H[b]	C-H	TV
Extraction	CRF[c]	FSM[d]	Viola-Jones	heuristic	SVM[e]
Face detection	Yes	No	No	No	Yes
Motion tracking	local features	states	skin detection Lucas-Kanade	skin detection	Kanade-Lucas-Tomasi
Training	Yes	NO	Yes	NO	Yes
Classification	classifier	templates	parameter analysis	parameter analysis	templates
Background	simple	complex	simple	simple	complex

[a]Trajectory vector; [b]Convex hull; [c]Conditional random fields; [d]Finite state machine; [e]Support vector machine

The overall success rate of the recent and the most relevant systems for gesture recognitions lie in the range of 70–96 % [1, 9, 11, 13]. As mentioned before, the gesture analysis methods as usual use skin recognition, which in case of presence of gloves would disable most of them [14]. Some methods process only static gestures applying complex algorithms that frequently do not provide fast recognition required for real-time applications. In this paper the proposed approach has been developed for high speed short gestures recognition in real-time using image acquisition and processing tool without any high quality requirements.

3   The Proposed Method for Gesture Detection

Commonly, an object detection process includes a differences frame implementation, background elimination and a method for the movement tracking. To achieve these several processes to work at real time we start making two copies of each frame in reduced different sizes: one with 50 × 50 pixels and another one with 100 × 80. The biggest image (the one with 130 × 95) is scanned for face detection using the Viola-Jones method [15]. When a face is detected the system grabs the frame t and the frame t-1 of the 50 × 50 frames and starts subtracting them in a loop until there is no face detection.

It this proposal we used a frame differences method due to its high speed to detect the motion in a video sequence. This is done with a dyadic pixel by pixel comparison and the difference calculation on both spatial axes x, y of the image [16]. Then a color reduction is done on each frame f on the pixel P at the t moment on the x, y position and its V value is obtained with the luminance calculation with its RGB values using the following equation (Eq. 1).

$$V = 0.21 * \text{red} + 0.72 * \text{green} + 0.07 * \text{blue} \tag{1}$$

Each of the frames $f(x, y, t - 1)$ and $f(x, y, t)$ are gray scale images and continuous from the real time input video sequence. Their difference is expressed as follows (Eq. 2) drawing desired pixels of detected objects in motion in black.

$$D_t(x, y) = -1|V(f(x, y, t - 1)) - V(f(x, y, t)| \tag{2}$$

Then the difference obtained from the matrices D_t is used to create a new binary image B_t by taking each D_t value and evaluate it at different thresholds. If the D_t value lies between *uMax* and *uMin*, then corresponding pixel in B_t would be *0*, otherwise the pixel is depicted as *1* (Eq. 3).

$$B_t(x, y) = val; val = 0, uMin > val < uMax$$
$$B_t(x, y) = val; val = 1, uMin < val > uMax \tag{3}$$

At the same time all of the gray values are evaluated to create the binary image. For this two histograms are established H_x and H_y each of them with 50 values. The *x histogram* (H_x) will have all the values of the columns of the binary image and the *y histogram* (H_y) all the ones that correspond to each row (see Fig. 1).

The motion tracking is achieved by crossing the maximum values in each of the histograms and the final classification is done by tracking the intersection of these two values of the histograms and comparing them to the heuristics (see Fig. 2).

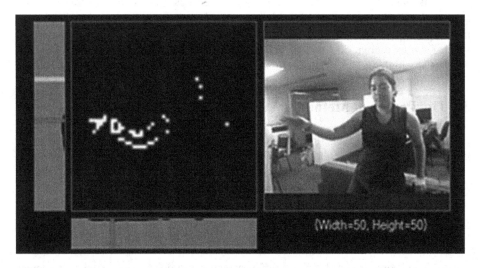

Fig. 1. Object in motion and its histogram. Despite of blurred object in motion due to low resolution camera, the approach satisfactory detects the region of interest.

The second part of the classification involves the face detection. It is used as a local reference parameter. The gesture detection and motion tracking start with a recognized face. At this moment we have new dynamic thresholds inside the image which are based on the position of the detected face.

We looked up for proportions of thresholds, where three gestures *(up, right, left)* could be recognized without collision reducing in this way false positives and true negatives. If both of the histograms are empty then it means there is no detected motion from the user. The block diagram of the procedure for fast gesture recognition and classification is shown in Fig. 3.

Applying fast feature extraction approaches with quite low computational complexity, the gesture detection is done effectively in real time as well as the used procedures are simple and easy to implement with low cost hardware.

Fig. 2. The motion tracking by crossing the maximum values in each of the histograms detecting the maximum motion in the image sequence.

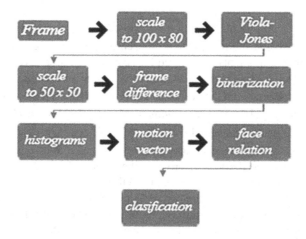

Fig. 3. Block diagram of the algorithm for gesture detection and recognition

4   The Proposed Algorithms for Recognition and Classification

In skeleton based classification, the gesture is determined by comparing the movement and the position of the wrist, elbow and shoulders from the detected body [17]. We start face recognition taking into account that it is important not only to determine if there is a user in motion, but also a face position must be detected. A face will be used as the reference of hand motion in relation to a human body without restricting its position.

This is used to determine three thresholds to create three different rules in order to find if the detected movement is *left, right* or *up* gesture. All the detected movements above the face are ignored. This allows the user to move in a natural way filtering several true negatives.

Three final thresholds related to a face have been proposed to use. The first one is measured as a center point of the height of the recognized face and it is used to detect the *up* gesture. If the face is detected at the starting point of the superior left side on the image $P(x, y)$ with a height H and a width W, then the center point would be calculated as it follows (see Eq. 4)

$$U_1 = P(x + \frac{W}{2}, y + \frac{H}{2}) \tag{4}$$

The second threshold is set to the right of the face at a distance localized in the face width multiplied by 2.5. This is the approximation of the hand position pointing towards right above its shoulder and elbow (see Eq. 5).

$$U_2 = P(x - W * 2.5, y + \frac{H}{2}) \tag{5}$$

The last threshold is at the end of the right side of the image, given by the same point $P(x, y)$ and the addition of the face width W (see Eq. 6).

$$U_3 = P(x + W, y) \tag{6}$$

On the Fig. 4 we can see the image with each of the mentioned thresholds. Additionally, there are two moments considered for gesture activation $M(t)$ and $M(t - 1)$. If a motion at the point $P(H_x(Max), H_y(Max))$ crosses one of the thresholds with two moments at the same time then the gesture is activated. With the histogram crossing the position of a movement is stored as a moment $M_{t-1}(P(H_x(Max), H_y(Max)))$ from the previous frame and one of the actual frame $M_t(P(H_x(Max), H_y(Max)))$. This generates a directional vector with the motion direction.

On Fig. 5 we can see the yellow cross of the center of face detection, done on a 50×50 image. The red lines indicate where the 3 thresholds are placed according to the position of a face. The white pixels are part of motion detection on the image and finally, with color circles we can see the motion vector direction. The red circle (closest to cross) represents where the motion came from and the green one (right most) shows the final position and defines the direction.

Fig. 4. Two examples of three thresholds related to a face used for gesture detection

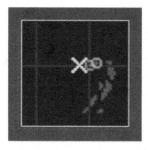

Fig. 5. Motion vector defined by red starting (left) circle and green final (right) circles. (Color figure online)

When the vector v completely crosses one of the thresholds, a gesture is detected and classified given by the direction of the vector $G(v)$ (see Eq. 7).

$$G(v) \begin{cases} UP, v > U_1 \\ LEFT, v < U_2 \\ RIGHT, v > U_3 \end{cases} \tag{7}$$

Additionally, the particular priority of gestures for classification is applied. The highest priority has the *up* gesture, then the outside threshold is analyzed specifying the *right gesture* and finally, the *left gesture* is evaluated analyzing presence of motion from center of the face to the left direction.

On the Fig. 6 we can see the histograms, the movement found by the histograms, the detected face *(yellow cross)* and the motion vector with the thresholds drawn in red accordingly with the face position.

Fig. 6. Final frame composed by the histograms for motion detection, the detected face (yellow cross) and the motion vector with the thresholds drawn in red (Color figure online)

The method compared with other proposals is the simplest in implementation and the easiest one from computational complexity point of view. It is not necessary to recognize skin color and the searching ace position is used only as the reference parameter to establish the thresholds for detection of the motion of user hands.

It is important to mention that the simplicity of each part contributes to low requirements for used hardware and implementation of the whole procedure.

5   Experimental Results and Evaluation

In Table 2 the results of several experiments are presented. The tests have been carried out with volunteers using the designed system based on the proposed method for gesture recognition. Every user had a session of interaction with a system making different gestures in a random sequence. All of the volunteers were sitting down, so the individual height of each user wouldn't varies too much to become an issue of adjustment every time at each test. Table 2 resumes the total number of attempts of each user for every produced gesture (up, right, left) and precision of recognition of gesture of this particular user by system. The recognition rates of each gesture for all users are resumed in the columns.

Table 2.   Results of gesture recognition by different users

Type of gesture					
	Attempts	*up/%*	*right/%*	*left/%*	Success rate
User 1	13	5	4	4	92.3 %
User 2	28	4	11	13	92.8 %
User 3	19	5	8	6	93.1 %
User 4	17	4	8	5	86.5 %
User 5	38	4	18	16	88.8 %
User 6	12	4	4	4	98.1 %
Total attempts:	127	26	53	48	
Total precision:		98.7 %	95.1 %	83.0 %	91.25 %

The average success recognition rate computed taking into account the number of attempts proportional to contribution of each user is about 91.25 %. That is a quite acceptable result for system that has no any particular requirements for high quality tool. For instance, the tests were made using webcam of low resolution (800×600) at the approximated distance of 1.5 m in ambient with soften indoor light.

The precision of recognition varies from 86.5 % (the case when user 4 has been illuminated with diffuse ambient light only) to 98.1 % (the case when user 6 has been illuminated with additional directional light from lamp directly to his face). Better illumination facilitates the face detection and increases precision of recognition.

Analyzing columns of Table 2, it is important to mention that the gestures *up* and *right* have been recognized with the highest precision (98.7 % and 95.1 %, respectively), while the recognition of left gesture (83 %) had errors due to more complex background of hand in motion to left for right handed users.

In Table 3 the final comparison of success rate of the proposed and recently used methods discussed in Sect. 2 is presented. Unfortunately, the most of reports use their proper non-standard video sequences or databases for performance evaluation of proposed and designed systems for gesture recognition. Therefore, the presented in Table 3 recognition rate may be considered only as related possibly achived efficiency of systems for gesture analysis in very specific controled environment described into each particular report.

Table 3. Recognition rate of well-known and the proposed systems for gesture recognition

	[9]	[10]	[11]	[12]	[13]	Proposed
Success rate	96 %	65.7 %	80 %	93.17 %	80 %	91.25 %

One of the main disadvantages of the proposed method is its sensitivity to the lighting conditions. Since the method was thought to work in airport information modules, the lighting of that ambient should be sufficient for satisfactory operation of the gesture recognition system. Besides this, the proposed low cost and simple procedure for gesture detection and recognition could be used for design several interactive systems with ease navigation. The real time gesture recognition during natural human-computer interaction may be considered as significant advantage of the proposal comparing it with other well-known systems.

6   Conclusions

We found that most recent methods for gesture recognition using artificial vision as whole or as part of their method depend on the skin recognition and some of them use specialized hardware. We proposed a simple and fast method to detect, track and recognize short gestures with high precision. It works with simple heuristics classifying three basic gestures in real time during natural human interaction with computers.

Compared to the high percentage of correct recognition of the gestures *up* and *right* we can assume that the heuristics of the lateral gesture *left* might not be the optimal still yet. In the future research we need to improve the performance of the approach, make adjustments to find new heuristics for gesture detection and recognition as well as increment the discrimination ability of the method in case of a the complex background and low or variable illumination.

References

1. Bondre, M.H.S., Pimple, J.: Survey on touch less computer control system using hand gesture recognition. Int. J. Recent Innov. Trends Comp. Commun. **3**(2), 209–213 (2015)
2. Mitra, S., Acharya, T.: Gesture recognition: a survey. IEEE Trans. Syst. Man Cybern. Part C Appl. Rev. **37**(3), 311–324 (2007)

3. Karam, M.: Ph.D. thesis: A framework for research and design of gesture-based human-computer interactions, University of Southampton (2006)
4. Athavale, S., Deshmukh, M.: Dynamic hand gesture recognition for human computer interaction, a comparative study. Int. J. Eng. Res. Gen. Sci. 2(2), 38–55 (2015)
5. Pu, Q., Gupta, S., Gollakota, S., Patel, S.: Whole-home gesture recognition using wireless signals. In: MobiCom 2013 Conference, Miami, FL, USA, pp. 1–12, 30 September 2013
6. Elmezain, M., Al-Hamadi, A., Michaelis, B.: A novel system for automatic hand gesture spotting and recognition in stereo color image sequences. J. WSCG 17, 89–96 (2009)
7. Chaudhary, A., Raheja, J.L., Raheja, S.: A vision based geometrical method to find fingers positions in real time hand gesture recognition. J. Softw. 7(4), 861–869 (2012)
8. Zabulis, X., Baltzakis, H., Argyros, A.: Vision-based hand gesture recognition for human-computer interaction. In: Universal Access Handbook, Chap. 34. Lawrence Erlbaum Associates, Inc. (LEA), pp. 1–30 (2009)
9. Bhuyan, M.K., Kumar, D.A., MacDorman, K.F., Iwahori, Y.: A novel set of features for continuous hand gesture recognition. J. Multimodal User Interfaces 8(4), 333–343 (2014)
10. Kshirsagar, K.P., Sahu, R.M., Bankar, S.M., Moje, R.K., Doye, D.D.: K one hand gesture recognition. Int. J. Innov. Res. Electr. Electron. Instrum. Control Eng. 1(7), 330–334 (2013)
11. Palochkin, V., Demidov, P.G., Alexander, M.J., Priorov, A.: Recognition of hand gestures on the video stream based on a statistical algorithm with pre-treatment. In: Open Innovations Association FRUCT Conference, St. Petersburg, Russia, pp. 105–111 (2014)
12. Rios-Soria, D.J., Schaeffer, S.E., Garza-Villarreal, S.E.: Hand-gesture recognition using computer-vision techniques (2013)
13. Lee, S.-H., Sohn, M.-K., Kim, D.-J., Kim, B., Kim, H.: Smart TV interaction system using face and hand gesture recognition. In: IEEE International Conference on Consumer Electronics, Mexico, pp. 173–174 (2013)
14. Brethes, L., Menezes, P., Lerasle, F., Hayet, J.: Face tracking and hand gesture recognition for human robot interaction. In: Proceedings of International Conference on Robotics and Automation, New Orleans, USA, vol. 2, pp. 1901–1906 (2004)
15. Viola, P., Jones, M.J.: Robust real-time face detection. Int. J. Comp. Vis. 57(2), 137–154 (2004)
16. Jun-Qin, W.: Moving object detection using the edge-prefetch frame difference method. Int. J. Adv. Comp. Technol. 5(5), 1139–1145 (2013)
17. Celebi, S., Aydin, A.S., Temiz, T.T., Arici, T.: Gesture recognition using skeleton data with weighted dynamic time warping. In: VISAPP, pp. 620–625 (2013)

Simultaneous Encryption and Compression of Digital Images Based on Secure-JPEG Encoding

Saqib Maqbool[1], Nisar Ahmad[1], Aslam Muhammad[1(✉)],
and A.M. Martinez Enriquez[2]

[1] Department of Computer Science and Engineering,
University of Engineering and Technology Lahore, Lahore, Pakistan
saqib_maqbool2003@yahoo.com, nisarahmedrana@yahoo.com,
maslam@uet.edu.pk
[2] Department of CS, CINVESTAV-IPN, Mexico, D.F., Mexico
ammartin@cinvestav.mx

Abstract. Confidentiality and efficient bandwidth utilization requires compression and encryption of digital images. Both of these parameters are necessary for most communication systems. Encryption and compression done separately sometimes result in decreased performance or reduced reconstruction quality. The paper presents a simultaneous encryption and compression scheme for digital images. It modifies standard JPEG compression in a way to encrypt data during compression. The encryption steps are based on a JPEG compressible image encryption scheme. The proposed Secure-JPEG algorithm provides the benefits of encryption along with the ability to provide lossless compression. This scheme results in improved performance and better reconstruction quality than existing schemes utilizing the similar approach.

Keywords: Multimedia security · Encryption · Compression · Simultaneous encryption and compression · JPEG

1 Introduction

The use of digital images and video applications has increased significantly due to availability of inexpensive capturing devices. Data compression is always desired due to limited storage or communication bandwidth. Wireless communication in particular requires low bit rate compression due to power and bandwidth constraints [1]. In contrast, encryption is required to protect the information for illicit use especially in wireless or public networks. Conventionally, compression is performed to reduce the data size and then it is encrypted using a suitable encryption algorithm. The decoder must perform this process in reverse order to obtain the actual data. The time consumed during encryption and decryption is a key tailback in real-time image communication and processing. Moreover, the processing time of compression and decompression of data pose another bottleneck. The computational cost incurred by encryption and decryption of data make it infeasible for many practical applications such as real-time embedded systems [2].

J.F. Martínez-Trinidad et al. (Eds.): MCPR 2016, LNCS 9703, pp. 145–154, 2016.
DOI: 10.1007/978-3-319-39393-3_15

The security of digital images is becoming increasingly important due to rapid evolution of Internet and digital technologies. Moreover, encryption of digital images is different from text data as it possesses high spatial correlation and redundancy. Therefore, traditional encryption schemes such as AES [3], and RSA [4] are not highly appropriate for image or other multimedia data. Degradation of visual content without achieving complete randomness can suffice the purpose of encryption for digital images. Many researchers have proposed encryption algorithms specifically for digital images [5–7]. Moreover, bulk-capacity and high-redundancy of image data require compression along with encryption.

The motivation of this research is acquired from the research of [8] which works on quantized Discrete Cosine Transform (DCT) coefficients which are produced during JPEG compression. In our research, we modify the DCT coefficients during JPEG compression process to produce an encrypted and compressed image which has visual quality nearly same as the original image and encrypted as well. Our Secure-JPEG technique has two benefits. One is that it allows achieving compression and encryption process in a single step. And the second is that it improves the image quality when decompressed and decrypted in the reverse process. Our results have been compared with standard JPEG algorithm and encryption scheme by [9] using Mean Squared Error (MSE), Peak-Signal to Noise Ratio (PSNR), Normalized Correction (NC), and Structural Similarity Index Matrix (SSIM).

The rest of the paper is organized as follows; Sect. 2 contains literature review of some contemporary encryption algorithms. Section 3 provides our proposed algorithm. Section 4 provides the results and discussions whereas the research is concluded in Sect. 5.

2   Literature Review

Several researchers [10–13] have focused only on compression of digital images where security aspects are not considered. Chaotic image encryption algorithms [3–7, 10] are gaining attention due to their inherent sensitivity to initial conditions, pseudo-randomness and ergodicity. They have good confusion and diffusion properties which satisfies the cryptographic requirements. However, these systems encompass security of images only and do not consider compression aspects. Consequently, the need for simultaneous encryption and compression of digital images is a necessary requirement. Several researchers follow this approach in their research and given attention to confidentiality along with data reduction [14–17].

There are two approaches followed by researchers while achieving encryption along with compression of digital images. In the first approach, encryption and compression are done at two different stages [18–20]. These two stages are completely independent of each other and sometime take more time while processing the image at two separate stages. In this scenario the adversary has to focus on cryptanalysis only to break the security of the algorithm without giving any consideration to compression algorithm. In the second approach, compression and encryption of digital images are performed simultaneously in a single stage [16, 21–23]. This combined encryption and compression result in reduced computational time and more security as the adversary has to consider the compression as well as the encryption algorithm while performing the cryptanalysis of the encryption algorithm.

[8] has presented a shared-key encryption algorithm for JPEG color images. The algorithm operates on DCT coefficients during quantization step. Their process is based on optical encryption by producing two random like shares of 8×8 blocks. Each 8×8 block is passed through the encryption process and random like shares are produced which are then fed to JPEG for further processing. The produced shares are of the same size as the original block so it does not result in increased size. Moreover, the share generation is lossless and the encryption does not add further error in the compressed image. During decryption process these shares are combined to obtain the original DCT-coefficients [8] and the original image is reproduced. They have also provided three extensions to their proposed schemes, one is intended to produce random looking pixel distribution, the second to produce asymmetric shares and the third one is to generate more than two shares. These additional extensions have their own limitations over the original proposed scheme.

In [9] a color image encryption scheme based on orthogonal basis vectors is proposed. The encryption scheme work in two phases: the first one divides the image into 8×8 blocks and then blocks are scrambled by means of Mersenne Twister [24]. This scrambled image is transformed to frequency domain by using DCT. In second phase, a random-number-matrix of the image size is generated using Mersenne Twister. Their proposed algorithms have demonstrated reasonable security but there was spatial correlation in horizontal direction which was explained in terms of orthogonal matrix. Moreover, their proposed scheme introduces intensity change due to grayscale stretching for several times. Although, the scheme was compression friendly and has shown significant compression ratio along with resistance to channel noise the scheme introduced redundant computation by performing DCT during encryption and then during JPEG compression. Moreover, encryption then compression introduces more quantization error as compared to performing the two steps in a single stage. So, it is always preferable to introduce less error by simultaneous encryption and compression.

3   Secure-JPEG

In our proposed Secure-JPEG scheme, a simultaneous compression and encryption algorithm is designed. The algorithm uses the JPEG compression algorithm and introduces encryption during the DCT quantization step. This encrypted sequence is further processed with JPEG compression steps of quantization and entropy coding to obtain an encrypted and compressed image. The security of encryption steps is demonstrated in [9] which is quite reasonable. However, the performance of that algorithm in terms of image quality was not good due to stretching of pixel values to complete grayscale range. The proposed methodology, therefore, provides all the security benefits of [9] algorithm along with its compression and also results in improvement of reproduced image quality.

Figure 1 provides the basic working of our Secure-JPEG scheme. Input plaintext image is fed to the Secure-JPEG algorithm which compresses and encodes the image based on secret key. The protected image data is transmitted on an insecure wireless channel where it is susceptible to different type of attacks. The decoder performs the decoding and finally outputs the decompressed plain image.

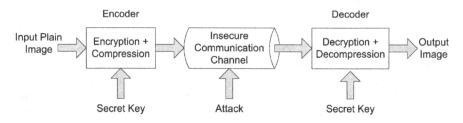

Fig. 1. Secure-JPEG

The lossy part of JPEG compression scheme uses DCT. The input plain image is separated into 8×8 non-overlapping blocks $(\beta_1, \beta_2, \cdots \beta_N)$, where N is equal to the total number of blocks. Zero-padding is used to convert the image matrix into multiple of 8. These blocks are scrambled based on permuted sequence ρ obtained from Pseudo-Random Number Generator (PRNG). We have used Mersenne Twister for PRNG but any cryptographically secure PRNG can be used for this purpose. These 8×8 pixels blocks are transformed using DCT into $(\Psi_1, \Psi_2, \ldots \Psi_N)$. Random matrices $(\gamma_1, \gamma_2, \ldots \gamma_N)$ of 8×8 are generated using the same PRNG separately for each DCT transformed image blocks. These randomly generated matrices are decomposed by SVD into U_i, Σ_i & V_i. The left singular vectors U_i are multiplied with DCT transformed image blocks to obtain encrypted matrix $(\mathcal{C}_1, \mathcal{C}_2, \cdots \mathcal{C}_N) = (\psi_1 \times U_1, \psi_{22} \times U_2, \ldots \psi_N \times U_N)$.

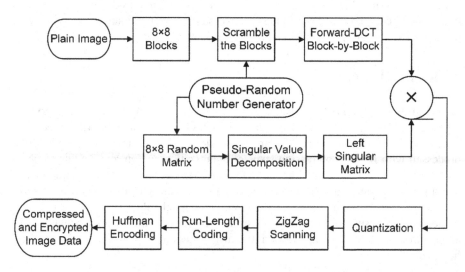

Fig. 2. Encryption process according to Secure-JPEG

This encrypted matrix is then quantized according to JPEG (see Fig. 2). Then run-length coding is performed by reading the 64 elements in zigzag scanning sequence. Then, variable length coding is performed using Huffman's algorithm [25] and compressed image data is obtained for transmission or storage.

The encryption key can be shared by using Diffie–Hellman key exchange or any other suitable algorithm. The decoding algorithm of the proposed scheme follows the same steps in reverse order. There could be two scenarios at decoding end which are provided below. In Scenario-I, the user treats the encoded image data as a JPEG compressed image and performs the decoding according to JPEG algorithm. The algorithm follows Huffman decoding of the compressed data which decodes the variable length code. Then, inverse run-length coding converts this data to its actual sequence. This data is then transformed from a vector sequence to 8 × 8 blocks. De-quantization retrieves the matrix values before quantization with error introduced by lossy compression factor. The image is formed after backward DCT transformation and combining the 8 × 8 blocks. The decoded image in this scenario represents the cipher image which is of no use to the intruder if he does not possess the knowledge of secret key and the encryption algorithm. The block diagram of this Scenario-I is shown in Fig. 3.

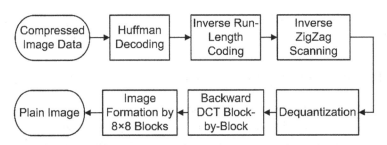

Fig. 3. Scenario-I; decoding according to standard JPEG

In scenario-II; the user possesses the information of the encryption algorithm and attempts to decode the image according to the algorithm. The compressed image data is decoded from variable length coded data using Huffman decoding. This decoded data is then transformed through inverse run-length coding to obtain the original data sequence. This sequence is transformed into 8 × 8 block through inverse Zigzag Scanning. De-quantization is done to retrieve the data before quantization with an error introduced by lossy compression. Whereas, 8 × 8 random matrices $\{\gamma_1, \gamma_2, \ldots \gamma_N\}$ are generated from the PRNG according to the secret key. These randomly generated matrices are decomposed using singular value decomposition to U_i, Σ_i & V_i. The left singular matrices are transposed to obtain its inverse as the left singular matrix is an orthogonal matrix and orthogonal matrix has inverse equal to its transpose. The transposed left singular vectors U_i^T are multiplied with de-quantized matrices to obtain decrypted matrices as $\left(\epsilon_1, \epsilon_2, \cdots \epsilon_N\right) = \left(\Psi_1 \times U_1^T, \Psi_2 \times U_2^T, \ldots \Psi_N \times U_N^T\right)$. These decrypted matrices are backward transformed using DCT to obtain the spatial image. Whereas, permutation sequence is generated same as during encryption. This permutation sequence is used to obtain inverse permutation sequence to perform the inverse

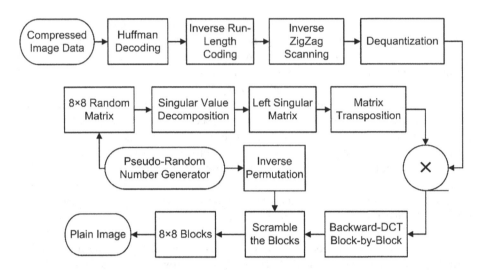

Fig. 4. Scenario-II; decoding according to Secure-JPEG

scrambling of blocks. These blocks are combined to form recovered image with dimensions of original image. The block diagram of this scenario-II is shown in Fig. 4.

4 Experimental Results and Discussions

The proposed scheme has been tested for a set of test images to obtain decoded images produced by standard JPEG and decoded by following Secure-JPEG. Figure 5(a) contains the test image used for encoding. Figure 5(b) contains the decoded image through standard JPEG which only decompress the image and display its output. This image has no visual information for the intruder. Figure 5(c) contains the decoded image through Secure-JPEG which is similar to the input plain image.

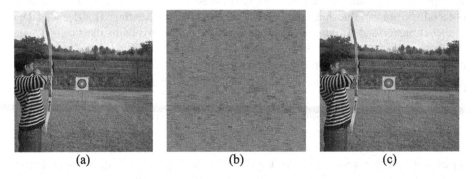

Fig. 5. (a) Original plain image-archer, (b) Decoded image through standard JPEG, (c) Decoded image through Secure-JPEG (Recovered deciphered image)

When the compression ratio is selected as 100 % at the time of encoding the decoded image is exact replica of the input image and only lossless compression is performed. When the image is provided less compression ratio such as for the images displayed in (c) where the compression ratio is selected as 90 % and the decoded image is visually similar but contains some error due to lossy compression during quantization. The change in decoded image due to quantization error is tested by employing Image quality metric [26].

4.1 Recovered Image Quality

Results in Tables 1, 2, and 3 provide the recovered image quality measured by MSE, PSNR, NC, and SSIM. These four metrics compare image quality based on different parameters except PSNR which is based on MSE. Any single matrix is not enough to measure the quality in all cases but they inspect the image similarity with different angles and a combined score of these parameters. Table 2 also provides image quality parameters with images compressed and decompressed with standard JPEG at 90 % quality factor. Table 3 provides the same results for scheme proposed by [9] and compressed by JPEG at 90 % quality factor. It can be seen from comparison of Table 1 and Table 3 that proposed scheme has better reconstruction quality then [9]. Moreover, the proposed scheme has better efficiency as it does not repeat the block formation, forward-DCT and quantization steps. Another significant improvement is reconstruction when it is encoded with 100 % quality factor which was not the same in [9] as they employed intensity scaling and de-scaling.

Table 1. Results of image quality metrics for proposed scheme at 90 % quality

Serial no.	Test images	MSE	PSNR	NC	SSIM
1	Archer	5.6577	37.8415	0.9880	0.9696
2	Cameraman	1.0807	39.8336	0.9981	0.9878
3	Flower	5.0009	38.1881	0.9774	0.9724
4	Glider	7.4606	38.8736	0.9872	0.9709
5	Kodim15	4.8454	38.7250	0.9993	0.9687
6	Lena	8.5593	37.0393	0.9717	0.9547
7	Mandrill	9.0010	35.9914	0.9805	0.9633
8	Peppers	3.0908	37.5445	0.9630	0.9476

Graphical forms are also used to provide comparison of quality parameters to see the reconstruction quality of three approaches. The plot of MSE for the 8 test images shows that the proposed scheme provides lower MSE than [9] followed by JPEG compression at 90 % quality factor. However, JPEG provide lower MSE then the proposed scheme for most of the cases. The graph for the comparison of PSNR shows that the proposed scheme has reasonably higher PSNR than [9].

The plot of the normalized correction shows that the similarity between original and reconstructed images. The proposed scheme has values near 0.98 which are good and

Table 2. Results of image quality metrics for Standard JPEG at 90 % quality

Serial no.	Test images	MSE	PSNR	NC	SSIM
1	Archer	5.3688	37.8027	1.0000	0.9737
2	Cameraman	0.8308	45.9048	0.9999	0.9966
3	Flower	3.9747	39.0987	0.9996	0.9773
4	Glider	4.4526	38.6481	1.0000	0.9750
5	Kodim15	4.9642	38.2276	0.9992	0.9571
6	Lena	4.0011	39.1199	0.9999	0.9672
7	Mandrill	7.3070	36.4888	1.0000	0.9767
8	Peppers	4.8888	38.2444	0.9998	0.9393

Table 3. Results of image quality metrics for [9] + JPEG compression at 90 % quality

Serial no.	Test images	MSE	PSNR	NC	SSIM
1	Archer	6.1243	31.8415	0.8880	0.8796
2	Cameraman	3.1656	32.8336	0.8981	0.8838
3	Flower	4.9373	30.5783	0.9089	0.8879
4	Glider	4.7440	31.0166	0.9186	0.8942
5	Kodim15	9.8341	29.1701	0.8386	0.8605
6	Lena	0.6267	28.7155	0.8937	0.8566
7	Mandrill	8.6986	29.1487	0.9555	0.7821
8	Peppers	5.0673	27.8639	0.8509	0.7450

can be regarded as perceptually similar and are much better than [9]. The graph for SSIM, which is a latest measure of image quality and it is claimed to be a full reference matrix, depicts improvements to MSE and PSNR values. The proposed scheme has higher value of SSIM than [9].

It is evident from the results of Tables 1, 2, and 3 that the proposed Secure-JPEG scheme has demonstrated reconstruction quality which is better than the previous method and the results are comparable to standard JPEG. The proposed Secure-JPEG scheme also provides improved performance due to reduced number of computational steps but the comparison of time consumed during computation on desktop computer cannot be provided as a reference as it is highly dependent on operating system, software environment, and other performance parameters. The Secure-JPEG scheme can be successfully used for encryption and compression of digital images. It also provides the ability to perform lossless operation as well as the lossy operation.

5   Conclusion

In this study, we intended to improve the performance and reconstruction quality of image encrypted by JPEG compressible image encryption scheme. As compression of digital images is vital to efficient bandwidth utilization but doing it separately may result in reduced performance. Moreover, in this scenario it was resulting in more error once while scaling of image pixels during encryption and the second while achieving lossy compression. Also the scheme was introducing error due to scaling even when the image was being compressed using lossless compression. In our proposed scheme, the advantages achieved are three fold. Firstly, the algorithm is resulting in increased performance due to reduced number of computational steps. Secondly, the error due to compression and encryption was reduced. Thirdly, the scheme resulted in errorless image construction when the compression quality is kept 100 %. Consequently, the algorithm is an improved version with similar security characteristics as of the JPEG compressible image encryption scheme and it can be used for lossless encryption and compression and lossy encryption and compression. In future, the similar mechanism of encryption and compression will be implemented in wavelet domain for JPEG 2000. Moreover, the proposed algorithm will be modified and used for encryption and compression of audio, image sequence, and other multimedia data.

References

1. Lu, Q., et al.: Low-complexity and energy efficient image compression scheme for wireless sensor networks. Comput. Netw. **52**(13), 2594–2603 (2008)
2. Lian, S., Kanellopoulos, D., Ruffo, G.: Recent advances in multimedia information system security. Informatica **33**(1) (2009)
3. Selent, D.: Advanced encryption standard. Rivier Acad. J. **6**(2), 1–14 (2010)
4. Smith, D.R., Palmer, J.T.: Universal fixed messages and the Rivest-Shamir-Adleman cryptosystem. Mathematika **26**(01), 44–52 (1979)
5. Lian, S.: Multimedia Content Encryption: Techniques and Applications. CRC Press, Boca Raton (2008)
6. Furht, B., Socek, D., Eskicioglu, A.M.: Fundamentals of multimedia encryption techniques. In: Multimedia Security Handbook, vol. **4** (2004)
7. Van Droogenbroeck, M., Benedett, R.: Techniques for a selective encryption of uncompressed and compressed images. In: Advanced Concepts for Intelligent Vision Systems (ACIVS) (2002)
8. Sudharsanan, S.: Shared key encryption of JPEG color images. IEEE Trans. Consum. Electron. **51**(4), 1204–1211 (2005)
9. Ahmed, N., et al.: A novel image encryption scheme based on orthogonal vectors. Nucleus **52**(2), 71–78 (2015)
10. Grigoras, V., Grigoras, C.: Chaos encryption method based on large signal modulation in additive nonlinear discrete-time systems. In: Proceedings of the 5th WSEAS International Conference on Non-linear Analysis, Non-linear Systems and Chaos. World Scientific and Engineering Academy and Society (WSEAS) (2006)
11. Philip, M., Das, A.: Survey: image encryption using chaotic cryptography schemes. IJCA, 1–4 (2011). Special Issue on "Computational Science-New Dimensions and Perspectives" NCCSE
12. Wei-bin, C., Xin, Z.: Image encryption algorithm based on Henon chaotic system. In: International Conference on Image Analysis and Signal Processing, 2009. IASP 2009. IEEE (2009)
13. Shum, H.-Y., Kang, S.B., Chan, S.-C.: Survey of image-based representations and compression techniques. IEEE Trans. Circuits Syst. Video Technol. **13**(11), 1020–1037 (2003)

14. Hossein, M., Mahmud, S., Biswas, N.: Image compression and encryption. Int. J. ElectroComput. World Knowl. Interface **1**(3) (2011)
15. Zhou, N., et al.: Image compression and encryption scheme based on 2D compressive sensing and fractional Mellin transform. Opt. Commun. **343**, 10–21 (2015)
16. Alfalou, A., Brosseau, C., Abdallah, N.: Simultaneous compression and encryption of color video images. Opt. Commun. **338**, 371–379 (2015)
17. Tong, X.-J., et al.: A new algorithm of the combination of image compression and encryption technology based on cross chaotic map. Nonlinear Dyn. **72**(1–2), 229–241 (2013)
18. Zhou, J., et al.: Designing an efficient image encryption-then-compression system via prediction error clustering and random permutation. IEEE Trans. Inf. Forensics Secur. **9**(1), 39–50 (2014)
19. Zhou, J., Liu, X., Au, O.C.: On the design of an efficient encryption-then-compression system. In: 2013 IEEE International Conference on Acoustics, Speech and Signal Processing (ICASSP). IEEE (2013)
20. Bansal, R., Sharma, M.R.: Designing an Efficient Image Encryption-Compression System Using a New Haar Wavelet (2014)
21. Zhu, H., Zhao, C., Zhang, X.: A novel image encryption–compression scheme using hyperchaos and Chinese remainder theorem. Sig. Process. Image Commun. **28**(6), 670–680 (2013)
22. Aldossari, M., Alfalou, A., Brosseau, C.: Simultaneous compression and encryption of closely resembling images: application to video sequences and polarimetric images. Opt. Express **22**(19), 22349–22368 (2014)
23. Zhou, N., et al.: Novel image compression–encryption hybrid algorithm based on key-controlled measurement matrix in compressive sensing. Opt. & Laser Technol. **62**, 152–160 (2014)
24. Matsumoto, M., Nishimura, T.: Mersenne twister: a 623-dimensionally equidistributed uniform Pseudo-random number generator. ACM Trans. Model. Comput. Simul. (TOMACS) **8**(1), 3–30 (1998)
25. Knuth, D.E.: Dynamic Huffman coding. J. Algorithms **6**(2), 163–180 (1985)
26. Naveed, A., et al.: Performance evaluation and watermark security assessment of digital watermarking techniques. Sci. Int. Lahore **27**(2), 6 (2015)

Pattern Recognition and Artificial Intelligent Techniques

Automatic Tuning of the Pulse-Coupled Neural Network Using Differential Evolution for Image Segmentation

Juanita Hernández and Wilfrido Gómez$^{(\boxtimes)}$

Information Technology Laboratory,
Center for Research and Advanced Studies of the National Polytechnic Institute,
Ciudad Victoria, Tamaulipas, Mexico
{jhernandez,wgomez}@tamps.cinvestav.mx

Abstract. The pulse-coupled neural network (PCNN) is based on the cortical model proposed by Eckhorn and is widely used in tasks such as image segmentation. The PCNN performance is particularly limited by adjusting its input parameters, where computational intelligence techniques have been used to solve the problem of PCNN tuning. However, most of these techniques use the entropy measure as a cost function, regardless of the relationship of inter-/intra-group dispersion of the pixels related to the objects of interest and their background. Therefore, in this paper, we propose using the differential evolution algorithm along with a cluster validity index as a cost function to quantify the segmentation quality in order to guide the search to the best PCNN parameters to get a proper segmentation of the input image.

Keywords: Automatic image segmentation · Pulse-coupled neural network · Differential evolution · Cluster validity index

1 Introduction

The pulse-coupled neural network (PCNN) is a bio-inspired model based on the cortical model proposed by Eckhorn in 1989. It is used in different applications of image segmentation, although its performance strongly dependents on the adequate tuning of its input parameters like decay constants, radio link, number of iterations, etc. [1].

The problem of tuning the PCNN parameters for image segmentation has been addressed by computational intelligence (CI) techniques because they are able to solve optimization problems in complex and changing environments in reasonable computation time [2]. Some relevant works that have optimized the segmentation performance of the PCNN include algorithms based on particle swarm optimization (PSO) [3], genetic algorithm (GA) [4], and differential evolution (DE) [5,6]. These CI-based methods require a proper choice of the cost function to quantify the quality of a potential solution to determine its survival

© Springer International Publishing Switzerland 2016
J.F. Martínez-Trinidad et al. (Eds.): MCPR 2016, LNCS 9703, pp. 157–166, 2016.
DOI: 10.1007/978-3-319-39393-3_16

in the population for ensuring an adequate and consistent segmentation of the input image.

The maximum entropy criterion has been widely used by CI-based techniques for tuning the PCNN parameters to quantify the segmentation quality generated by a potential solution. The maximum entropy criterion only provides a measure of overlap between the intensity probability distributions related to the objects and their back ground, but it does not consider the inter-/intra-group dispersion, that is, how similar are the intensity levels of the objects and how dissimilar are relative to the background intensities. This inter-/intra-group dispersion could be measured by a cluster validity index (CVI).

A CVI is an internal validation index usually used by clustering algorithms to evaluate the quality of a candidate grouping. Currently, using a CVI as a cost function by segmentation algorithms based on CI techniques has not been explored, although it is feasible to be applied because the process of image segmentation is basically a grouping process of pixels with similar intensities. In this context, a CVI quantifies the segmentation quality considering the inter-/intra-group ratio, that is, minimizing the intra-group dispersion while maximizing the inter-group dispersion simultaneously.

Hence, the purpose of this study is to demonstrate that the automatic tuning of the PCNN parameters by means the DE algorithm guided by a CVI improves the performance of image segmentation than solely maximizing the entropy criterion.

The organization of this paper is divided into five sections. Section 2 describes the problem of image segmentation as an optimization problem. Section 3 presents the materials and methods used in this study as well as the description of the proposed approach. Section 4 summarizes the experimental results. Finally, Sect. 5 gives the conclusion and future work.

2   Problem Statement

The problem of image segmentation can be considered as a clustering problem, where an input image $\mathbf{R}$ is partitioned into two groups, c_1 and c_2, containing the pixels that belong to the objects of interest and their background, respectively, to form a grouping denoted by $\mathbf{C} = \{c_1, c_2\}$ that should satisfy the following three conditions:

1. $c_i \neq \emptyset$ for $i = 1, 2$;
2. $c_1 \cup c_2 = \mathbf{R}$;
3. $c_1 \cap c_2 = \emptyset$.

On the other hand, let $\mathbf{x} = [x_1, \ldots, x_d]$ be the vector containing the d input parameters of a PCNN that generates a grouping (or segmentation) $\mathbf{C}$ given $\mathbf{R}$. Let $\mathbf{X} = \{\mathbf{x}_1, \ldots, \mathbf{x}_N\}$ be the set of vectors of PCNN parameters that generates N feasible groupings of $\mathbf{R}$. Then, the problem of finding the best clustering can be formulated as an optimization problem, where $\Omega = \{\mathbf{C}^{\mathbf{x}_1}, \ldots, \mathbf{C}^{\mathbf{x}_N}\}$ is the set of candidate groupings of the pixels in $\mathbf{R}$ given the set $\mathbf{X}$, so the optimal grouping $\mathbf{C}^* \in \Omega$ should satisfy

$$\forall \mathbf{C} \in \Omega : f(\mathbf{C}^*) < f(\mathbf{C}) \tag{1}$$

where $f(\cdot)$ is a cost function given in terms of a CVI, which measures the dispersion intra-/inter-group between c_1 and c_2. Note that $f(\cdot)$ is minimized without loss of generality.

3 Materials and Methods

3.1 PCNN Model

The PCNN is a bidimensional single layer, laterally connected network of integrate-and-fire neurons, with a 1-to-1 correspondence between the image pixels and network neurons as illustrated in Fig. 1.

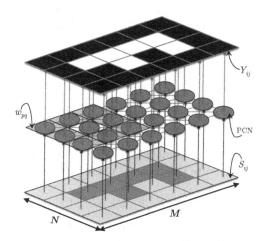

Fig. 1. PCNN configuration, where M and N are the width and height of the input image, S_{ij} is the intensity level, w_{pq} is a synaptic value, and Y_{ij} is the output of a single neuron.

A single pulse-coupled neuron (PCN) has two input channels named feeding and linking, whose responses are combined to regulate the internal neuron activity, which is further compared with a trigger threshold to generate a pulse. Hence, a PCN consists of three main parts: input field, modulation field, and pulse generator, as shown in Fig. 2 [7].

The input field can be seen as an integrator of leaks simulating the dendritic part of the biological neuron, in which each neuron $(N_{i,j})$ receives signals from external sources, in the form of stimuli $(S_{i,j})$ that represents the pixel intensity

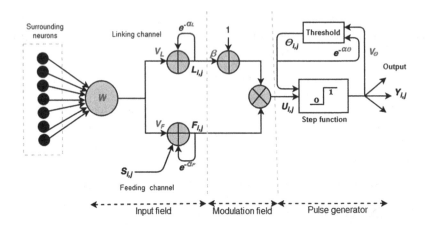

Fig. 2. Typical model of a single PCN. In red, the input parameters.

in the input image, and internal sources, which are the responses of neighboring neurons within a specified radius linked by synaptic weights ($W_{k,l}$). At iteration t, these input signals reach the neuron via the feeding ($F_{i,j}$) and linking ($L_{i,j}$) channels expressed by

$$F_{i,j}[t] = e^{-\alpha_F} F_{i,j}[t-1] + V_F \sum_{(k,l)\in N(i,j)} W_{i,j,k,l} Y_{i,j}[t-1] + S_{i,j} \qquad (2)$$

$$L_{i,j}[t] = e^{-\alpha_L} L_{i,j}[t-1] + V_L \sum_{(k,l)\in N(i,j)} W_{i,j,k,l} Y_{i,j}[t-1] \qquad (3)$$

In the modulation field, the signals from $F_{i,j}$ and $L_{i,j}$ channels are combined in a nonlinear way to generate the internal neuron activity expressed by

$$U_{i,j}[t] = F_{i,j}[t](1 + \beta L_{i,j}[t]) \qquad (4)$$

where β is a connection factor that regulates the internal activity, which simulates the electrical potential generated in the biological neuron.

An adaptative threshold $\theta_{i,j}$ (Eq. 5) is used for the pulse generator, that operates as a step function, which controls the trigger event $Y_{i,j}$ (Eq. 6). This process simulates the action of polarization and repolarization generated in biological neurons, obviously considering a refractory period dependent on a time interval.

$$\theta_{i,j}[t] = e^{-\alpha_\theta} \theta_{i,j}[t-1] + V_\theta Y_{i,j}[t] \qquad (5)$$

$$Y_{i,j}[t] = \begin{cases} 1 & \text{if } U_{i,j}[t] > \theta_{i,j}[t] \\ 0 & \text{otherwise} \end{cases} \qquad (6)$$

3.2   Differential Evolution Algorithm

The DE algorithm is inspired by the natural evolution of individuals within a population, that is, the survival of the fittest. DE maintains a population of potential solutions that mutate and recombine to produce new individuals, which are further evaluated and selected based on their fitness measured by a cost function. The DE process involves the following basic steps:

1. **Initialization**: the population with N individuals is denoted by the set $\mathbf{X} = \{\mathbf{x}_1, \ldots, \mathbf{x}_N\}$. For the ith individual, a d-dimensional vector is defined by $\mathbf{x}_i = [x_{i,1}, \ldots, x_{i,d}]$, where each variable is randomly initialized in the range $[\mathrm{LL}, \mathrm{UL}]$ representing the lower and upper limits, respectively, of the search space.
2. **Mutation**: for the ith target vector in generation g, $\mathbf{x}_{i,g}$, a mutant vector, $\mathbf{v}_{i,g}$, is created, which combines three members of the population, the current best individual, $\mathbf{x}_{\mathrm{best},g}$, and two individual randomly chosen from the current population, $\mathbf{x}_{r1,g}$ and $\mathbf{x}_{r2,g}$, such that $r1 \neq r2 \neq i$. The mutant vector is generated by using the current-to-best strategy as [8]

$$\mathbf{v}_{i,g} = \mathbf{x}_{i,g} + F \cdot (\mathbf{x}_{\mathrm{best},g} - \mathbf{x}_{i,g}) + F \cdot (\mathbf{x}_{r1,g} - \mathbf{x}_{r2,g}) \tag{7}$$

   where $F = 0.8$ is the scaling factor that controls the amplification of the vector differences.
3. **Crossover**: a test vector, $\mathbf{u}_{i,g}$, is created by exchanging the elements of the target vector $\mathbf{x}_{i,g}$ and the mutant vector $\mathbf{v}_{i,g}$, which is performed by the binomial crossover as

$$u_{i,j} = \begin{cases} v_{i,j} & \text{if } \mathrm{rand}(0,1) < CR \\ x_{i,j} & \text{otherwise} \end{cases} \tag{8}$$

   where $j = 1, \ldots, d$ and $CR = 0.9$ is the crossover factor that controls the amount of information that is copied from the mutant to the test vector.
4. **Penalty**: in order to prevent the solution falling outside the search space limits $[\mathrm{LL}, \mathrm{UL}]$, the bounce-back strategy [9] is used to reset out-of-bound test variables by selecting a new value that lies between the target variable value and the bound being violated.
5. **Selection**: if the fitness of the test vector $f(\mathbf{u}_{i,g})$ is better than the fitness of the target vector $f(\mathbf{x}_{i,g})$, then $\mathbf{u}_{i,g}$ replaces $\mathbf{x}_{i,g}$ in the next generation, which is expressed by

$$\mathbf{x}_{i,g+1} = \begin{cases} \mathbf{u}_{i,g} & \text{if } f(\mathbf{u}_{i,g}) < f(\mathbf{x}_{i,g}) \\ \mathbf{x}_{i,g} & \text{otherwise} \end{cases} \tag{9}$$

where $f(\cdot)$ is a cost function, which is minimized without loss of generality.

3.3   Proposed Segmentation Approach

The pseudo-code of the proposed segmentation method based on DE and PCNN is shown in Algorithm 1. Note that each individual in the population codifies the nine PCNN parameters summarized in Table 1, which are randomly initialized using their respective lower and upper limit values. Also, $N = 20$ individuals are considered in the population, which evolves during $G_{\mathrm{max}} = 50$ generations. As mentioned previously, a CVI represents the cost function used by the DE algorithm to evaluate the segmentation quality produced by the PCNN given a potential solution. Here, four CVIs are considered: Calinski-Harabasz (CH) [10], PBM (PBM) [11], Davies-Bouldin (DB) [12] and Xie-Beni (XB) [13]. Depending on the CVI type used by the proposed approach, four algorithm variants are defined: DE-PCNN-CH, DE-PCNN-DB, DE-PCNN-PBM, and DE-PCNN-XB. Besides, for performance comparison purposes, the CVI in the proposed algorithm is replaced by the entropy criterion (ENT) [14] to define the DE-PCNN-ENT algorithm.

Table 1. PCNN parameters and their limit values.

Parameter	Description	Lower limit	Upper limit
α_L	Link attenuation factor	0.01	2.50
α_F	Feed attenuation factor	0.01	2.50
α_θ	Impulse attenuation factor	0.01	2.50
V_L	Link potential	0.01	2.50
V_F	Feed potential	0.01	2.50
V_θ	Impulse potential	1	25
β	Connection factor	0.01	2.50
W	Radius of synaptic weights	1	5
n	Number of iterations	2	10

3.4   Performance Evaluation

For evaluating the proposed approach, an image data set containing 30 natural gray scale images is considered. Every image includes three reference segmentations defined manually by three different persons [15]. The image data set is public and it can be downloaded from http://www.wisdom.weizmann.ac.il/~vision/Seg_Evaluation_DB/2obj/index.html.

The output of a segmentation algorithm (S_A) is compared with a reference segmentation (S_R) by using the Jaccard index defined by

$$J(S_A, S_R) = \frac{|S_A \cap S_R|}{|S_A \cup S_R|} \tag{10}$$

Algorithm 1. Proposed segmentation algorithm based on PCNN tuned by DE.

Require: grayscale image ($\mathbf{R}$), population size (N), number of generations ($G_{\max}$), CVI type (f)
Ensure: segmented image ($\mathbf{C}^*$), best PCNN parameters ($\mathbf{x}^*$)
    Initialize population randomly: $\mathbf{X}_0 = \{\mathbf{x}_{1,0}, \ldots, \mathbf{x}_{N,0}\}$
    Segment $\mathbf{R}$ with $\mathbf{X}_0$ using PCNN: $\Omega_0 = \{\mathbf{C}^{\mathbf{x}_{1,0}}, \ldots, \mathbf{C}^{\mathbf{x}_{N,0}}\}$
    Evaluate initial population: $f(\Omega_0)$
    for $g = 1$ **to** $G_{\max}$ **do**
        for $i = 1$ **to** N **do**
            Apply mutation strategy (current-to-best): $\mathbf{v}_{i,g}$
            Apply binomial crossover: $\mathbf{u}_{i,g}$
            Apply bounce-back strategy to $\mathbf{u}_{i,g}$
            Segment $\mathbf{R}$ with $\mathbf{u}_{i,g}$ using PCNN: $\mathbf{C}^{\mathbf{u}_{i,g}}$
            if $f(\mathbf{C}^{\mathbf{u}_{i,g}}) < f(\mathbf{C}^{\mathbf{x}_{i,g}})$ **then**
                Replace target vector with test vector: $\mathbf{x}_{i,g} \leftarrow \mathbf{u}_{i,g}$
            else
                Keep the target vector in the population
            end if
        end for
    end for
    Get best individual $\mathbf{x}^*$ and its associated segmentation $\mathbf{C}^*$ that satisfies Eq. 1.

This index returns a value in the range [0,1], where '1' indicates perfect similarity between both segmentations and '0' indicates total disagreement.

In order to statistically determine the segmentation performance of the proposed approach, 31 runs are considered for each algorithm variant. From the Jaccard index results, the median (MED) and the median absolute deviation (MAD) are calculated to determine the central tendency and the dispersion, respectively. These estimators are chosen because they are capable of coping with outliers and non-normal distributions. Additionally, statistical significance analysis is conducted by the Kruskal-Wallis test ($\alpha = 0.05$) to evaluate whether the median values between groups are different under the assumption that the shapes of the underlying distributions are the same. Finally, the wall clock time in seconds is also measured.

The testing plataform employed a Linux-based computer with 16 cores at 2.67 GHz (Intel Xeon) and 32 GB of RAM. All the algorithms were developed in MATLAB 2015a (The MathWorks, Natick, MA, USA).

4 Results

The experimental results in terms of the Jaccard index are shown in Table 2. The Kruskal-Wallis test indicated that all groups are statistically significant different ($p < 0.001$), where the DE-PCNN-XB variant attained the best segmentation performance compared to its counterparts, whereas the DE-PCNN-ENT obtained the worst performance.

Table 2. Performance segmentation in terms of the Jaccard index. The results are sorted from the best to the worst performance.

Algorithm	MED	MAD
DE-PCNN-XB	0.738	0.154
DE-PCNN-PBM	0.694	0.201
DE-PCNN-CH	0.660	0.246
DE-PCNN-DB	0.368	0.336
DE-PCNN-ENT	0.251	0.140

Figure 3 shows the computation time of the five algorithm variants, whose median values are in the range 5−7 s. Besides, the DE-PCNN-XB variant obtained the lowest MAD value with 0.42 s, whereas the DE-PCNN-ENT variant reached the largest MAD value with 0.69 s.

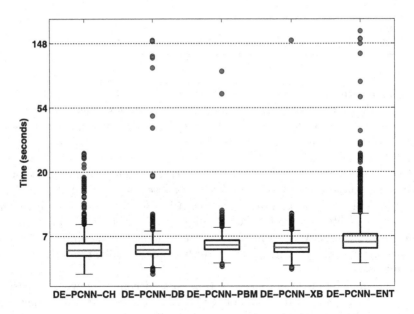

Fig. 3. Boxplots of the computation time regarding the five algorithm variants. For illustrative purposes, the vertical axis is log-compressed.

Figure 4 illustrates a subjective comparison among the outputs of the five segmentation algorithm variants considering four different images (flowers, moth, helicopters, and iceland) from the data set. Notice that all segmentations obtained with the DE-PCNN-XB variant are quite close to their respective reference images. Also, the DE-PCNN-PBM variant is capable to adequately segment three images (flowers, moth, and iceland), although for images where the objects

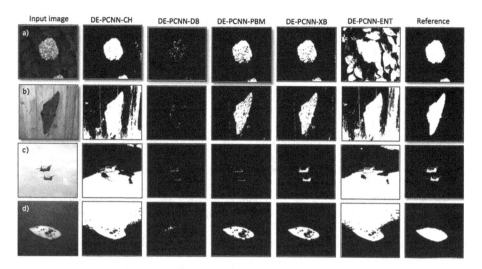

Fig. 4. Segmentation comparison using the five algorithm variants including the reference segmentation. Four images from the data set are considered: (a) flowers, (b) moth, (c) helicopters, and (d) iceland.

are small (such as helicopters) the segmentation is not satisfactory. Finally, DE-PCNN-CH, DE-PCNN-DB and DE-PCNN-ENT failed to properly segment all the images: DE-PCNN-DB tends to under-segment the input image, whereas DE-PCNN-CH and DE-PCNN-ENT tend to over-segment the objects.

5   Conclusion and Future Work

In this paper, a segmentation method based on PCNN tuned by DE algorithm was presented. Five cost functions were evaluated: CH index, PBM index, DB index, XB index, and the maximum entropy criterion. These cost functions quantified the segmentation quality to guide the DE algorithm to find a set of PCNN parameters.

The experimental results pointed out that the DE-PCNN-XB variant obtained the best segmentation performance with low dispersion for distinct algorithm runs. In terms of the Jaccard index, the MED/MAD values were 0.738/0.154. These findings indicated that using a CVI as a cost function is suitable to obtain adequate and consistent segmentations.

Therefore, it was demonstrated that using the XB index instead of the entropy criterion is appropriate to find adequate PCNN parameters to obtain satisfactory segmentation results. This is because a CVI quantifies the relationship between the intensities of the objects and their background, that is, how similar are the intensity levels of the objects and how dissimilar are relative to the background intensities.

Future work involves evaluating other variants of PCNN (e.g., simplified models) as well as other CI-based techniques such as PSO and GA.

Acknowledgments. The authors would like to thanks to CONACyT Mexico for the financial support received through a scholarship to pursue Masters studies at Center for Research and Advanced Studies of the National Polytechnic Institute, Information Technology Laboratory.

References

1. Lindblad, T., Kinser, J.: Image processing using pulse-coupled neural networks. Springer Verlag, Heidelberg (2005)
2. Engelbrecht, A.P.: Computational Intelligence: An Introduction, 2nd edn. Wiley Publishing, Hoboken (2007)
3. Xu, X., Ding, S., Shi, Z., Zhu, H., Zhao, Z.: Particle swarm optimization for automatic parameters determination of pulse coupled neural network. J. Comput. **6**, 1546–1553 (2011)
4. Using a genetic algorithm to find an optimized pulse coupled neural network solution. vol. 6979 (2008)
5. A Self-Adapting Pulse-Coupled Neural Network Based on Modified Differential Evolution Algorithm and Its Application on Image Segmentation. vol. 6 (2012)
6. Gómez, W., Pereira, W., Infantosi, A.: Evolutionary pulse-coupled neural network for segmenting breast lesions on ultrasonography. Neurocomputing **129**, 216–224 (2015)
7. Wang, Z., Ma, Y., Cheng, F., Yang, L.: Review of pulse-coupled neural networks. Image Vis. Comput. **28**, 5–13 (2010)
8. Zhang, J., Sanderson, A.: Jade: Adaptive differential evolution with optional external archive. Evol. Comput. IEEE Trans. **13**, 945–958 (2009)
9. Price, K., Storn, R., Lampinen, J.: Differential Evolution: A Practical Approach to Global Optimization. Natural Computing Series. Springer, Heidelberg (2005)
10. Caliski, T., Harabasz, J.: A dendrite method for cluster analysis. Commun. Stat. **3**, 1–27 (1974)
11. Maulik, U., Bandyopadhyay, S.: Performance evaluation of some clustering algorithms and validity indices. Pattern Anal. Mach. Intell. IEEE Trans. **24**, 1650–1654 (2002)
12. Davies, D.L., Bouldin, D.W.: A cluster separation measure. Pattern Anal. Mach. Intell. IEEE Trans. **PAMI–1**, 224–227 (1979)
13. Xie, X.L., Beni, G.: A validity measure for fuzzy clustering. Pattern Anal. Mach. Intell. IEEE Trans. **13**, 841–847 (1991)
14. Ma, Y., Qi, C.: Study of automated pcnn system based on genetic algorithm. J. Syst. Simul. **18**, 722–725 (2006)
15. Unnikrishnan, R., Pantofaru, C., Hebert, M.: Toward objective evaluation of image segmentation algorithms. Pattern Anal. Mach. Intell. IEEE Trans. **29**, 929–944 (2007)

Efficient Counting of the Number of Independent Sets on Polygonal Trees

Guillermo De Ita[1(✉)], Pedro Bello[1], Meliza Contreras[1],
and Juan C. Catana-Salazar[2]

[1] Facultad de Ciencias de la Computación,
Benemérita Universidad Autónoma de Puebla, Puebla, Mexico
{deita,pbello,mcontreras}@cs.buap.mx
[2] Posgrado en Ciencia e Ingeniería de la Computación,
Universidad Nacional Autónoma de México, Mexico City, Mexico
j.catanas@uxmcc2.iimas.unam.mx

Abstract. We present a method to compute the number of independent sets for two basic graph patterns: simple cycles and trees. We show how to extend this initial method for processing efficiently more complex graph topologies.

We consider polygonal array graphs that are a graphical representation of molecular compounds. We show how our method processes polygonal trees based on a 2-treewidth decomposition of the input graph.

A pair of *macros* is associated to each basic graphic pattern in this decomposition, allowing us to represent and perform a series of repetitive operations while the same pattern graph is found. This results in a linear-time algorithm for counting its number of independent sets.

Keywords: Recognition of graph patterns · Counting the number of independent sets · Symbolic variables · Programming macros

1. Introduction

Counting problems are not only mathematically interesting, also they arise in many applications. Regarding hard counting problems, the computation of the number of independent sets of a graph G, denoted as $NI(G)$, has been a key in determining the frontier between efficient counting and intractable counting algorithms.

The recognition of structural patterns lying on graphs has been helpful to design efficient algorithms for counting its number of independent sets. For example, the linear-time Okamoto's algorithm [2] computes $NI(G)$ when G is a chordal graph, and where the decomposition of G in its clique tree gives the possibility of applying dynamic programming in an efficient way. Other case, is the Zhao's algorithm for computing $NI(G)$ on regular graphs [9]. However, for some kind of graphical patterns there is not an efficient method known, at least up to today, to compute $NI(G)$.

J.F. Martínez-Trinidad et al. (Eds.): MCPR 2016, LNCS 9703, pp. 167–176, 2016.
DOI: 10.1007/978-3-319-39393-3_17

Polygonal array graphs have been widely investigated, and they represent a relevant area of interest in mathematical chemistry because they have been used to study intrinsic properties of molecular graphs. In addition, it is also of great importance to recognize substructures of those compounds and learn messages from the graphic model by clear elucidation of their structures and properties [4]. There are several works analyzing extremal values for the number of independent sets (known in mathematical chemistry area as the Merrifield-Simmons index) on hexagonal chain graphs, but none of those works have presented the methods applied to compute that index [4,5].

The use of *macros* (common structural patterns) that are basic tools used in planning community, has been a common tool in AI (Artificial Intelligence), in order to represent accumulative series of basic operations into a plan action [6]. A main property of the macros is the possibility to represent accumulative preconditions and effects by making macros indistinguishable from individual operators, and allowing them to perform efficiently, a series of repetitive operations while the same pattern graph is found, as it occurs in the case of polygonal tree graphs.

Decompositions of graphs such as clique separators, treewidth decomposition and clique decomposition are often used to design efficient graph algorithms. There are even wonderful general results stating that a variety of NP-complete graph problems can be solved in linear time for graphs of bounded treewidth and bounded clique-width, respectively [7]. In order to obtain an efficient algorithm using this approach, the input graphs have to be restricted to a graph class nicely treewidth decomposable.

We show here, that a polygonal tree G has a 2-treewidth decomposition allowing the application of macros for computing $NI(G)$ efficiently. Our algorithm could be adapted as a computational tool for mathematical chemistry researchers, to contribute in the analysis of intrinsic properties on those molecular graphs. In fact, our algorithm can be adapted to compute also other intrinsic properties on molecular graphs; for example, for counting the number of matching edges of a chemical compound, known as the Hosoya index.

2.   Notation

Let $G = (V, E)$ be an undirected graph with vertex set (or node set) V and set of edges E. The *neighborhood* for $x \in V$ is $N(x) = \{y \in V : \{x, y\} \in E\}$, and its *closed neighborhood* is $N(x) \cup \{x\}$ which is denoted by $N[x]$. We denote the cardinality of a set A, by $|A|$. The degree of a vertex x, denoted by $\delta(x)$, is $|N(x)|$, and the degree of G is $\Delta(G) = max\{\delta(x) : x \in V\}$. The size of the neighborhood of x, $\delta(N(x))$, is $\delta(N(x)) = \sum_{y \in N(x)} \delta(y)$.

A path from v to w is a sequence of edges: $v_0v_1, v_1v_2, \ldots, v_{n-1}v_n$ such that $v = v_0$ and $v_n = w$ and v_k is adjacent to v_{k+1}, for $0 \leq k < n$. The length of the path is n. A simple path is a path where $v_0, v_1, \ldots, v_{n-1}, v_n$ are all distinct. A cycle is a non-empty path such that the first and last vertices are identical, and a simple cycle is a cycle in which no vertex is repeated, except that the first and last vertices are identical. A graph G is acyclic if it has no cycles. P_n, C_n, K_n, denote respectively, a path graph, a simple cycle and the complete graph, all of those graphs have n vertices.

Given a graph $G = (V, E)$, let $G' = (V', E')$ be a subgraph of G if $V' \subseteq V$ and E' contains edges $\{v, w\} \in E$ such that $v \in V'$ and $w \in V'$. If E' contains every edge $\{v, w\} \in E$ where $v \in V'$ and $w \in V'$ then G' is called the *induced graph* of G. A *connected component* of G is a maximal induced subgraph of G, that is, a connected component is not a proper subgraph of any other connected subgraph of G. If an acyclic graph is also connected, then it is called a *free tree*.

Given a graph $G = (V, E)$, $S \subseteq V$ is an independent set in G if for every two vertices v_1, v_2 in S, $\{v_1, v_2\} \notin E$. Let $I(G)$ denote the set of all independent sets of G. An independent set $S \in I(G)$ is *maximal* if it is not a subset of any larger independent set and, it is *maximum* if it has the largest size among all independent sets in $I(G)$. The determination of the maximum independent set has received much attention since it is an NP-complete problem.

The corresponding counting problem on independent sets, denoted by $NI(G)$, consists of counting the number of independent sets of a graph G. $NI(G)$ is a #P-complete problem for graphs G where $\Delta(G) \geq 3$.

3.   Counting Independent Sets on Basic Topology Graphs

In this section we present two algorithms to compute $NI(G)$ for two basic graph patterns: cycles and trees.

Since $NI(G) = \prod_{i=1}^{k} NI(G_i)$ where $G_i, i = 1, \ldots, k$, are the connected components of G [3], then the total time complexity for computing $NI(G)$, denoted as $T(NI(G))$, is given by the maximum rule as $T(NI(G)) = max\{T(NI(G_i)) : G_i$ is a connected component of $G\}$. Thus, a first helpful decomposition of the graph is done via its connected components, and from here on, we consider as an input graph only one connected component.

It turns out that the combinatorial meaning of the Fibonacci numbers are closely related to the number of independent sets of some kind of basic graphs patterns. In fact, it is shown in [8] that F_{n+2} is equal to the number of subsets (including the empty set) in $\{1, 2, ..., n\}$, such that no two elements are adjacent, i.e. there is no two consecutive integers in any subset. If we think of $\{1, 2, ..., n\}$ as the vertex set of a path graph, say P_n, where there exists an edge $e_i = \{i, i+1\}$, $i = 1, \ldots, n-1$, for each pair of sequential nodes, then F_{n+2} is equal to $NI(P_n)$.

On the other hand, if we consider that the n-th element in $\{1, 2, ..., n\}$ is adjacent to the first vertex, then P_n turns into a simple cycle C_n, and the number of subsets (including the empty set) with no two adjacent elements is characterized by the n-th Lucas number. Therefore, L_n is equal to $NI(C_n)$ [8].

3.1.   Cycles

Let $C_n = (V, E)$ be a simple cycle graph, and $|V| = n = |E| = m$, i.e. every node in V has degree two.

Theorem 1. *The number of independent sets in C_n is equal to $F_{n+1}+F_n-F_{n-2}$.*

Proof. We decompose the cycle C_n as: $G \cup \{c_m\}$, where $G = (V, E')$, $E' = \{c_1, ..., c_{m-1}\}$. G is a path of n nodes, and $c_m = \{v_m, v_1\}$ is called the *back edge* of the cycle.

We build the family $\mathcal{F}_i = \{G_i\}, i = 1, ..., n$ where $G_i = (V_i, E_i)$ is the induced graph of G formed by just the first i nodes of V.

We associate to each node $v_i \in V$ a pair (α_i, β_i), where α_i represents the number of sets in $I(G_i)$ where v_i does not appear, while β_i conveys the number of sets in $I(G_i)$ where v_i appears, thus $NI(G_i) = \alpha_i + \beta_i$. The first pair (α_1, β_1) is $(1, 1)$ since for the induced subgraph $G_1 = \{v_1\}$, $I(G_1) = \{\emptyset, \{v_1\}\}$. It is not hard to see that a new pair $(\alpha_{i+1}, \beta_{i+1})$ is built from the previous one by a Fibonacci sequence, as shown in Eq. 1.

$$(\alpha_{i+1}, \beta_{i+1}) = (\alpha_i + \beta_i, \alpha_i) \qquad (1)$$

Note that every independent set in G is an independent set in C_n, except for the sets $S \in I(G)$ where $v_1 \in S$ and $v_m \in S$. In order to eliminate those conflicting sets, we use two computing threads denoted by $\alpha\beta-$pairs to compute $NI(C_n)$, one of those for computing $NI(G)$ and the other one: (α'_i, β'_i), for computing $|\{S \in I(G) : v_1 \in S \wedge v_m \in S\}|$.

The second thread $\alpha\beta-$pair is done by starting $(\alpha'_1, \beta'_1) = (0, 1)$, and considering only the independent sets of $I(G)$ where v_1 appears $(0, \beta_1)$.

Expressing the computation of $NI(C_m)$ in terms of Fibonacci numbers, we have $(\alpha'_1, \beta'_1) = (0, 1) = (F_0, F_1) \rightarrow (\alpha'_2, \beta'_2) = (1, 0) = (F_1, F_0) \rightarrow (\alpha'_3, \beta'_3) = (1, 1) = (F_2, F_1), ..., (\alpha'_n, \beta'_n) = (F_{n-1}, F_{n-2})$, and the value for the final pair is $(0, F_{n-2})$, then $|\{S \in I(G') : v_1 \in S \wedge v_n \in S\}| = 0 + \beta_n = F_{n-2}$. Then, the last pair associated to the computation of $NI(C_n)$ is $(F_{n+1}, F_n - F_{n-2}) = (F_{n+1}, F_{n-1})$. Then, $NI(C_n) = F_{n+1} - F_{n-1}$, obtaining a well known identity, the n-th Lucas number. □

Algorithm 1. Count_Ind_Sets_cycles(C_n)

Input: Given a simple cycle graphs C_n, where $G \cup \{c_m\}$, where $G = (V, E')$, $E' = \{c_1, ..., c_{m-1}\}$.
Output: $NI(C_n)$
  Compute the NI(G) of the simple path G
  Compute $|\{S \in I(G) : v_1 \in S \wedge v_m \in S\}|$
  return $\alpha_{v_n} + \beta_{v_n} - \beta'_{n-2}$

3.2. Trees

Let $T = (V, E)$ be a rooted tree at a vertex $v_r \in V$. We denote with (α_v, β_v) the pair associated with the node v ($v \in V(T)$). We compute $NI(T)$ while we are traversing by T in post-order.

Algorithm 2. Count_Ind_Sets_trees(T)

Input: A tree T

Output: NI(T)

    Traversing T in post-order, and when a node $v \in T$ is left, assign:

    if v is a leaf node in T **then**

        $(\alpha_v, \beta_v) = (1, 1)$

    else if v is a parent node with a list of child nodes associated, i.e., $u_1, u_2, ..., u_k$ are the child nodes of v, as we have already visited all child nodes, then each pair $(\alpha_{u_j}, \beta_{u_j})$ $j = 1, ..., k$ has been determined based on recurrence (1) **then**

        Let $\alpha_v = \prod_{j=1}^{k} \alpha_{v_j}$ and $\beta_v = \prod_{j=1}^{k} \beta_{v_j}$. Notice that this step includes the case when v has just one child node.

    else if v_r is the root node of G **then**

        return $\alpha_{v_r} + \beta_{v_r}$

    end if

This algorithm returns the number of independent sets of G in time $O(n+m)$ which is the necessary time for traversing G in post-order.

We call *Linear_NI* to the algorithm that computes $NI(G)$ when G is a simple cycle or a tree. *Linear_NI* will be applied to process any acyclic graph or simple cycles that we find as part of a more complex graph.

4.   Polygonal System Graphs

Let $G = (V, E)$ be a molecular graph. Denote by $n(G, k)$ the number of ways in which k mutually independent vertices can be selected in G. By definition, $n(G, 0) = 1$ for all graphs, and $n(G, 1) = |V(G)|$. Then $\sigma(G) = \sum_{k \geq 0} n(G, k)$ will be the *Merrifield-Simmons index* of G, that is, exactly the number of independent sets of G. Merrifield and Simmons showed the correlation between $NI(G)$ and boiling points on polygonal chain graphs representing chemical molecules [4,5]. The Merrifield-Simmons index is a typical example of a graph invariant used in mathematical chemistry for quantifying relevant details of molecular structure.

A polygonal chain is a graph $P_{k,t}$ obtained by identifying a finite number of t congruent regular polygons, and such that each basic polygon, except the first and the last one, is adjacent to exactly two basic polygons. When each polygon in $P_{k,t}$ has the same number of k nodes, then $P_{k,t}$ is a linear array of t k-gons.

The way that two adjacent k-gons are joined, via a common vertex or via a common edge, defines different classes of polygonal chemical compounds. A special class of polygonal chains is the class of hexagonal chains, chains formed by n 6-gons. Hexagonal systems play an important role in mathematical chemistry as natural representations of catacondensed benzenoid hydrocarbons.

Let $T_P = H_1 H_2 \cdots H_n$ be a hexagonal chain with n hexagons, where each H_i and H_{i+1} have a common edge for each $i = 1, 2, \ldots, n-1$. A hexagonal chain with at least two hexagons has two end-hexagons: H_1 and H_n, while $H_2, \ldots, H_{n-1}$ are the internal hexagons of the chain. If the array of hexagons follows the structure of a tree where instead of nodes we have hexagons, and any two consecutive hexagons share exactly one edge, then we call to that graph a hexagonal tree (see Fig. 1). And when the nodes in the tree are represented by any k-gon then we call that graph a polygonal tree.

The propensity of carbon atoms to form compounds, made of hexagonal arrays fused along the edges gives a relevant importance to the study of chemical properties of benzenoid hydrocarbons. Thus, it has been paralleled by the study of its corresponding graphs, the so-called polygonal tree graphs. Those graphs have been widely investigated and represent a relevant area of interest in mathematical chemistry, since it is used for quantifying relevant details of the molecular structure of the benzenoid hydrocarbons [1,4]. We show now, how to count the number of independent sets for polygonal trees based on the use of macros.

4.1. Counting Independent Sets on Polygonal Trees

A simple cycle graph of length n, C_n, represents a polygon of n sides. Algorithm 1 allow us to compute $NI(C_n)$ for any polygon C_n. But, if we apply Algorithm 1 using symbolic variables (α, β) as the associated pair to each node in the polygon, then we can compute the Merrifield-Simmon index for any chain of polygons at the same time that we visit each edge of the array once.

Let us show how to use symbolic variables during the computation of $NI(P_4)$. The two computing threads and its associated pairs are expressed by the symbolic variables: α and β, as shown in Eq. 2.

$$
\begin{array}{l}
L_p : \ (\alpha, \beta) \to (\alpha + \beta, \alpha) \to (2\alpha + \beta, \alpha + \beta) \to (3\alpha + 2\beta, 2\alpha + \beta) \\
L_c : \ (0, \beta) \to (\beta, 0) \to \quad\ (\beta, \beta) \to \quad\quad (2\beta, \beta) \\
\hline
\quad\quad\quad\quad\quad\quad\quad\quad\quad\quad\quad\quad\quad\quad\quad\quad (3\alpha + 2\beta, 2\alpha)
\end{array}
\tag{2}
$$

Thus we obtain Eq. 3:

$$
NI(P_4) = 3\alpha + 2\beta + 2\alpha = 5\alpha + 3\beta \tag{3}
$$

In fact, applying Algorithm 1 and using symbolic variables for computing $NI(P_k)$, where P_k is a polygon of k nodes, we obtain a last pair: $(F_k\alpha + F_{k-1}\beta, F_{k-1}\alpha)$ which it is a system of linear equations, where F_k is the k-nth Fibonacci number. Thus, using symbolic variables α and β, we obtain Eq. 4.

$$
NI(P_k) = (F_k + F_{k-1})\alpha + F_{k-1}\beta \tag{4}
$$

We can codify the last pair $(F_k\alpha + F_{k-1}\beta, F_{k-1}\alpha)$ obtained from the computation of $NI(P_k)$ as a *macro*, that in terms of the initial symbolic variables α, β can be written as shown in Eq. 5.

$$
\alpha' = F_k\alpha + F_{k-1}\beta, \ \beta' = F_{k-1}\alpha \tag{5}
$$

This part of forming the pair of linear equations, is called the *formation of the macro*. Each symbolic variable in the macro: (α', β') represents a linear system on two symbolic initial variables: α and β. This macro would be associated to a common edge between any two polygons.

For example, let $P_1 \circ P_2$ be two contiguous hexagons with common edge $\{x, y\}$. Algorithm 1 computes $NI(P_1)$ beginning at the common node (between both polygons) x, and at the end of the algorithm, the pair $(F_k\alpha + F_{k-1}\beta, F_{k-1}\alpha)$ is obtained after visiting node y. Such pair of linear equations is associated to the common edge $\{x, y\}$, indicating that it does not matter the initial values for any α and β, they can be substituted by a current pair of values in order to obtain a current final pair of linear equations and such that in those new equations, the value $NI(P_1)$ has been considered as part of its accumulative operations.

The *expansion* of the macro (α', β') consists in considering the linear equations that they represent and substitute the symbolic variables α and β by its current values represented by new variables. This process of expansion is well-defined since no macro appears in its own expansion.

When the computation of $NI(P_2)$ is started by Algorithm 1, it begins in the node v_1 of P_2, and when the node x is visited, a current pair of symbolic variables (χ, δ) has been computed in the main line L_P and other pair (φ, γ) has been obtained in the secondary line L_{P_2}. To visit and process the common edge $\{x, y\}$ for P_2 implies to substitute the variables α and β in the macro $(F_k\alpha + F_{k-1}\beta, F_{k-1}\alpha)$ by χ and δ in the main thread L_P, and by φ and γ in the secondary thread L_{P_2}. Algorithm 1 continues with those new pairs of linear equations in both threads, and for the application of the Fibonacci recurrence, such pairs will be updated until arrive to the last edge of P_2, where v_1 is one of its end-points.

At the end of processing a new polygon, we obtain again a new pair of linear equations $(a \cdot \chi + b \cdot \delta, c \cdot \chi + d \cdot \delta)$ as a result for $NI(P_1 \circ P_2)$. For using symbolic variables in the last pair obtained through the processing of polygons, it allows us to substitute such macro with current variables α and β, updating the macro associated to the common edge. And in this way we can process any polygonal chain with any number of polygons in linear time, on the number of edges in the array.

Algorithm 3. Count_Ind_Sets_Polygonal_trees(G)

Traversing T_P in post-order, and consider all vertex forming the current polygon P_i that is being visited

repeat

    1) Form $I_P \subseteq I$ be the indices such that $v \in X(i), i \in I_P$ iff $x \in V(P_i)$

    2) Let $X_a, X_b \in X(i), i \in I_P$ be initial nodes of T containing a common edge $\{x, y\}$ between P_i and its father polygon in T_P. Assume X_a the father node of X_b

    3) Apply algorithm 2 on each $X(i), i \in I_P$ in post-order. The resulting macro is associated to the common edge $\{x, y\} \in X_a$

    /* If there is a macro (α_i, β_i) in any $X(i), i \in I_P$ then a macro-expansion is performed */

    4) Eliminate $X(i), i \in I_P$ from T, with exception of X_a

until the root node in T_G is evaluated

Substitute $\alpha = 1, \beta = 1$ in the last macro obtained, resulting a pair of integers (a, b)

Returns $NI(T_P) = a + b$.

---

5.   A 2-Treewidth Decomposition for Polygonal Trees

Treewidth is one of the most basic parameters in graph algorithms. There is a well established theory on the design of polynomial (or even linear) time algorithms for many intractable problems when the input is restricted to graphs of bounded treewidth. What is more important, there are many problems on graphs with n vertices and treewidth at most k that can be solved in time $O(c^k \cdot n^{O(1)})$, where c is some problem dependent constant [7].

For example, a maximum independent set (a MIS) of a graph can be found in time $O(2^k \cdot n)$ given a tree decomposition of width at most k. So, a quite natural approach to solve the $NI(G)$ problem would be to find a treewidth T_G of G and to determine how to join the partial results on the nodes of T_G. However, given a general graph G, finding its treewidth is an NP-complete problem.

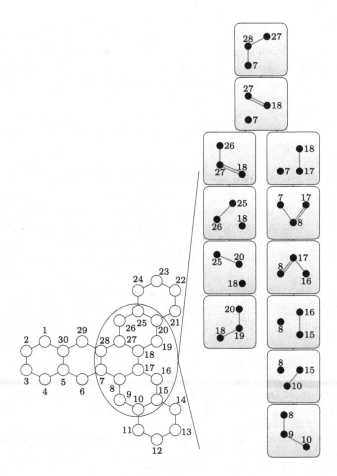

Fig. 1. A 2-treewidth decomposition for Trinaftileno

The treewidth decomposition of T_P-polygonal tree graph is based in the well-known 2-treewidth decomposition of any simple cycle (a polygon); and as any two adjacent polygons have just one common edge, then it is enough to join the 2-treewidth decomposition of the two contiguous polygons with the nodes containing the common edge, as it is shown in Fig. 1.

Let $T_G = (T, F)$ be the 2-tree decomposition of $T_P = (V, E)$, where $T = (I, F)$ is a tree, I is an index set and X is a function, $X : I \to 2^V$, satisfying the tree constraints of any k-tree decomposition [7]. We refer to $x \in V(G)$ as vertex and $X(i) \in T$ as nodes of T. A vertex x is associated with a node $i \in I$, or vice versa, whenever $v \in X(i)$. Our treewidth decomposition of T_P keeps the structure of a tree, and then we can apply the Algorithms 2 and 3 for computing $NI(T_P)$.

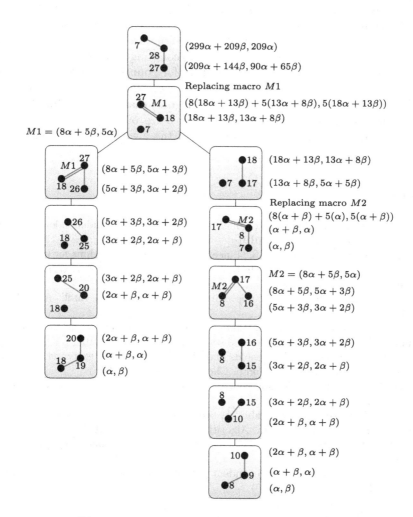

Fig. 2. Processing the 2-treewidth decomposition

5.1. Processing Polygonal Trees

In Figs. 1 and 2, we show the 2-treewidth decomposition of a polygonal tree graph, as well as the application of macros for counting independent sets on the nodes of the tree. Notice that all common edges are visited twice; the first one, a macro is formed, and in the second one, an expansion of the macro is performed. This provides a linear-time algorithm that traverses by all node and edge on the treewidth.

6. Conclusions

We have proposed a 2-treewidth decomposition for any polygonal tree graph T_P, where all common edge between two adjacent polygons will appear exactly in two consecutive nodes of the treewidth. This structure for the treewidth allows the efficient application of macros for solving counting problems on T_P.

We present a novel linear-time algorithm for counting independent sets on T_P. Our method exploits the use of macros for performing repetitive operations appearing on the basic common patterns (polygons) that form T_P. Our algorithm can be adapted to solve other intrinsic properties on polygonal tree graphs, impacting directly on the time complexity of the algorithms for solving those problems.

References

1. Došlić, T., Måløy, F.: Chain hexagonal cacti: matchings and independent sets. Discrete Math. **310**, 1676–1690 (2010)
2. Okamoto, Y., Uno, T., Uehara, R.: Linear-time counting algorithms for independent sets in chordal graphs. In: Kratsch, D. (ed.) WG 2005. LNCS, vol. 3787, pp. 433–444. Springer, Heidelberg (2005)
3. Roth, D.: On the hardness of approximate reasoning. Artif. Intell. **82**, 273–302 (1996)
4. Wagner, S., Gutman, I.: Maxima and minima of the Hosoya index and the Merrifield-Simmons index. Acta Applicandae Math. **112**(3), 323–346 (2010)
5. Deng, H.: Catacondensed benzenoids and phenylenes with the extremal third-order Randic Index. Comm. Math. Comp. Chem. **64**, 471–496 (2010)
6. Bäckström, C., Jonsson, A., Jonsson, P.: Automaton plans. J. Artif. Intell. Res. **51**, 255–291 (2014)
7. Fomin, F.V., Gaspers, S., Saurabh, S., Stepanov, A.A.: On two techniques of combining branching and treewidth. Algorithmica **54**, 181–207 (2009)
8. Prodinger, H., Tichy, R.F.: Fibonacci numbers of graphs. Fibonacci Q. **20**(1), 16–21 (1982)
9. Zhao, Y.: The number of independent sets in a regular graph. Comb. Probab. Comput. **19**(02), 315–320 (2010)

SMOTE-D a Deterministic Version of SMOTE

Fredy Rodríguez Torres[(✉)], Jesús A. Carrasco-Ochoa,
and José Fco. Martínez-Trinidad

Instituto Nacional de Astrofísica, Óptica y Electrónica, Puebla, Mexico
{frodriguez,ariel,fmartine}@inaoep.mx

Abstract. Imbalanced data is a problem of current research interest. This problem arises when the number of objects in a class is much lower than in other classes. In order to address this problem several methods for oversampling the minority class have been proposed. Oversampling methods generate synthetic objects for the minority class in order to balance the amount of objects between classes, among them, SMOTE is one of the most successful and well-known methods. In this paper, we introduce a modification of SMOTE which deterministically generates synthetic objects for the minority class. Our proposed method eliminates the random component of SMOTE and generates different amount of synthetic objects for each object of the minority class. An experimental comparison of the proposed method against SMOTE in standard imbalanced datasets is provided. The experimental results show an improvement of our proposed method regarding SMOTE, in terms of *F-measure*.

Keywords: Imbalanced datasets · Oversampling · Supervised classification

1 Introduction

In supervised classification, imbalanced data arises when in some classes the number of objects is much lower than in other classes. Usually, the minority class (the class with the lowest amount of objects) is the class of interest in a class imbalance problem. However, when working on imbalanced datasets, classifiers tend to have a bias towards the majority class (the class with the largest amount of objects), resulting in poor classification performance for the minority class. This behavior is more notorious when the imbalance among classes is higher.

Some researchers have tackled the class imbalance problem [12,13] through oversampling methods [4–10]. Oversampling methods have the advantage of being independent of the classification method to be used, since they generate synthetic objects based only on the training set. SMOTE [2] is one of the most used and well known oversampling methods, which generates synthetic objects along the line segments joining objects in the minority class with some of their nearest neighbors. Thus, by increasing the amount of objects of the minority class, SMOTE tries to balance the amount of objects for all the classes. SMOTE

© Springer International Publishing Switzerland 2016
J.F. Martínez-Trinidad et al. (Eds.): MCPR 2016, LNCS 9703, pp. 177–188, 2016.
DOI: 10.1007/978-3-319-39393-3_18

has a random component for generating synthetic objects, producing a different result each time it is applied. Therefore, whether SMOTE is applied several times, choosing the best result becomes an issue.

In this paper, we propose a new oversampling method based on SMOTE, which computes in a deterministic way how many new synthetic objects should be generated from each object of the minority class and where these new objects should be placed. According to our experiments the proposed method performs better than SMOTE in different datasets with different imbalance level.

The rest of this document is organized as follows: in the Sect. 2, some related works are described; in the Sect. 3, the proposed method is introduced; in the Sect. 4, the experimental setup for the experiments is described; in the Sect. 5, the experimental results are shown; and in the Sect. 6, our conclusions and some future work directions are discussed.

2   Related Work

In the literature there are two types of extensions of SMOTE, those which combine SMOTE with other methods like noise filters (SMOTE-IPF) [6], subsampling methods (SMOTE-RSB*) [5] or feature selectors (E-SMOTE) [7]; and those that modify SMOTE like Bordeline-SMOTE [4], Safe-Level-SMOTE [8], SMOTE-OUT [9], SMOTE-COSINE [9] or Random-SMOTE [10]. Our work belongs to these last kind of methods. Thus we briefly describe some methods that modify SMOTE:

Borderline-SMOTE only oversamples the objects in the borderline of the minority class. First, it finds out the borderline objects of the minority class; then, synthetic objects are generated from these objects and they are added to the original training set. Borderline-SMOTE works as follows:

- For each object in the minority class its k nearest neighbors, from the whole training set, are calculated.
- If the k nearest neighbors contain objects from the minority and majority classes and the amount of its nearest neighbors in the majority class is larger than the amount of its nearest neighbors in the minority class, the object is considered as a borderline object.

For each borderline object, Borderline-SMOTE calculates its k nearest neighbors in the minority class and generates synthetic objects in the same way as SMOTE.

Safe-Level-Synthetic Minority Oversampling technique, assigns to each object in the minority class its safe level before generating synthetic objects. Each synthetic object is positioned closer to the largest safe level object, in this way all the synthetic objects are generated in safe regions. The safe level of an object is defined as the number of minority class objects among it's k nearest neighbors.

SMOTE-OUT randomly generates synthetic objects along the outside line of attributes between an object of the minority class and its nearest neighbor of the majority class. Once a synthetic object has been generated then SMOTE-OUT finds out its nearest object in the minority class and randomly generates another synthetic object along the line of attributes between them.

SMOTE-Cosine works as SMOTE but it computes the k-nearest neighbors by voting using two distance metrics (Euclidean and Cosine).

Random-SMOTE generates temporally synthetic objects along the line of attributes between two objects of the minority class which are selected randomly. After, synthetic objects along the line of attributes between each temporal synthetic object and one object of the minority class are generated.

3   Proposed Method

The proposed method takes into account the distances between each object of the minority class and its k-nearest neighbors in the same class in order to determine how many synthetic objects should be generated from each object. For each object in the minority class, the higher its distance dispersion against its k-nearest neighbors, the higher the number of synthetic objects to generate. Additionally, the distance between an object and each one of its k-nearest neighbors is also taken into account individually to determine how many objects should be generated between each pair of objects. The larger the distance between a pairs of objects allows generating a greater amount of synthetic objects between them. The new synthetic objects are generated by dividing the difference in attributes between two objects by the number of objects to be generated between them. In this way, the synthetic objects will be created in a deterministic and uniform way.

For evaluating the dispersion of objects around each $object_i$ of the minority class we propose to use the standard deviation (σ_i) of the distances between the $object_i$ and its k nearest neighbors ($object_{ij}$ $j = 1, ..k$). We propose generating an amount of synthetic objects around each $object_i$ in the minority class such that it be proportional to the fraction of the standard deviation of the distances (σ_i) with respect to the sum of all the standard deviations of the distances computed for all the objects in the minority class ($\sum_{i=1}^{m} \sigma_i$, where m is the amount of objects in the minority class). Then, around objects of the minority class with higher standard deviation of distances, with respect to its k-nearest, neighbors more objects will be created.

After determining the proportion (p_i) of synthetic objects to generate from each object of the minority class, we proceed to calculate the proportion (p_{ij}) of synthetic objects to generate between each object of the minority class and each one of its k nearest neighbors. This proportion of synthetic objects is calculated as the fraction that represents the distance between the object and each nearest neighbor regarding to the sum of all the distances between the object and all its k nearest neighbors.

In the Fig. 1, we can see an example with three objects of the minority class and its 3-nearest neighbors (into the minority class), where $\sum_{i=1}^{m} \sigma_i = 2$. Here, the fraction ($p_1$) of σ_1 is $p_1 = \sigma_1 / \sum_{i=1}^{m} \sigma_i = 0.5$, therefore 50 % of the synthetic objects to be generated will be generated from $object_1$. For σ_2 the fraction is 30 % and 20 % for σ_3, If we have to generate a total of 10 synthetic objects, 5 would be generated around $object_1$, 3 around $object_2$, and 2 around $object_3$.

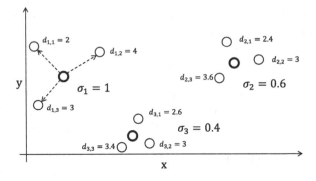

Fig. 1. Three objects of a minority class with distance values and standard deviation of distances for 3-nearest neighbors.

In order to determine the number of objects to generate ($s_{ij} = p_{ij} * p_i * n$) between an $object_i$ and each one of its nearest neighbors ($object_{ij}$) we take into account the amount of objects to generate for the minority class (n), this amount is given by the difference between the number of objects in the minority and majority class ($n = (M - m) * R$, where M is the number of objects in majority class, m is the number of objects in the minority class and R is a parameter defining the proportion of the difference to be reached).

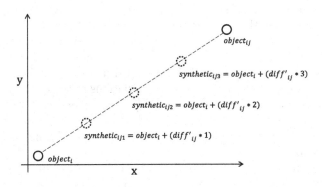

Fig. 2. Generation of 3 synthetic objects between an $object_i$ of the minority class and one of its nearest neighbors $object_{ij}$.

After calculating the amount of synthetic objects s_{ij} to generate between each $object_i$ an each one of its nearest neighbors ($object_{ij}, j = 1, \ldots, k$), attribute differences $diff_{ij}$ ($object_i, object_{ij}$) between $object_i$ and $object_{ij}$ are calculated. These differences are divided by the amount of synthetic objects to generate plus one, obtaining $diff'_{ij} = diff_{ij}/(s_{ij} + 1)$. This difference is added s_{ij} times to $object_i$; with each addition we obtain a new synthetic object and all the new synthetic objects are added to the original minority class. In the Fig. 2 we can see an example of the generation of three synthetic objects between an object of the minority class $object_i$ and one of its k-nearest neighbors $object_{ij}$.

If for an object in the minority class the amount of objects to be generated regarding to the proportion representing the fraction of the standard deviation of the distances with respect to the sum of all standard deviations of distances represents less than 1 object, SMOTE-D does not generate synthetic objects from this object. The same happens if the proportion that represents the distance between the object and one of its nearest neighbors regarding to the sum of all the distances between the object and all its k nearest neighbors represents less than the 1 %.

Given a training set with M objects in the majority class and m objects in the minority class. The detailed procedure of SMOTE-D is as follows:

- Calculate the amount of objects to be generated for the minority class ($n = (M - m) * R$) according to a parameter $R \in [0, 1]$.
- Calculate the distances (d_{ij}) between each $object_i$ in the minority class and its k nearest neighbors, $j = 1, \ldots, k$ (k is a parameter).
- Calculate the standard deviation (σ_i) of the distances between each $object_i$ and its k-nearest neighbors.
- Calculate for each $object_i$ the fraction (p_i) of its standard deviation (σ_i) from the total sum of all standard deviations as $p_i = \sigma_i / \sum_{i=1}^{m} \sigma_i$.
- Calculate the fraction (p_{ij}) of each distance (d_{ij}) with respect to the sum of distances of each $object_i$ and its k nearest neighbors as $p_{ij} = d_{ij} / \sum_{j=1}^{k} d_{ij}$.
- Calculate the number of objects (s_{ij}) to generate between an $object_i$ and one of its nearest neighbors $object_{ij}$ as $s_{ij} = p_i * p_{ij} * n$.
- Get the attribute difference $diff_{ij}$ between an $object_i$ and each one of its k nearest neighbors $object_{ij}$ as $diff_{ij} = object_{ij} - object_i; j = 1, \ldots, k$.

    • Divide the difference between an object and each one of its neighbors by the amount of synthetic objects to be generated from this pair plus 1. ($diff'_{ij} = diff_{ij} / (s_{ij} + 1)$)
    • Add the difference $diff'_{ij}$ to the object of the minority class as many times as objects to generate (s_{ij}).

- Add the generated synthetic objects to the minority class.

4 Experimental Setup

For evaluating the proposed method, we use 66 datasets taken from the repository KEEL [1], we used 5-fold cross validation. In order to measure the degree of imbalance in a dataset, in the literature, the imbalance ratio (IR) is commonly used. The IR of a dataset with M objects in the majority class and m objects in the minority class is computed as $IR = M/m$. All the datasets used in our experiments are from binary problems with numeric attributes, and IR ranging from 1.82 to 129.44 (see Table 1).

Into the KEEL repository these 66 datasets are provided together with the results of applying SMOTE with $k = 5$ as the number of nearest neighbors,

Table 1. Sumary of the datasets used in our experiments.

Name	# attributes	# objects	IR	Name	# attributes	# objects	IR
glass1	9	214	1.82	glass-0-4_vs_5	9	92	9.22
ecoli-0_vs_1	7	220	1.86	ecoli-0-3-4-6_vs_5	7	205	9.25
wisconsin	9	683	1.86	ecoli-0-3-4-7_vs_5-6	7	257	9.28
pima	8	768	1.87	yeast-0-5-6-7-9_vs_4	8	528	9.35
iris0	4	150	2.00	vowel0	13	988	9.98
glass0	9	214	2.06	ecoli-0-6-7_vs_5	6	220	10.00
yeast1	8	1484	2.46	glass-0-1-6_vs_2	9	192	10.29
haberman	3	306	2.78	ecoli-0-1-4-7_vs_2-3-5-6	7	336	10.59
vehicle2	18	846	2.88	led7digit-0-2-4-5-6-7-8-9_vs_1	7	443	10.97
vehicle1	18	846	2.90	glass-0-6_vs_5	9	108	11.00
vehicle3	18	846	2.99	ecoli-0-1_vs_5	6	240	11.00
glass-0-1-2-3_vs_4-5-6	9	214	3.20	glass-0-1-4-6_vs_2	9	205	11.06
vehicle0	18	846	3.25	glass2	9	214	11.59
ecoli1	7	336	3.36	ecoli-0-1-4-7_vs_5-6	6	332	12.28
new-thyroid1	5	215	5.14	cleveland-0_vs_4	13	177	12.62
new-thyroid2	5	215	5.14	ecoli-0-1-4-6_vs_5	6	280	13.00
ecoli2	7	336	5.46	shuttle-c0-vs-c4	9	1829	13.87
segment0	19	2308	6.02	yeast-1_vs_7	7	459	14.30
glass6	9	214	6.38	glass4	9	214	15.47
yeast3	8	1484	8.10	ecoli4	7	336	15.80
ecoli3	7	336	8.60	page-blocks-1-3_vs_4	10	472	15.86
page-blocks0	10	5472	8.79	abalone9-18	8	731	16.40
ecoli-0-3-4_vs_5	7	200	9.00	glass-0-1-6_vs_5	9	184	19.44
yeast-2_vs_4	8	514	9.08	shuttle-c2-vs-c4	9	129	20.50
ecoli-0-6-7_vs_3-5	7	222	9.09	yeast-1-4-5-8_vs_7	8	693	22.10
ecoli-0-2-3-4_vs_5	7	202	9.10	glass5	9	214	22.78
glass-0-1-5_vs_2	9	172	9.12	yeast-2_vs_8	8	482	23.10
yeast-0-3-5-9_vs_7-8	8	506	9.12	yeast4	8	1484	28.10
yeast-0-2-5-7-9_vs_3-6-8	8	1004	9.14	yeast-1-2-8-9_vs_7	8	947	30.57
yeast-0-2-5-6_vs_3-7-8-9	8	1004	9.14	yeast5	8	1484	32.73
ecoli-0-4-6_vs_5	6	203	9.15	ecoli-0-1-3-7_vs_2-6	7	281	39.14
ecoli-0-1_vs_2-3-5	7	244	9.17	yeast6	8	1484	41.40
ecoli-0-2-6-7_vs_3-5	7	224	9.18	abalone19	8	4174	129.44

HVDM [3] as distance function and an oversampling rate N such that $IR = 0.0$. Thus, for our experiments SMOTE-D was configured in the same way ($k = 5$, HVDM and $R = 1$ in order to get $IR = 0.0$).

For comparing the results of SMOTE-D and SMOTE, decision trees, support vector machines (SVM) and KNN with $k = 5$ were used. We applied SMOTE-D in all datasets and compare the classification results against those obtained by using SMOTE.

One of the measures commonly used for assessing the quality of classifiers on imbalanced datasets is the *F-measure* [11]. Therefore, in our experiments we used this measure to asses our results, additionally we applied a t-test with statistical significant difference at the 5 % of significance level between the results of SMOTE and SMOTE-D.

5   Experimental Results

The results of comparing SMOTE-D and SMOTE with different classifiers in terms of *F-measure* are shown in Tables 2 and 3. In both tables, the first column shows the name of the databases, the following columns show the classification

Table 2. Results of applying the evaluated classifiers over the results of SMOTE and SMOTE-D over the tested datasets using HVDM metric distance.

Dataset name	Decision tree		KNN		SVM	
	SMOTE	SMOTE-D	SMOTE	SMOTE-D	SMOTE	SMOTE-D
glass1	0.6910	**0.6967**	0.7502	**0.7575**	0.5693	**0.5746**
ecoli-0_vs_1	0.9744	**0.9822**	**0.9715**	0.9675	**0.9861**	**0.9861**
wisconsin	0.9315	**0.9444**	**0.9501**	0.9456	**0.9590**	0.9527
pima	**0.6073**	0.5533	**0.5692**	0.5378	**0.6699**	0.6664
iris0	**1.0000**	**1.0000**	**1.0000**	**1.0000**	**1.0000**	**1.0000**
glass0	**0.6904**	0.5935	**0.7073**	0.6861	0.6464	**0.6691**
yeast1	0.5226	**0.5278**	**0.5240**	0.4833	0.5859	**0.5885**
haberman	0.3670	**0.3755**	**0.4026**	0.3723	**0.4506**	0.4147
vehicle2	**0.9343**	0.8946	0.8605	**0.8712**	0.9212	**0.9329**
vehicle1	**0.5591**	0.5183	0.4265	**0.4385**	0.6581	**0.6651**
vehicle3	0.5283	**0.5630**	0.4555	**0.4701**	**0.6267**	0.6257
glass-0-1-2-3_vs_4-5-6	**0.8549**	0.8496	0.8650	**0.9197**	0.8378	**0.8688**
vehicle0	0.8295	**0.8812**	0.8434	**0.8658**	**0.9220**	0.9167
ecoli1	**0.7501**	0.7369	**0.7366**	0.7282	**0.7788**	0.7761
new-thyroid1	**0.9448**	0.8959	**0.9581**	**0.9581**	**0.9733**	0.9713
new-thyroid2	**0.9617**	0.8694	0.9600	**1.0000**	0.9617	**0.9713**
ecoli2	0.6610	**0.7394**	**0.8623**	0.8124	0.7192	**0.7402**
segment0	**0.9786**	0.9728	0.9701	**0.9715**	0.9822	**0.9850**
glass6	**0.7719**	0.7178	0.8469	**0.8621**	0.7872	**0.8256**
yeast3	0.7320	**0.7447**	0.6700	**0.6893**	0.6585	**0.7076**
ecoli3	0.5078	**0.6084**	0.5837	**0.6054**	0.5916	**0.6872**
page-blocks0	0.7927	**0.8303**	0.7653	**0.7982**	**0.1747**	0.1603
ecoli-0-3-4_vs_5	0.7597	**0.7921**	0.6678	**0.8000**	0.6854	**0.7351**
yeast-2_vs_4	0.6642	**0.7371**	0.7092	**0.7343**	0.6999	**0.7423**
ecoli-0-6-7_vs_3-5	0.5044	**0.6521**	0.6033	**0.7460**	0.4835	**0.5442**
ecoli-0-2-3-4_vs_5	0.6732	**0.8100**	0.6946	**0.7937**	0.6745	**0.7254**
glass-0-1-5_vs_2	**0.3464**	0.2244	**0.2956**	0.2817	**0.1489**	0.0000
yeast-0-3-5-9_vs_7-8	**0.3314**	0.3188	**0.4597**	0.4146	**0.3606**	0.3360
yeast-0-2-5-7-9_vs_3-6-8	0.4249	**0.4673**	0.4676	**0.4964**	0.5197	**0.5941**
yeast-0-2-5-6_vs_3-7-8-9	**0.7654**	0.6747	**0.6954**	0.6648	0.7004	**0.7766**
ecoli-0-4-6_vs_5	0.6057	**0.6797**	0.7461	**0.7784**	0.6844	**0.7695**
ecoli-0-1_vs_2-3-5	0.6005	**0.6523**	0.5571	**0.6587**	0.5690	**0.6945**
ecoli-0-2-6-7_vs_3-5	**0.6255**	0.6094	**0.7034**	0.6714	0.5710	**0.7052**

Table 2. (*Continued*)

Dataset name	Decision tree		KNN		SVM	
	SMOTE	SMOTE-D	SMOTE	SMOTE-D	SMOTE	SMOTE-D
glass-0-4_vs_5	0.8800	**0.9333**	0.8743	**1.0000**	0.7886	**0.8933**
ecoli-0-3-4-6_vs_5	0.7429	**0.7562**	0.7621	**0.8548**	0.7168	**0.7529**
ecoli-0-3-4-7_vs_5-6	0.6582	**0.7740**	0.6472	**0.7321**	0.6524	**0.7349**
yeast-0-5-6-7-9_vs_4	0.4268	**0.4624**	**0.4760**	0.4061	0.4509	**0.5086**
vowel0	**0.8989**	0.8959	**1.0000**	1.0000	**0.8131**	0.7696
ecoli-0-6-7_vs_5	0.5600	**0.6871**	0.6793	**0.6976**	0.5814	**0.7611**
glass-0-1-6_vs_2	0.3174	**0.3333**	0.2256	**0.3665**	**0.1382**	0.0000
ecoli-0-1-4-7_vs_2-3-5-6	0.6229	**0.7140**	0.6349	**0.7228**	0.5956	**0.6845**
led7digit-0-2-4-5-6-7-8-9_vs_1	0.7450	**0.7766**	0.3622	**0.5133**	0.5363	**0.6542**
glass-0-6_vs_5	0.6167	**0.7714**	0.6760	**0.7792**	0.5797	**0.6358**
ecoli-0-1_vs_5	0.7933	**0.9333**	0.8600	**0.9200**	0.6489	**0.7200**
glass-0-1-4-6_vs_2	0.3456	**0.3837**	0.2734	**0.2939**	**0.1931**	0.1435
glass2	0.3825	**0.4089**	**0.3262**	0.2876	**0.1806**	0.0500
ecoli-0-1-4-7_vs_5-6	0.5765	**0.6737**	0.6781	**0.7766**	0.5235	**0.7065**
cleveland-0_vs_4	0.4544	**0.5410**	**0.1653**	0.1500	0.5367	**0.6324**
ecoli-0-1-4-6_vs_5	0.6106	**0.6133**	0.6368	**0.8159**	0.6240	**0.6444**
shuttle-c0-vs-c4	**1.0000**	0.9884	**0.9959**	0.9841	**0.9959**	0.9959
yeast-1_vs_7	0.2685	**0.2900**	0.3047	**0.3574**	0.2939	**0.3036**
glass4	**0.6276**	0.5743	0.7833	**0.7881**	0.6308	**0.6558**
ecoli4	0.6543	**0.7790**	0.8148	**0.8492**	0.6328	**0.8047**
page-blocks-1-3_vs_4	0.9110	**0.9667**	0.7231	**0.8013**	0.3535	**0.4013**
abalone9-18	0.2573	**0.4583**	0.3176	**0.4992**	0.3684	**0.6419**
glass-0-1-6_vs_5	0.5619	**0.6800**	0.6933	0.5943	0.4870	**0.5743**
shuttle-c2-vs-c4	1.0000	0.9333	1.0000	1.0000	1.0000	1.0000
yeast-1-4-5-8_vs_7	0.1393	**0.2179**	**0.1880**	0.1791	0.1276	**0.1490**
glass5	**0.7867**	0.7276	**0.6933**	0.6076	0.4918	**0.7156**
yeast-2_vs_8	0.4653	**0.5418**	**0.4879**	0.4599	0.5610	**0.6681**
yeast4	0.2915	**0.3118**	**0.3446**	0.3190	0.2812	**0.3117**
yeast-1-2-8-9_vs_7	**0.1678**	0.1473	0.1748	**0.2012**	0.1428	**0.1726**
yeast5	0.5546	**0.6338**	0.6953	**0.7120**	0.4709	**0.5003**
ecoli-0-1-3-7_vs_2-6	0.2833	**0.5400**	0.3667	**0.5333**	0.2183	**0.3994**
yeast6	0.3706	**0.4041**	**0.4743**	0.4324	0.2647	**0.3465**
abalone19	**0.0515**	0.0235	0.0293	**0.0353**	0.0466	**0.0587**
Average	0.6199	**0.6513***	0.6310	**0.6583***	0.5831	**0.6257***

Table 3. Results of applying the evaluated classifiers over the results of SMOTE and SMOTE-D over the tested datasets using euclidean metric distance.

Dataset name	Decision tree		KNN		SVM	
	SMOTE	SMOTE-D	SMOTE	SMOTE-D	SMOTE	SMOTE-D
glass1	0.6469	**0.6511**	**0.7642**	0.6511	0.5688	**0.6511**
ecoli-0_vs_1	0.9618	**0.9862**	0.9474	**0.9862**	0.9659	**0.9862**
wisconsin	0.9253	**0.9338**	**0.9475**	0.9338	**0.9572**	0.9338
pima	**0.5908**	0.5762	0.5552	**0.5762**	**0.6658**	0.5762
iris0	**1.0000**	**1.0000**	**1.0000**	**1.0000**	**1.0000**	**1.0000**
glass0	0.6838	**0.7161**	0.7156	**0.7161**	0.6533	**0.7161**
yeast1	**0.5280**	0.5269	0.5042	**0.5269**	**0.5790**	0.5269
haberman	0.3707	**0.3807**	0.3645	**0.3807**	**0.4349**	0.3807
vehicle2	0.9167	**0.9193**	0.8672	**0.9193**	0.9186	**0.9193**
vehicle1	**0.5339**	0.5175	0.4399	**0.5175**	**0.6656**	0.5175
vehicle3	0.5443	**0.5608**	0.4470	**0.5608**	**0.6268**	0.5608
glass-0-1-2-3_vs_4-5-6	0.8370	**0.8402**	**0.9114**	0.8402	**0.8423**	0.8402
vehicle0	**0.8626**	0.8508	**0.8591**	0.8508	**0.9292**	0.8508
ecoli1	0.7658	**0.7700**	0.7488	**0.7700**	**0.7759**	0.7700
new-thyroid1	**0.9228**	0.8979	**0.9713**	0.8979	**0.9713**	0.8979
new-thyroid2	**0.9027**	0.8739	**0.9920**	0.8739	**0.9713**	0.8739
ecoli2	0.7470	**0.7645**	**0.7980**	0.7645	0.7259	**0.7645**
segment0	**0.9816**	0.9743	**0.9759**	0.9743	**0.9894**	0.9743
glass6	0.7314	**0.8221**	**0.8517**	0.8221	**0.8315**	0.8221
yeast3	**0.7300**	0.7195	0.6957	**0.7195**	0.6755	**0.7195**
ecoli3	**0.5287**	0.5086	**0.6249**	0.5086	**0.5862**	0.5086
page-blocks0	0.8173	**0.8474**	0.7904	**0.8474**	0.2182	**0.8474**
ecoli-0-3-4_vs_5	**0.7626**	0.7042	**0.7871**	0.7042	0.6815	**0.7042**
yeast-2_vs_4	**0.6963**	0.6934	**0.7361**	0.6934	**0.7011**	0.6934
ecoli-0-6-7_vs_3-5	**0.6051**	0.5535	**0.7040**	0.5535	**0.5937**	0.5535
ecoli-0-2-3-4_vs_5	**0.7254**	**0.7254**	**0.7725**	0.7254	0.7058	**0.7254**
glass-0-1-5_vs_2	0.3150	**0.4103**	0.3091	**0.4103**	0.1648	**0.4103**
yeast-0-3-5-9_vs_7-8	0.3031	**0.3903**	0.3375	**0.3903**	0.3641	**0.3903**
yeast-0-2-5-7-9_vs_3-6-8	0.4661	**0.4996**	0.4747	**0.4996**	**0,5223**	0.4996
yeast-0-2-5-6_vs_3-7-8-9	0.7641	**0.7690**	0.7180	**0.7690**	0.6997	**0.7690**
ecoli-0-4-6_vs_5	**0.7176**	0.6478	**0.7837**	0.6478	**0.7664**	0.6478
ecoli-0-1_vs_2-3-5	0.6160	**0.6796**	**0.6887**	0.6796	0.6738	**0.6796**
ecoli-0-2-6-7_vs_3-5	**0.6340**	0.6157	**0.7284**	0.6157	**0.6418**	0.6157
glass-0-4_vs_5	**0.9333**	**0.9333**	**1.0000**	0.9333	0.8793	**0.9333**
ecoli-0-3-4-6_vs_5	0.7235	**0.7862**	**0.8498**	0.7862	**0.8014**	0.7862

Table 3. (*Continued*)

Dataset name	Decision tree		KNN		SVM	
	SMOTE	SMOTE-D	SMOTE	SMOTE-D	SMOTE	SMOTE-D
ecoli-0-3-4-7_vs_5-6	0.6427	**0.6567**	**0.7390**	0.6567	**0.6845**	0.6567
yeast-0-5-6-7-9_vs_4	0.4379	**0.4812**	0.4579	**0.4812**	0.4485	**0.4812**
vowel0	**0.8902**	0.8577	**1.0000**	0.8577	0.8095	**0.8577**
ecoli-0-6-7_vs_5	0.6841	**0.7743**	0.7245	**0.7743**	0.7060	**0.7743**
glass-0-1-6_vs_2	**0.2728**	0.1602	**0.2564**	0.1602	**0.1673**	0.1602
ecoli-0-1-4-7_vs_2-3-5-6	0.6514	**0.7260**	0.6895	**0.7260**	0.6189	**0.7260**
led7digit-0-2-4-5-6-7-8-9_vs_1	**0.7625**	0.7614	0.5207	**0.7614**	0.6838	**0.7614**
glass-0-6_vs_5	0.6969	**0.7702**	0.7685	**0.7702**	0.6989	**0.7702**
ecoli-0-1_vs_5	0.8667	**0.9333**	0.9200	**0.9333**	0.8200	**0.9333**
glass-0-1-4-6_vs_2	**0.4101**	0.3417	**0.3445**	0.3417	0.1926	**0.3417**
glass2	**0.3580**	0.2416	**0.2876**	0.2416	0.1844	**0.2416**
ecoli-0-1-4-7_vs_5-6	0.6811	**0.6911**	**0.7468**	0.6911	0.6066	**0.6911**
cleveland-0_vs_4	**0.6593**	0.6181	0.1479	**0.6181**	0.5657	**0.6181**
ecoli-0-1-4-6_vs_5	**0.5986**	0.5739	**0.7833**	0.5739	**0.6080**	0.5739
shuttle-c0-vs-c4	**1.0000**	0.9920	**0.9955**	0.9920	**1.0000**	0.9920
yeast-1_vs_7	**0.2823**	0.2624	0.2613	**0.2624**	**0.2804**	0.2624
glass4	**0.7019**	0.5500	**0.8155**	0.5500	**0.6027**	0.5500
ecoli4	0.7221	**0.7312**	**0.8068**	0.7312	**0.7801**	0.7312
page-blocks-1-3_vs_4	**0.9667**	0.9624	0.7951	**0.9624**	0.3344	**0.9624**
abalone9-18	0.4217	**0.4974**	**0.5117**	0.4974	**0.5074**	0.4974
glass-0-1-6_vs_5	0.6600	**0.7200**	**0.7533**	0.7200	0.6953	**0.7200**
shuttle-c2-vs-c4	**0.9333**	**0.9333**	**1.0000**	0.9333	**1.0000**	0.9333
yeast-1-4-5-8_vs_7	0.0984	**0.1314**	0.1089	**0.1314**	**0.1328**	0.1314
glass5	0.7600	**0.7733**	0.6476	**0.7733**	0.7538	**0.7733**
yeast-2_vs_8	**0.4217**	0.3800	0.3424	**0.3800**	**0.5369**	0.3800
yeast4	0.3273	**0.3329**	0.3152	**0.3329**	0.2891	**0.3329**
yeast-1-2-8-9_vs_7	0.1673	**0.1753**	0.1092	**0.1753**	0.1386	**0.1753**
yeast5	**0.7110**	0.6051	**0.7208**	0.6051	0.4720	**0.6051**
ecoli-0-1-3-7_vs_2-6	0.5054	**0.5889**	0.5133	**0.5889**	0.2974	**0.5889**
yeast6	**0.4023**	0.3911	0.3613	**0.3911**	0.2996	**0.3911**
abalone19	0.0434	**0.0451**	**0.0582**	0.0451	**0.0511**	0.0451
Average	0.6444	**0.6470**	**0.6540**	0.6470	0.6169	**0.6470**

results, in terms of *F-measure*, for decision tree, KNN and SVM classifiers, respectively. In the last row, the average over the 66 datasets for each classifier is shown. The best results of *F-measure* appear boldfaced. Datasets with (*) are those where the t-test showed statistical significant difference at the 5 % of significance level.

In the results that appear in the Table 2, the proposed method obtains a better performance in 67 % of the datasets using decision trees, in 61 % using KNN, and 73 % using SVM. In average the results show a statistical improvement of 3.14 %, 2.73 % and 4.26 % in terms of *F-measure* for decision trees, KNN and SVM respectively. Considering only those datasets with an IR greater than 15.8 the results obtained by all the classifiers when SMOTE-D is applied are statistical significant better for all used classifiers, these databases can be seen with the names in bold.

In the results that appear in the Table 3, the proposed method obtains a better performance in 50 % of the datasets using decision trees, in 46 % using KNN, and 50 % using SVM. In average the results do not show a statistical difference for all classifiers. Considering only those datasets with an IR greater than 10.59 the results obtained by SVM classifier when SMOTE-D is applied are statistical significant better for SVM classifier, these databases can be seen with the names in bold

6   Conclusions

This paper introduces a new oversampling method, SMOTE-D, which is a deterministic version of SMOTE. Comparisons against SMOTE in terms of *F-measure* using decision trees, KNN and SVM classifiers show that SMOTE-D get better results than SMOTE. These results give evidence that estimating the dispersion of the objects of the minority class (based on the standard deviation of distances) to determine how many objects should be generated around each object in the minority class and how many objects should be created between each object and its nearest neighbors, together with a deterministic and uniform creation of the synthetic objects, allows an oversampling of the minority class such that better results than SMOTE can be obtained.

From our experiments, we can conclude that when a dataset has an imbalance ratio higher than 10.0, then SMOTE-D,using either Euclidean or HVD distance, performs better than SMOTE.

As future work, we are going to extend the proposed method for working with nominal attributes.

Acknowledgment. This work was partly supported by National Council of Science and Technology of Mexico under the scholarship grant 627301.

References

1. Alcalá-Fdez, J., Fernandez, A., Luengo, J., Derrac, J., García, S., Sánchez, L., Herrera, F.: KEEL data-mining software tool: data set repository, integration of algorithms and experimental analysis framework. J. Multiple-Valued Logic Soft Comput. **17**(2–3), 255–287 (2011)
2. Chawla, N.V., et al.: SMOTE: synthetic minority over-sampling technique. J. Artif. Intell. Res. **16**, 321–357 (2002)
3. Wilson, D., Randall Martinez, T.R.: Improved heterogeneous distance functions. J. Artif. Intell. Res. **6**, 1–34 (1997)
4. Han, H., Wang, W.-Y., Mao, B.-H.: Borderline-SMOTE: a new over-sampling method in imbalanced data sets learning. In: Huang, D.-S., Zhang, X.-P., Huang, G.-B. (eds.) ICIC 2005. LNCS, vol. 3644, pp. 878–887. Springer, Heidelberg (2005)
5. Ramentol, E., et al.: SMOTE-RSB*: a hybrid preprocessing approach based on over-sampling and undersampling for high imbalanced data-sets using SMOTE and rough sets theory. Knowl. Inf. Syst. **33**(2), 245–265 (2012)
6. Sáez, J.A., et al.: SMOTE IPF: addressing the noisy and borderline examples problem in imbalanced classification by a re-sampling method with filtering. Inf. Sci. **291**, 184–203 (2015)
7. Deepa, T., Punithavalli, M.: An E-SMOTE technique for feature selection in high-dimensional imbalanced dataset. In: 2011 3rd International Conference on Electronics Computer Technology (ICECT), vol. 2. IEEE (2011)
8. Bunkhumpornpat, C., Sinapiromsaran, K., Lursinsap, C.: Safe-level-SMOTE: safe-level-synthetic minority over-sampling technique for handling the class imbalanced problem. In: Theeramunkong, T., Kijsirikul, B., Cercone, N., Ho, T.-B. (eds.) PAKDD 2009. LNCS, vol. 5476, pp. 475–482. Springer, Heidelberg (2009)
9. Koto, F.: SMOTE-OUT, SMOTE-COSINE, and selected-SMOTE: an enhancement strategy to handle imbalance in data level. In: 2014 International Conference on Advanced Computer Science and Information Systems (ICACSIS). IEEE (2014)
10. Dong, Y., Wang, X.: A new over-sampling approach: random-SMOTE for learning from imbalanced data sets. In: Xiong, H., Lee, W.B. (eds.) KSEM 2011. LNCS, vol. 7091, pp. 343–352. Springer, Heidelberg (2011)
11. Larsen, B., Aone, C.: Fast and effective text mining using linear-time document clustering. In: Proceedings of the Fifth ACM SIGKDD International Conference on Knowledge Discovery and Data Mining. ACM (1999)
12. Shakiba, N., Rueda, L.: MicroRNA identification using linear dimensionality reduction with explicit feature mapping. In: BMC Proceedings. BioMed Central (2013)
13. Batuwita, R., Palade, V.: Adjusted geometric-mean: a novel performance measure for imbalanced bioinformatics datasets learning. J. Bioinf. Comput. Biol. **10**(04), 1250003 (2012)

A Glance to the Goldman's Testors from the Point of View of Rough Set Theory

Manuel S. Lazo-Cortés[✉], José Francisco Martínez-Trinidad, and Jesús Ariel Carrasco-Ochoa

Instituto Nacional de Astrofísica Óptica y Electrónica, San Andrés Cholula, Mexico {mlazo,fmartine,ariel}@inaoep.mx

Abstract. In this article, we revisit the concept of Goldman's fuzzy testor and re-study it from the perspective of the conceptual approach to attribute reduct introduced by Y.Y. Yao in the framework of Rough Set Theory. We reformulate the original concept of Goldman's fuzzy testor and we introduce the Goldman's fuzzy reducts. Additionally, in order to show the usefulness of the Goldman's fuzzy reducts, we build a rule based classifier and evaluate its performance in a case of study.

1 Introduction

The concept of testor (initially *test*) was created by S.V. Yablonskii as a tool for solving problems of control and diagnosis of faults in contact networks and combinatorial circuits. From these early works [2,22] derived a research line that is primarily concerned with this kind of problems. In the middle of the sixties of the past century, the methods of Testor Theory were extended to prediction problems in such domains as geology and medicine [9]. Later Testor Theory has been developed in both directions, tools and applications.

The primary concept of testor (and typical testor) has had numerous generalizations and adaptations to different environments [17]. In 1980, R.S. Goldman published an article in Russian introducing a type of testor which he called fuzzy testor [14]. The main characteristic of this kind of testor is that comparisons between two values of the same attribute take real values in the interval [0,1], in such a way that 0 is interpreted as equality (or minimal difference) and 1 is interpreted as the maximum possible difference. An exhaustive review of the literature published since then allows us to ensure that this fuzzy testor is not well known and the study of its properties and possibilities has been very limited.

Undoubtedly, when one refers to fuzziness, there is unanimity in understanding that one refers to the fuzzy set theory and the derived logic introduced by Zadeh in 1965 [25]. However, in many problems, sources of fuzziness are varied and even more so their treatment. Classification problems are no exception.

In the rough set theory, the fuzzy approach has also been studied; so for example, to address the issues of feature selection and attribute reduction, there are some works on fuzzy reducts. Since the publication of the work of

© Springer International Publishing Switzerland 2016
J.F. Martínez-Trinidad et al. (Eds.): MCPR 2016, LNCS 9703, pp. 189–197, 2016.
DOI: 10.1007/978-3-319-39393-3_19

Dubois and Prade [10], there have been many studies regarding basic concepts of Rough Set Theory from the perspective of fuzziness. Fuzzy attributes [5,27], fuzzy positive region [6,7], fuzzy decision trees [11,12], fuzzy rules [19,20], fuzzy discernibility matrix [27], fuzzy decision reduct to certain degree [8] and fuzzy subset of attributes [26] are some examples that show the variety of studies that have been made related to fuzziness within the context of Rough Set Theory.

The notion of reduct plays an essential role in rough set analysis. Previously, the concepts of testor and reduct have been related [3,15,16]. In this paper, we revisit the old concept of Goldman's fuzzy testor re-studying it from the Rough Set Theory point of view and particularly we analyze its relation with the concept of reduct from the conceptual approach developed by Y.Y. Yao in [23], as well as its application to supervised classification problems.

This document is organized as follows: In Sect. 2, we present some basic concepts regarding Goldman's fuzzy testor and the conceptual definition of reduct. In Sect. 3, we show that a Goldman's typical fuzzy testor can be defined as a reduct. In Sect. 4, we exemplify a practical use of Goldman's typical fuzzy testors to build a rule based classifier and additionally, we illustrate their practical usefulness through a case study. Our concluding remarks are summarized in Sect. 5.

2   Basic Concepts

The basic representation of data in Rough Set Theory is an information system, which is a table with rows representing objects while columns specify attributes. Formally, an information system is defined as a 4-tuple $IS = (U, A_t^* = A_t \cup \{d\}, \{V_a \mid a \in A_t^*\}, \{I_a \mid a \in A_t^*\})$, where U is a finite non-empty set of objects, A_t^* is a finite non-empty set of attributes, d denotes the decision attribute, V_a is a non-empty set of values of $a \in A_t^*$, and $I_a : U \to V_a$ is an information function that maps an object of U to exactly one value in V_a.

Let us define for each attribute a in A_t a real valued dissimilarity function $\varphi_a : V_a \times V_a \to [0,1]$ in such a way that 0 is interpreted as equality (or minimal difference) and 1 is interpreted as the maximum possible difference.

Applying these comparison functions to all possible pairs of objects belonging to different classes in IS, a [0,1]-pairwise discernibility matrix can be built. We will denote such discernibility matrix as DM. We assume that IS is consistent, that is, there is not a pair of indiscernible objects belonging to different classes, this means that there will be no complete row of zeros in DM.

Example 1 illustrates an information system and its corresponding [0,1]-pairwise discernibility matrix, applying the heterogeneous distance function used by Giraud-Carrier et al. [13] which is normally associated to the Heterogeneous Euclidean-Overlap Metric function HEOM [21]. This function defines the distance between two values x and y of a given attribute a as:

$$\varphi_a(x,y) = \begin{cases} 1 & \textit{if } x \textit{ or } y \textit{ is unknown, else} \\ overlap_a(x,y) & \textit{if } a \textit{ is nominal, else} \\ rn_{diff_a}(x,y) & \textit{otherwise} \end{cases} \tag{1}$$

being

$$overlap_a(x, y) = \begin{cases} 0 & if \ x = y \\ 1 & otherwise \end{cases}$$

and

$$rn_{diff_a}(x, y) = \frac{|x - y|}{max_a - min_a}$$

Example 1. M shows an example of an information table, d represents the decision attribute. DM is the corresponding [0,1]-pairwise discernibility matrix. DM is obtained applying $\varphi_a(x, y)$ in (1), attribute by attribute, to each pair of objects in M belonging to different classes.

$$M = \begin{pmatrix} a_1 & a_2 & a_3 & a_4 & d \\ 1 & 36 & white & 13.6 & 1 \\ 1 & 29 & yellow & 12.9 & 1 \\ 0 & 42 & yellow & 11.8 & 1 \\ 0 & 17 & red & 31.1 & 0 \\ 0 & 11 & white & 29.4 & 0 \\ 1 & 21 & red & 40.3 & 0 \end{pmatrix}$$

$$DM = \begin{pmatrix} a_1 & a_2 & a_3 & a_4 \\ 1 & 0.61 & 1 & 0.61 \\ 1 & 0.81 & 0 & 0.55 \\ 0 & 0.48 & 1 & 0.94 \\ 1 & 0.39 & 1 & 0.64 \\ 1 & 0.58 & 1 & 0.58 \\ 0 & 0.26 & 1 & 0.96 \\ 0 & 0.81 & 1 & 0.68 \\ 0 & 1.00 & 1 & 0.62 \\ 1 & 0.68 & 1 & 1.00 \end{pmatrix}$$

From the second row of DM, for example, we can state that there exists a pair of objects (belonging to different classes in M) which are discernible in grade 0.81 regarding a_2. The same pair of objects are indiscernible regarding a_3. These objects are the corresponding to the first and third rows of M.

From the example, we can also affirm that all objects belonging to different classes in M are discernible in a certain grade regarding attributes a_2 (at least 0.26) and a_4 (at least 0.55). Although a_1 and a_3 do not fulfill the same property.

We will introduce the notation $\mu_{f_i}(a_j)$ to refer to the value corresponding to row f_i and to the column associated to attribute a_j in DM. So, for example $\mu_{f_4}(a_2) = 0.39$, $\mu_{f_6}(a_1) = 0$, $\mu_{f_1}(a_3) = 1$.

Definition 1 *(Goldman's fuzzy testor).* Let $A_t = \{a_1, a_2, ..., a_n\}$ and let $T = \{a_{r_1}|\mu_{r_1}, a_{r_2}|\mu_{r_2}, ..., a_{r_s}|\mu_{r_s}\}$ be a fuzzy subset of A_t such that $\forall p \in \{1, 2, ..., s\}$ $\mu_{r_p} \neq 0$. T is a Goldman's fuzzy testor with respect to IS if $\forall f_i \in DM$ (being f_i the i-th row in DM) $\exists a_{r_p}|\mu_{r_p} \in T$ such that $\mu_{r_p} \leq \mu_{f_i}(a_{r_p})$.

We will denote the set of all Goldman's fuzzy testors of an information system by Ψ.

According to the above definition, a Goldman's fuzzy testor is a fuzzy subset of attributes such that the set of attributes belonging to it, with the corresponding membership degrees, are able to discern all pairs of objects belonging to different classes.

Definition 2 *(Goldman's typical fuzzy testor).* $T = \{a_{r_1}|\mu_{r_1}, a_{r_2}|\mu_{r_2}, ..., a_{r_s}|\mu_{r_s}\}$, *($\mu_{r_p} \neq 0$; $p \in \{1, 2, ..., s\}$) is a Goldman's typical fuzzy testor with respect to IS if:*

(i) T *is a Goldman's fuzzy testor with respect to IS.*
(ii) $\forall p \in \{1, 2, ..., s\}, T \setminus \{a_{r_p}|\mu_{r_p}\}$ *is not a Goldman's fuzzy testor with respect to IS.*
(iii) $\forall T'$ *such that* $T \subset T'$ *and* $supp(T) = supp(T')$ *(it means that* $\forall p \in \{1, 2, ..., s\}$ $\mu_{r_p} \leq \mu'_{r_p}$ *and for at least one index the inequality is strict)* T' *is not a Goldman's fuzzy testor with respect to IS.*

We will denote the set of all Goldman's typical fuzzy testors of an information system by Ψ^*.

Condition *(ii)* means that if a fuzzy singleton $\{a_{r_p}|\mu_{r_p}\}$ is eliminated from a Goldman's typical fuzzy testor, the resulting subset is no longer a Goldman's fuzzy testor. Condition (iii) means that if the membership degree of some attribute to T is increased, then T stops being a Goldman's fuzzy testor.

Example 2. Regarding M in Example 1, $\{a_4|0.55\}$, $\{a_2|0.81, a_4|0.58\}$ and $\{a_1|1.0, a_2|0.81, a_4|0.94\}$ are examples of Goldman's typical fuzzy testors.

$\{a_1|1.0, a_2|0.48, a_3|1.0, a_4|0.96\}$ and $\{a_1|1.0, a_2|0.39, a_4|0.96\}$ are both Goldman's fuzzy testors but not typical, because $\{a_1|1.0, a_2|0.48, a_4|0.96\}$ is a Goldman's typical testor.

In the classic formulation, typical testors are minimal by inclusion, this is also quite common for different extensions of the primary concept of testor, see for example [17]. However, Goldman's fuzzy testors do not keep the property of minimality by inclusion. Therefore, in order to define Goldman's typical fuzzy testors in a similar way, we introduce the following partial order.

Definition 3 *(partial order $\preceq$).* *Let us consider the following binary relation over the set* $\mathfrak{P}(A)$ *of all fuzzy subsets of A. Let* t_1, $t_2 \in \mathfrak{P}(A)$, *then*
$t_1 \preceq t_2 \Leftrightarrow (t_1 \cap t_2) \cup ((supp(t_1) \setminus supp(t_2)) \cap t_1) \cup ((supp(t_2) \setminus supp(t_1)) \cap t_2) = t_2$

It is not difficult to prove that $\preceq$ is a partial order. For saving space, we omit the proof.

Proposition 1. $T \in \Psi$ *is a Goldman's typical fuzzy testor with respect to IS if* T *is a minimal element for the relation* $\preceq$ *defined over* Ψ.

Proof. *Let* T *be a minimal element in* Ψ *for the relation* $\preceq$ *and let us suppose that* T *is not a Goldman's typical fuzzy testor, then either (a) we can eliminate any fuzzy singleton from* T *or (b) we can increase the membership degree of any attribute belonging to* $supp(T)$ *(or both) and* T *continues being a Goldman's fuzzy testor. Let us suppose (a) is fulfilled. Let* $T = \{a_{r_1}|\mu_{r_1}, a_{r_2}|\mu_{r_2}, ..., a_{r_s}|\mu_{r_s}\} \in \Psi$ *and for simplicity suppose that* $T' = \{a_{r_2}|\mu_{r_2}, ..., a_{r_s}|\mu_{r_s}\}$ *is also a Goldman's fuzzy testor,* $T' \in \Psi$. *Then,* $(T' \cap T) = T'$, $((supp(T) \setminus supp(T')) \cap T) = \emptyset$ *and* $(supp(T') \setminus supp(T) = \{a_{r_1}\}$ *then* $((supp(T') \setminus supp(T)) \cap T') = \{a_{r_1}|\mu_{r_1}\}$ *and hence* $(T' \cap T) \cup ((supp(T') \setminus supp(T)) \cap T') \cup ((supp(T) \setminus supp(T')) \cap T) = T$, *which means that* $T' \preceq T$ *which contradicts that* T *is a minimal element in* Ψ.

Now, let us suppose (b) is fulfilled. Let $T = \{x_{r_1}|\mu_{r_1}, x_{r_2}|\mu_{r_2}, ..., x_{r_s}|\mu_{r_s}\} \in \Psi$ and for simplicity suppose that $T' = \{x_{r_1}|\nu_{r_1}, x_{r_2}|\mu_{r_2}, ..., x_{r_s}|\mu_{r_s}\}$ with $\nu_{r_1} > \mu_{r_1}$, is also a Goldman's fuzzy testor, $T' \in \Psi$. In this case $(T' \cap T) = T$, and $supp(T') = supp(T)$ hence, $(T' \cap T) \cup ((supp(T') \setminus supp(T)) \cap T') \cup ((supp(T) \setminus supp(T')) \cap T) = T$, which once again contradicts that T is a minimal element in Ψ.

3   Goldman's Fuzzy Reducts

In this section, we follow Y.Y. Yao's approach to the conceptual formulation of the Rough Sets Theory, namely the conceptual formulation of the concept of reduct [23]. The notion of reduct plays an essential role in rough set analysis. In order to formulate an in-depth conceptual understanding of reducts, Y.Y. Yao searched for an explanation and interpretation of the reduct concept in a wider context. Given a set of attributes, the question is whether there exists a subset that serves for the same purpose as that of the entire set. Such a subset may be considered as a reduct of the original set of attributes. In [24], a conceptual definition of a reduct of a set of attributes is developed based on this intuitive understanding.

Suppose A is a finite set and $\mathfrak{p}(A)$ is the power set of A. Let $\mathbb{P}$ denote a unary predicate on subsets of A. For $S \in \mathfrak{p}(A)$, $\mathbb{P}(S)$ stands for the statement that subset S has property $\mathbb{P}$. The values of $\mathbb{P}$ are computed by an evaluation e with reference to certain available data, for example, an information system. For a subset $S \in \mathfrak{p}(A)$, $\mathbb{P}(S)$ is true if S has property $\mathbb{P}$, otherwise, it is false. In this way, a conceptual definition of reduct is given based on an evaluation e as follows.

Definition 4 *(Subset-based definition). Given an evaluation e of $\mathbb{P}$, if a subset $R \subseteq A$ fulfills the following conditions:*

(a) existence: $\mathbb{P}_e(A)$;
(b) sufficiency: $\mathbb{P}_e(R)$;
(c) minimization: $\forall B \subseteq R(\neg \mathbb{P}_e(B))$;

we call R a reduct of A.

These three conditions reflect the fundamental characteristics of a reduct. Condition of existence (a) ensures that a reduct of S exists, in the great majority of the studies it is explicitly assumed that the whole set A must have the property $\mathbb{P}$, and then A itself is a candidate to be a reduct. Condition of sufficiency (b) expresses that a reduct R of A is sufficient for preserving the property $\mathbb{P}$ of A. Condition of minimization (c) expresses that a reduct is a minimal subset of A having property $\mathbb{P}$ in the sense that none of the proper subsets of R has the property. Since it is needed to check all the subsets of R to verify the Definition 4, Y.Y. Yao called this definition a subset-based definition [23].

For our convenience, we consider $\mathfrak{P}(A)$ as the set of all subsets of A, including fuzzy subsets, instead of the classical power set $\mathfrak{p}(A)$. Besides, we consider the partial order $\preceq$ previously defined instead of the classic inclusion.

Let $A = \{a_1, a_2, ..., a_n\}$ be as before the set of condition attributes in IS, and $S = \{a_{r_1}|\mu_{r_1}, a_{r_2}|\mu_{r_2}, ..., a_{r_s}|\mu_{r_s}\}$ a fuzzy subset of A, i.e. $\{a_{r_1}, a_{r_2}, ..., a_{r_s}\} \subseteq A$ and $0 < \mu_{r_p} \leq 1$, $p = \{1, 2, ...s\}$. Let $\mu_i^o = min\{\varphi_i(u, v) \neq 0\}$ for all pair of objects u and v belonging to different classes in IS, $1 \leq i \leq n$, and let $A^o = \{a_1|\mu_1^o, a_2|\mu_2^o, ..., a_n|\mu_n^o\}$. A^o is a fuzzy subset of A built considering the minimum among the nonzero values of each column of the $[0,1]$-pairwise discernibility matrix DM associated to IS.

Example 3. For the information table in Example 1, $A^o = \{a_1|1.0, a_2|0.26, a_3|1.0, a_4|0.55\}$.

Let us consider the following predicate $\mathbb{P}$:

$\mathbb{P}(S) \equiv \forall u, v \in IS[I_d(u) \neq I_d(v)] \rightarrow \exists x_{r_p}|\mu_{r_p} \in S$ such that $\mu_{r_p} \leq \varphi_{r_p}(u, v)$).

Notice that A^o has the property $\mathbb{P}$ since by construction $\forall u, v \in IS$ with $I_d(u) \neq I_d(v)$ $[\mu_{r_p^o} \leq \varphi_{r_p}(u, v)]$ unless $min\{\varphi_{r_p}(u, v)\} = 0$, $\forall p \in \{1, ..., n\}$, but this is not possible since we have assumed that IS is consistent. Then, we have $\mathbb{P}_e(A^o)$.

On the other hand, let $T = \{a_{r_1}|\mu_{r_1}, a_{r_2}|\mu_{r_2}, ..., a_{r_s}|\mu_{r_s}\}$ be a Goldman's fuzzy testor, then from Definition 1 it follows that T also has the property $\mathbb{P}$, i.e. $\mathbb{P}_e(T)$. In fact, T has the property $\mathbb{P}$ iff T is a Goldman's fuzzy testor. Finally, taking into account that minimal elements by $\preceq$ in Ψ are Goldman's typical fuzzy testors, it follows that $\forall B \prec T [\neg \mathbb{P}_e(B)]$.

Now, we can introduce the following definition.

Definition 5. *Let A^o, $\mathbb{P}$ and $\preceq$ as above. Let T be a Goldman's typical fuzzy testor. Considering that:*

(a) existence: $\mathbb{P}_e(A^o)$;
(b) sufficiency: $\mathbb{P}_e(T)$;
(c) minimization: $\forall B \prec T[\neg \mathbb{P}_e(B)]$;

T satisfies Definition 4 so we can say that T is a Goldman's fuzzy reduct.

4   Goldman's Fuzzy Reducts for Supervised Classification

Goldman's fuzzy reducts would be just a theoretical curiosity if they lacked practical use. In this section, we present an example of their application to a problem of supervised classification. Let IS be an information system, as before. Let $u_1, u_2, ..., u_m$ be the objects in U and $a_1, a_2, ..., a_n$ the attributes in A_t. In a supervised classification problem, U represents the training sample. Let d be the decision attribute, and $c_1, c_2, ..., c_s$ be the values that d takes, i.e., $c_1, c_2, ..., c_s$ are the class labels, and $I_d(u)$ denotes the class u belongs to. Let v be an object to be classified and Ψ^* be the set of all Goldman's fuzzy reducts of IS. We define a rule based classifier based on the set of Goldman's fuzzy reducts as follows:

Proposed classifier
v: object to be classified
$supp_1 = supp_2 = ...supp_s = 0$ (initializing the support counter for each class)
    for each $T = \{a_{r_1}|\mu_{r_1}, a_{r_2}|\mu_{r_2}, ..., a_{r_s}|\mu_{r_s}\} \in \Psi^*$
        for each $u_i \in U$
           if $[\varphi_{a_{r_1}}(v, u_i) \leq \mu_{r_1}] \vee [\varphi_{a_{r_2}}(v, u_i) \leq \mu_{r_2}] \vee ...[\varphi_{a_{r_s}}(v, u_i) \leq \mu_{r_s}]$
           and $I_d(u_i) = c_k$ then $supp_k = supp_k + 1$
    if $supp_k > supp_w$ for all $w \neq k$ then v is assigned to class c_k.

This algorithm behaves like a rule based classifier, where for each Goldman's fuzzy reduct and each object in the training sample a rule is built and evaluated. If the rule built for an object in the i-th class is fulfilled, the support for this class is increased. In a general case one should take into account issues such as if the training sample is imbalanced. To show the performance of the algorithm we will avoid this issue by considering a balanced training sample.

For evaluating the practical usefulness of this classification algorithm based on Goldman's fuzzy reducts, we study its behavior over the Iris database [18]. As it is widely known, the Iris database consists of 3 classes with 50 objects each, described in terms of 4 real-valued attributes. We built five training sets by randomly splitting the data, taking 70 % of the objects for training and the remaining 30 % for testing. Table 1(a) contains the confusion average matrix; for our proposed classifier, the mean percentage of correct classification was 94.67 %. It is important to comment that unlike other rule based classifiers, our classifier can work directly with the original data.

For comparing our results, in Table 1(b), we present the average classification results over the same five partitions using the RSES software [1], which is a well known system within the community of Rough Set Theory. For each case, classic reducts and rules based on them were calculated and evaluated.

Another experiment was done discretizing the continuous data to crisp data. For this, we used the discretization proposed by F. Coenen [4]. Then we computed the new set of reducts and rules using RSES. Results of this third experiment are shown in Table 1(c).

As it can be seen, both classification results based on classic reducts, with and without discretization, were worse than the classification results obtained by the proposed algorithm based on Goldman's fuzzy reducts.

Table 1. Confusion matrices for Iris data applying three methods

Classes	(a) Goldman's fuzzy reducts				(b) Classic reducts without discretization				(c) Classic reducts with discretization			
	Set	Vers	Virg	%	Set	Vers	Virg	%	Set	Vers	Virg	%
I. Setosa	15	0	0	100	15	0	0	100	15	0	0	100
I. Versicolor	0.2	13.6	1.2	90.7	0	13.8	1.2	92	0.42	13.2	1.38	88
I. Virginica	0	1	14	93.3	0	3.4	11.6	77.3	0.06	2.04	12.9	86
				94.67				89.78				91.33

5 Conclusions

In this paper, we introduced a new type of fuzzy reducts inspired by the Goldman's typical fuzzy testors. These reducts are fuzzy in the sense that they are fuzzy subsets of the set of attributes in an information system. Theoretically we show that this new type of reducts can be interpreted as reducts in the sense of the subset-based conceptual definition of reducts given by Y.Y. Yao [23].

Based on our proposed conceptualization of Goldman's fuzzy reducts, we introduce a new classifier based on rules. An advantage of the proposed method is that we can directly work with the original data without preprocessing continuous data by fuzzification or discretization.

The results achieved in our case study suggest that, as future work, it makes sense to deepen the study of these new reducts both theoretically and experimentally. For example, it would be interesting to study the case when data are described in terms of different types of attributes (nominal, Boolean, real, etc.). Likewise, it would be interesting to study the behavior of the classifier if only a proper subset of the Goldman's fuzzy reducts are used instead of using all of them, and how to select this subset. The problem of finding efficient algorithms for calculating Goldman's fuzzy reducts is also a research line.

We conjecture that, using Goldman's fuzzy reducts, we could obtain robust classifiers for datasets with real-valued attributes, but this point remains to be confirmed by further research.

References

1. Bazan, J., Szczuka, M.S.: The rough set exploration system. In: Peters, J.F., Skowron, A. (eds.) Transactions on Rough Sets III. LNCS, vol. 3400, pp. 37–56. Springer, Heidelberg (2005)
2. Cheguis, I.A., Yablonskii, S.V.: Logical methods of control of work of electric schemes. Trudy Mat. Inst. Steklov. **51**, 270–360 (1958). (in Russian)
3. Chikalov, I., Lozin, V., Lozina, I., Moshkov, M., Nguyen, H.S., Skowron, A., Zielosko, B.: Three Approaches to Data Analysis. Intelligent Systems Reference Library, vol. 41. Springer, Heidelberg (2013)
4. Coenen, F.: The LUCS-KDD Discretised/normalised ARM and CARM DataLibrary. Department of Computer Science, The University of Liverpool, UK (2003). http://www.csc.liv.ac.uk/~frans/KDD/Software/LUCS_KDD_DN/
5. Cornejo, M.E., Medina-Moreno, J., Ramírez, E.: On the classification of fuzzy-attributes in multi-adjoint concept lattices. In: Rojas, I., Joya, G., Cabestany, J. (eds.) IWANN 2013, Part II. LNCS, vol. 7903, pp. 266–277. Springer, Heidelberg (2013)
6. Cornelis, C., Jensen, R.: A noise-tolerant approach to fuzzy-rough feature selection. In: IEEE World Congress on Computational Intelligence FUZZ-IEEE 2008, pp. 1598–1605 (2008)
7. Cornelis, C., Jensen, R., Hurtado, G., Ślzak, D.: Attribute selection with fuzzy decision reducts. Inf. Sci. **180**(2), 209–224 (2010)
8. Cornelis, C., Martín, G.H., Jensen, R., Ślęzak, D.: Feature selection with fuzzy decision reducts. In: Wang, G., Li, T., Grzymala-Busse, J.W., Miao, D., Skowron, A., Yao, Y. (eds.) RSKT 2008. LNCS (LNAI), vol. 5009, pp. 284–291. Springer, Heidelberg (2008)

9. Dmitriev, A.N., Zhuravlev, Y.I., Krendelev, F.P.: On mathematical principles for classifications of objects and phenomena. Diskretnyi Analiz **7**, 3–11 (1966). (in Russian)

10. Dubois, D., Prade, H.: Rough fuzzy sets and fuzzy rough sets. Int. J. Gen. Syst. **17**, 191–209 (1990)

11. Elashiri, M.A., Hefny, H.A., Elwahab, A.H.A.: Induction of fuzzy decision trees based on fuzzy rough set techniques. In: 2011 International Conference on IEEE Computer Engineering & Systems (ICCES), pp. 134–139 (2011)

12. Elashiri, M.A., Hefny, H.A., Abd Elwhab, A.H.: Construct fuzzy decision trees based on roughness measures. In: Das, V.V., Stephen, J. (eds.) CNC 2012. LNICST, vol. 108, pp. 199–207. Springer, Heidelberg (2012)

13. Giraud-Carrier, C., Martinez, T.: An efficient metric for heterogeneous inductive learning applications in the attribute-value language. In: Intelligent Systems, pp. 341–350 (1995)

14. Goldman, R.S.: Problems of fuzzy test theory. Avtomat. Telemech. **10**, 146–153 (1980)

15. Lazo-Cortés, M.S., Martínez-Trinidad, J.F., Carrasco-Ochoa, J.A., Sanchez-Diaz, G.: Are reducts and typical testors the same? In: Bayro-Corrochano, E., Hancock, E. (eds.) CIARP 2014. LNCS, vol. 8827, pp. 294–301. Springer, Heidelberg (2014)

16. Lazo-Cortes, M.S., Martinez-Trinidad, J.F., Carrasco-Ochoa, J.A., Sanchez-Diaz, G.: On the relation between rough set reducts and typical testors. Inf. Sci. **294**, 152–163 (2015)

17. Lazo-Cortes, M., Ruiz-Shulcloper, J., Alba-Cabrera, E.: An overview of the evolution of the concept of testor. Pattern Recogn. **34**(4), 753–762 (2001)

18. Lichman, M.: UCI Machine Learning Repository Irvine. University of California, School of Information and Computer Science, CA (2013). http://archive.ics.uci.edu/ml

19. Tsang, E.C.C., Zhao, S., Yeung, D.S., Lee, J.W.T.: Learning from an incomplete information system with continuous-valued attributes by a rough set technique. In: Yeung, D.S., Liu, Z.-Q., Wang, X.-Z., Yan, H. (eds.) ICMLC 2005. LNCS (LNAI), vol. 3930, pp. 568–577. Springer, Heidelberg (2006)

20. Wang, X., Tsang, E.C., Zhao, S., Chen, D., Yeung, D.S.: Learning fuzzy rules from fuzzy samples based on rough set technique. Inf. Sci. **177**(20), 4493–4514 (2007)

21. Wilson, D.R., Martinez, T.R.: Improved heterogeneous distance functions. J. Artif. Intell. Res. **11**, 134 (1997)

22. Yablonskii, S.V., Cheguis, I.A.: On tests for electric circuits. Uspekhi Mat. Nauk **10**(4), 182–184 (1955). (in Russian)

23. Yao, Y.Y.: The two sides of the theory of rough sets. Knowl.-Based Syst. **80**, 67–77 (2015)

24. Yao, Y., Fu, R.: The concept of reducts in pawlak three-step rough set analysis. In: Peters, J.F., Skowron, A., Ramanna, S., Suraj, Z., Wang, X. (eds.) Transactions on Rough Sets XVI. LNCS, vol. 7736, pp. 53–72. Springer, Heidelberg (2013)

25. Zadeh, L.A.: Information and control. Fuzzy Sets **8**(3), 338–353 (1965)

26. Zhao, S., Chen, H., Li, C., Zhai, M., Du, X.: RFRR: Robust fuzzy rough reduction. IEEE Trans. Fuzzy Syst. **21**(5), 825–841 (2013)

27. Zhao, S.Y., Tsang, E.C., Wang, X.Z., Chen, D.G., Yeung, D.S.: Fuzzy matrix computation for fuzzy information system to reduce attributes. In: 2006 IEEE International Conference on Machine Learning and Cybernetics, pp. 2300–2304 (2006)

Automatic Construction of Radial-Basis Function Networks Through an Adaptive Partition Algorithm

Ricardo Ocampo-Vega[1], Gildardo Sanchez-Ante[1]([✉]), Luis E. Falcon-Morales[1], and Humberto Sossa[2]

[1] Tecnologico de Monterrey, Campus Guadalajara, Av. Gral Ramon Corona 2514, 45201 Zapopan, Jal, Mexico
gildardo@itesm.mx
[2] Instituto Politecnico Nacional-CIC, Av. Juan de Dios Batiz S/N, Gustavo A. Madero, 07738 Distrito Federal, Mexico

Abstract. Radial-Basis Function Neural Networks (RBFN) are a well known formulation to solve classification problems. In this approach, a feedforward neural network is built, with one input layer, one hidden layer and one output layer. The processing is performed in the hidden and output layers. To adjust the network for any given problem, certain parameters have to be set. The parameters are: the centers of the radial functions associated to the hidden layer and the weights of the connections to the output layer. Most of the methods either require a lot of experimentation or may demand a lot of computational time. In this paper we present a novel method based on a partition algorithm to automatically compute the amount and location of the centers of the radial-basis functions. Our results, obtained by running it in seven public databases, are comparable and even better than some other approaches.

Keywords: Radial-basis functions · Neural networks · Adaptive parameter adjustment · Classification

1 Introduction

We humans have the ability to quickly identify the face of a friend among many people, to distinguish oranges from apples in the market and to remember the name of a song playing in the radio. All those tasks could be considered as instances of what is called the *classification* problem. Solving it implies assigning a certain label -the class- to a set of features. There are quite a few methods to solve classification tasks. Those methods can be broadly grouped in supervised or unsupervised. The *supervised* approach tries to find an approximate model for the classes, whose error is smaller than a certain threshold. In order to do that, it is required to have a set of solved instances of the problem. Those instances are used to improve an initial model through a *training algorithm* [1].

© Springer International Publishing Switzerland 2016
J.F. Martínez-Trinidad et al. (Eds.): MCPR 2016, LNCS 9703, pp. 198–207, 2016.
DOI: 10.1007/978-3-319-39393-3_20

One of the most well known supervised methods is the Artificial Neural Networks (ANN). Current ANN models are used to solve complex problems, such as image processing in medical applications [2], the control of devices [3], as well as face [4] and speech recognition [5], to mention just a few [6]. There are different types of ANNs. One of them is called the Radial-Basis Function (RBF) Neural Networks.

Radial-Basis Function Neural Networks (RBFNN) require three layers of artificial neurons: the first layer connects the network with the environment, the second layer is the only hidden layer, usually implemented by radial activated functions. The third layer is the output, which is implemented as a weighted sum of the outputs of the hidden units (that is, a linear function) [7]. Figure 1 illustrates a RBF NN with three inputs, four hidden units and one output. RBFs are important given their capability to model non-linear systems with only two layers. Under certain conditions on the shape of the activation functions, RBFNN are universal approximators [8]. That means that such networks can approximate any continuous function with arbitrary precision, if enough hidden neurons are provided. Other ANN models, like the Multi-Layer Perceptron, can also approximate non-linear functions, but they may require multiple intermediary layers [7].

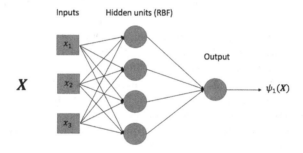

Fig. 1. Architecture of a radial basis function network.

When RBF networks are used in classification problems, the inputs are associated to the features that characterize the problem, and the outputs represent the classes or labels. The hidden units correspond to subclasses. Several different functions have been considered as activation functions for the hidden neurons [9]. Undoubtedly, the Gaussians are the most common ones. Equation 1 presents the definition of a Gaussian function.

$$\phi_j(\mathbf{X}) = \exp[-(\mathbf{X} - \boldsymbol{\mu}_j)^T \Sigma_j^{-1}(\mathbf{X} - \boldsymbol{\mu}_j)] \tag{1}$$

for $j = 1, \ldots, L$, where $\mathbf{X}$ is the feature vector, L is the number of hidden units, μ_j and Σ_j are the mean and the covariance matrix of the j-th Gaussian function.

As for the output layer, it is computed through a weighted sum, defined in Eq. 2:

$$\boldsymbol{\Psi}_k(\mathbf{X}) = \sum_{j=1}^{L} \lambda_{jk}\phi_j(\mathbf{X}) \tag{2}$$

for $k = 1, \ldots, M$, where λ_{jk} are the output weights, M is the number of output units, and ϕ_j represents node j in the hidden layer. It is common that for pattern classification problems the output is limited to the interval $(0, 1)$ by a sigmoidal function.

The training of the RBF neural networks is often performed in two steps: the first step is to find the number of neurons in the hidden layer, which means defining the corresponding radial-basis functions. The second step consists in finding the weights for the connections to the output layer. Despite the importance of RBF Networks, it turns out that creating one to solve a particular problem is an issue that may imply a good amount of experimental work. One traditional method to perform those steps consists in first deciding on the number L of choosing hidden neurons. Then choosing arbitrarily L data points as centers. Such procedure often causes ill-conditioning, but it has been shown that for large training sets, this approach gives reasonable results [1]. Another way to determine the number of neurons in the hidden layer is performed by using clustering (K-means). Still, it is up to the person solving the problem to decide the number L of neurons. Once that has been done, the K-means algorithm finds the centers of each cluster. This is an iterative process that implies solving an optimization problem. From that, is possible to define the radial-basis functions $\phi_j(\mathbf{X}); \forall j \in \{1, 2, \ldots, L\}$. Then, the weights for the output layer are computed, usually through Least-Means Squares (LMS) or Recursive-Means Squares (RMS). Since the number of neurons (centers) is defined arbitrarily, usually several values are tested to determine with which the RBFN gives the best results.

In this work, we propose a new method to automatically determine the number of neurons on the hidden layer. Moreover, our method adjusts the number of neurons so that we get the best performance in the RBFN for a given training dataset. Our contribution is to provide a simple algorithm to construct the best RBFNN for a given problem, requiring no intervention from the user. The experimental results show that our approach is competitive in practice compared with RBFNN generated with the traditional trial and error process. The remainder of the paper is organized as follows: Sect. 2 describes other works related with ours. Section 3 presents our method. Section 4 describe the experiments and results and Sect. 5 discusses possible future work.

2   Related Works

There is a lot of research in neural networks and their applications. In particular, some researchers have attempted to develop methods to automatically choose some of the parameters that control the functioning of the RBFNs. For instance,

Moody et al. [10] analyzed three versions of a learning algorithm, using nearest neighbor prediction, adaptive processing units with LMS and self-organizing adaptive units. According to their results, their proposal allows to train faster than with backpropagation.

In another approach, Poggio et al. [11] introduce a supervised method based on regularization to find the centers of the radial-basis functions and to update the weights. In a way, their work leads to a generalization of RBFNs. In this work, the authors focus on the mathematical formulation of their approach and no information is given regarding the applicability of it to practical problems.

Chen et al. [9] on the other hand, consider the application of Orthogonal Least-Squares (OLS). Under this approach, the functions are chosen one at a time until an appropriate network is built. The center of each Gaussian is given by one of the data points. The points are chosen by constructing a set of orthogonal vectors. The authors show several applications and mention as further work a comparison with Moody's algorithm [10].

Kurban et al. [12] introduces a comparison of training algorithms of radial basis function (RBF) neural networks for classification purposes. The authors compare: gradient descent (GD), genetic algorithms (GA), Kalman filtering (KF) and Artificial Bee Colonies (ABC). They use datasets from UCI to test the algorithms. In their results, ABC offered a good performance, so they used that implementation to solve a problem related with the classification of terrains given information from inertial sensors. The main issue with the approach is that in general the convergence time is bigger than for methods such as GD and KF.

Chun-Tao et al. [13] introduce a methodology based on Particle Swarm Optimization (PSO) and simulation (Resource allocation), to train RBF Networks. Their results seems to suggest to use the method with caution since it got trapped in local minima, though. Oh et al. [14] extended this work and use fuzzy logic to help in the process. Their results are comparable to previous works.

In a very recent work, Reiner [15] propose two algorithms to construct RBF networks. One of them is based on an Incremental Extreme Machine Learning added with a simplex optimization. The second one uses a Levenberg-Marquardt algorithm to find the locations and configurations of the Radial-Basis functions. What the author finds is that doing this allows to define more compact networks with smaller errors, compared with ten different algorithms.

3   Methodology

The classic training method for RBFN is hybrid, and it is comprised by two steps. In the first step, the parameters of the hidden nodes are set. Afterwards, the weights of the output layer are adjusted until a halt criteria is satisfied. For this method, the k-means algorithm is used to set the parameters of the Gaussian nodes. The k-means algorithm find L clusters with minimum internal variation, but with maximum variation among clusters. The number of clusters (L) must be manually configured, and the method does not guarantee halting in a global minimum solution. When k-means finishes, a network is created considering each cluster as a hidden layer node, and a single output layer node.

The proposed method is also hybrid, and it is performed in two steps as well. However, we use the SMLP-P algorithm [16] to set the number of clusters L and the radial basis functions parameters. We adjust the weights of the output layer employing the Recursive Least Squares (RLS) algorithm. We call our method RBFN-SLMP.

3.1   Estimation of Hidden Layer Parameters

In the hidden layer, it is required to set three kinds of parameters: (1) number of Gaussian nodes L, (2) center of each Gaussian function $\mathbf{x}_j$ and (3) amplitude σ of the Gaussian functions $\varphi_j(\mathbf{x}_j, \sigma)$. It is proposed to use SLMP-P [16] to estimate the latter. Originally, this algorithm is used to train the Lattice Neural Networks with Dendrite Processing (LNNDP) [17]. With this approach, we can automatically calculate the L Gaussian nodes, and their parameters $\mathbf{x}_j$ and σ. Although, the amplitude of the Gaussian nodes is not uniform for all of them, we calculate a general parameter as Haykin suggests in [7]. The proposed algorithm of training guarantees to stop in a global minimum.

Given p classes, C^k, $k = 1, 2, 3..., p$, each of them with n features, the training algorithm is the following:

1. A hypercube HC^n is created. It contains all the cases of the training set. Looseness is added to each of the HC sides in order to add tolerance to noise.
2. HC^n is divided into 2^n equal sized, smaller hypercubes.
3. For each HC, it is verified if there exist at least two observations with different classes. If it is the case, it returns to step 2; otherwise, it goes to step 4.
4. For each HC, $\mathbf{x}_j$ is calculated. Also, σ is calculated.
5. A RBFN is created with L hidden nodes, where L is the number of HC obtained.

The functions for σ and center($\mathbf{x}_j$), based on [7] are:

$$\sigma = \frac{d_{max}}{\sqrt{2L}} \tag{3}$$

$$\text{center} (\mathbf{x}_j) = \frac{1}{|S_j|} \sum_i \mathbf{x}_{j_i} \tag{4}$$

where d_{max} is the maximum Euclidean distance among the center of the clusters and $|S_j|$ is the cardinality of cluster S_j that contains all observations $\mathbf{x}_{j_i}$.

3.2   Estimation of Output Layer Weights

The hidden layer's training is recursive, therefore, Haykin [7] suggests that the computation of the weights in the output layer should be recursive too. Thus, we use *Recursive Least Squares* (RLS) to adjust the weights. The training algorithm is the following:

1. Begin with $\hat{\mathbf{w}}(0) = \mathbf{0}$ and $\mathbf{P}(0) = \lambda^{-1}\mathbf{I}$, where $\mathbf{I}$ is the identity matrix. In this case, we use an initialization to zero for the weights, as in [7,18].

2. Given a sample $\{\boldsymbol{\phi}(i), d(i)\}_{i=1}^{N}$ calculate the following for $n = 1, 2, ..., N$:

$$\mathbf{P}(n) = \mathbf{P}(n-1) - \frac{\mathbf{P}(n-1)\boldsymbol{\phi}(n)\boldsymbol{\phi}^T(n)\mathbf{P}(n-1)}{1+\boldsymbol{\phi}^T(n)\mathbf{P}(n-1)\boldsymbol{\phi}(n)}$$

$$\mathbf{g}(n) = \mathbf{P}(n)\boldsymbol{\phi}(n)$$

$$\alpha(n) = d(n) - \hat{\mathbf{w}}^T(n-1)\boldsymbol{\phi}(n)$$

$$\hat{\mathbf{w}}(n) = \hat{\mathbf{w}}^T(n-1) + \mathbf{g}(n) * \alpha(n)$$

3. When $MSE < \theta$ halt.

where θ is manually defined by the user, λ is a positive constant as small as possible, and $d(i)$ is the class of the sample i. The vector $\boldsymbol{\phi}(i)$, results after the transformation in the hidden layer. It is defined as,

$$\boldsymbol{\phi}(i) = [\varphi(\mathbf{x}_i, \boldsymbol{\mu}_1), \varphi(\mathbf{x}_i, \boldsymbol{\mu}_2), ..., \varphi(\mathbf{x}_i, \boldsymbol{\mu}_L)]^T \qquad (5)$$

$\boldsymbol{\mu}_j$ is the hidden node center j. The Gaussian function is expressed as follows:

$$\varphi(\mathbf{x}_i, \boldsymbol{\mu}_j) = exp\left(-\frac{1}{2\sigma_j^2}||\mathbf{x}_i - \boldsymbol{\mu}_j||^2\right), \; j = 1, 2, ..., L \qquad (6)$$

4   Experiments and Results

The experiments were run in a server with processors Xeon E5-2650 (20M Cache, 2.00 GHz, 8.00 GT/s Intel QPI) with a shared memory of 100GB. The OS was CentOS 6.5. The code is programmed in Python 3.3.3 with embedded code in R 3.0.3. We tried with 25, 50, 75, 100 and 300 iterations for RLS algorithm. The λ was selected as small as possible.

4.1   Databases

We used the benchmarks published on the Center for Machine Learning and Intelligent Systems (UCI) [19]. The benchmarks were selected following the recommendation of Babu and Suresh in [20]. Table 1 shows for each dataset the amount of features it has, the number of classes and the number of samples used for training and testing. In total, we apply our method to seven different datasets comprising up to a few dozens of features and a few hundreds of samples.

4.2   Comparison

We compare the performance of two classifiers. One is our proposed RBFN-SLMP and the second is a Lattice Neural Network with Dendritic Processing (RNMPD). The metric used is the percentage of accuracy, defined as:

$$Acc = \frac{N_{correct}}{N_{Total}} \times 100$$

Table 1. Description of the benchmarks used.

Number of samples				
Databases	Features	Classes	Training	Test
LD	6	2	200	145
PIMA	8	2	400	368
BC	9	2	300	206
HEART	13	2	70	200
ION	34	2	100	251
IRIS	4	3	45	105
WINE	13	3	60	118

where $N_{correct}$ is the number of correct predictions and N_{Total} is the total number of data points.

Table 2, shows the results of the experiments. As it is possible to see, RBFN-SLMP outperforms RNMPD in all cases. In case that the number of features exceeded 13, we extracted new features using Principal Component Analysis (PCA), and we used the first 10 components. The test conditions are the same for both methods. Besides, we tested the algorithms with the databases standardized, and also using the raw data. We ran the experiments 20 times and calculated the average of the results.

Table 2. Comparison of the accuracy achieved by RBFN-SLMP and RNMPD.

	RBFN-SLMP	RNMPD
LD	68.73 %	57.59 %
PIMA	76.40 %	68.42 %
HEART	82.65 %	71.90 %
ION	91.62 %	85.88 %
BC	87.87 %	93.41 %
IRIS	95.33 %	89.81 %
WINE	94.58 %	86.44 %

Figure 2 shows how smaller values for sigma (the amplitude of the Gaussian) imply worse results for all databases. Also, it shows that changes in the number of iterations do not have a big impact in the performance.

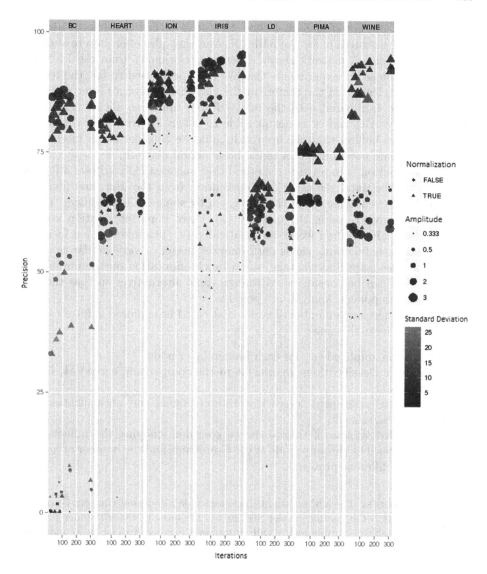

Fig. 2. The image shows the impact on the performance of the RBFNN, when modifying: iteration number, value of sigma and normalization.

5   Conclusion and Future Work

One of the main purposes of this paper was to test whether the algorithm that was originally developed to train a Lattice Neural Networks with Dendritic Processing (LNNDP) could be applied to the automatic selection of the parameters associated with the hidden layer of a Radial-Basis Function Neural Network (RBFNN). Also, it was our interest to compare the performance of the automatically tuned RBFNN with a LNNDP.

The results presented in Sect. 4 show that it is in fact possible to use the LNNDP training algorithm for the RBFNN. And they also offer data to support the claim that such a network shows a better performance than the LNNDP. According to the set of experiments that we have performed so far, the results show that the selection of the centers of the Gaussians allow a better accuracy in the classifier. Compared with the RNMDP, our approach always achieves a higher accuracy. Although it is not possible to generalize from only one set of experiments, we believe that the approach could be consider as an interesting option to alleviate the tuning of a radial-basis neural network.

Our contribution is to provide a method that automatically adjusts a RBFN depending on the data, requiring no intervention from the user, and giving a good accuracy, even better than the one given by some other classifiers. There are two main advantages in the method: (1) There is no intervention needed from the user. The method works automatically. (2) The amount and location of the radial-basis functions yields performances better or very similar to other approaches. The main disadvantage would be the limitation on the number of attributes that can be considered. Although some alternatives such as the use of Principal Component Analysis are possible.

Of course, there are a number of issues that could be explored in future work. We identify at least this ones:

- Calculate the amplitude σ_j for each hidden layer node.
- Include more classifications algorithms in the comparison, e.g. SVM, ELM, MLP, Bayes.
- Run a design experiment to see if iterations, sigma and normalization are statistically significant.
- In a extended version of this work, we consider studying the influence of parameters such as the amplitude of the Gaussians, and the number of hidden neurons for different problems.

Acknowledgments. The authors thank Tecnológico de Monterrey, Campus Guadalajara, as well as IPN-CIC under project SIP 20161126, and CONACYT under project 155014 and 65 within the framework of call: Frontiers of Science 2015 for the economical support to carry out this research.

References

1. Bishop, C.M., et al.: Pattern Recognition and Machine Learning. Information Science and Statistics, vol. 1. Springer, New York (2006)
2. Miller, A., Blott, B., et al.: Review of neural network applications in medical imaging and signal processing. Med. Biol. Eng. Comput. **30**(5), 449–464 (1992)
3. Willis, M., Montague, G., Di Massimo, C., Tham, M., Morris, A.: Artificial neural network based predictive control. In: Advanced Control of Chemical Processes (ADCHEM 1991): Selected Papers from the IFAC Symposium, p. 261, 14–16 October 1991. Toulouse, France (2014)
4. Le, T.H.: Applying artificial neural networks for face recognition. Adv. Artif. Neural Syst. **2011**, 16 (2011). doi:10.1155/2011/673016. Article ID 673016

5. Siniscalchi, S.M., Svendsen, T., Lee, C.H.: An artificial neural network approach to automatic speech processing. Neurocomputing **140**, 326–338 (2014)
6. Maren, A.J., Harston, C.T., Pap, R.M.: Handbook of Neural Computing Applications. Academic Press, New York (2014)
7. Haykin, S.: Neural Networks and Learning Machines. Number v. 10 in Neural networks and learning machines. Prentice Hall, Upper Saddle River (2009)
8. Park, J., Sandberg, I.W.: Universal approximation using radial-basis-function networks. Neural Comput. **3**(2), 246–257 (1991)
9. Chen, S., Cowan, C.F., Grant, P.M.: Orthogonal least squares learning algorithm for radial basis function networks. IEEE Trans. Neural Networks **2**(2), 302–309 (1991)
10. Moody, J., Darken, C.J.: Fast learning in networks of locally-tuned processing units. Neural Comput. **1**(2), 281–294 (1989)
11. Poggio, T., Girosi, F.: Networks for approximation and learning. Proc. IEEE **78**(9), 1481–1497 (1990)
12. Kurban, T., Beşdok, E.: A comparison of RBF neural network training algorithms for inertial sensor based terrain classification. Sensors **9**(8), 6312–6329 (2009)
13. Chun-Tao, M., Kun, W., Li-yong, Z.: A new training algorithm for RBF neural network based on PSO and simulation study. In: WRI World Congress on Computer Science and Information Engineering, vol. 4, pp. 641–645 (2009)
14. Oh, S.K., Kim, W.D., Pedrycz, W., Park, B.J.: Polynomial-based radial basis function neural networks (P-RBF NNs) realized with the aid of particle swarm optimization. Fuzzy Sets Syst. **163**(1), 54–77 (2011)
15. Reiner, P.D.: Algorithms for Optimal Construction and Training of Radial Basis Function Neural Networks. Ph.D. thesis, Auburn University (2015)
16. Sossa, H., Guevara, E.: Efficient training for dendrite morphological neural networks. Neurocomputing **131**, 132–142 (2014)
17. Ritter, G.X., Iancu, L., Urcid, G.: Morphological perceptrons with dendritic structure. In: Proceedings of the IEEE International Conference on Fuzzy Systems, vol. 2, pp. 1296–1301. IEEE (2003)
18. Fun, M.H., Hagan, M.T.: Recursive orthogonal least squares learning with automatic weight selection for Gaussian neural networks. In: Proceedings of the International Joint Conference on Neural Networks, vol. 3, pp. 1496–1500 (1999)
19. Bache, K., Lichman, M.: UCI machine learning repository (2013)
20. Babu, G.S., Suresh, S.: Sequential projection-based metacognitive learning in a radial basis function network for classification problems. IEEE Trans. Neural Netw. Learn. Syst. **24**(2), 194–206 (2013)

Feature Selection Using Genetic Algorithms for Hand Posture Recognition

Uriel H. Hernandez-Belmonte and Victor Ayala-Ramirez$^{(\boxtimes)}$

División de Ingenierías, Campus Irapuato- Salamanca DICIS,
Universidad de Guanajuato DICIS, Carr. Salamanca-Valle Km. 3.5+1.8,
Palo Blanco, 36700 Salamanca, Mexico
hailehb@laviria.org, ayalav@ugto.mx

Abstract. In this work, we propose a feature selection algorithm to perform hand posture recognition. The hand posture recognition is an important task to perform the human-computer interaction. The hand is a complex object to detect and recognize. That is because the hand morphology varies from human to human. The object recognition community has developed several approaches to recognize hand gestures, but still, there are not a perfect system to recognize hand gestures under diverse conditions and scenarios. We propose a method to perform the hand recognition based on feature selection. The feature selection is performed by a genetic algorithm that combines several features to build a descriptor. The evolved descriptor is used to train a perceptron, which is used as a weak classifier. Each weak learner is used in the AdaBoost algorithm to build a strong classifier. To test our approach, we use a standard image dataset and the full image evaluation methodology. The results were compared with a state of the art algorithm. Our approach demonstrated to be comparable with this algorithm and improve its performance in the some of the cases.

1 Introduction

Hand gesture recognition is an important task to perform interactions with computers and robots. This is because the hand is the body part most used in the interaction between humans. The hand has proven to be the most challenging body part to be recognized.

Several approaches have been developed to recognize the hand gestures. There are approaches which are based on wearable sensors, where the user needs to use gloves or markers to perform the hand recognition. Other non-intrusive methods use 3D sensors and computer vision to perform this task. In this work, we focus on a computer vision based method to recognize the hand in video sequences.

There are many ways to recognize the hand using computer vision. The most used approaches in the literature to classify hand gestures is based on hand segmentation. That is because the hand can be segmented by using color information or 3D information.

© Springer International Publishing Switzerland 2016
J.F. Martínez-Trinidad et al. (Eds.): MCPR 2016, LNCS 9703, pp. 208–218, 2016.
DOI: 10.1007/978-3-319-39393-3_21

Some of the problems that arise in the methods based on hand segmentation are their sensitivity to illumination changes, a need for initialization step, the problems of the sensors (like Kinect) to work well in outdoor locations and a need for controlling the scenario to exhibit good performance. Conversely, object detection methods have proven to be an excellent alternative to detect objects in several types of scenarios (both indoor and outdoor environments).

The feature selection step is a core element in the object detection frameworks. Viola and Jones object detection framework [14] performs an exhaustive search in each round to find iteratively good features to classify the objects. To select a feature, they test a set of M predefined features in several sizes and positions in the image. The ranking of a feature is based on its discriminative power.

This feature selection approach has proven to be a powerful strategy to build a good classifier [7]. Because of the good results of this type of feature selection strategy, several works have been focused in proposing new methodologies to improve the selection.

To avoid the exhaustive evaluation of features, a common strategy is to define a fixed number of features to be tested [3]. Each feature is created randomly. The number of features is defined in most cases experimentally. Hidaka and Kurita [9] proposed the use of the Particle Swarm Optimization to perform the search of the features and reduce the learning time.

In contrast, Dalal and Triggs [5] proposed the use of the Histogram of Oriented Gradients (HOG) and a Support Vector Machine (SVM) to perform the feature selection in a pedestrian detection task.

They designed a descriptor manually by dividing the pattern into cells. Each cell is a rectangle of equal dimensions. The cells can also have an overlapping with others cells. The HOG is computed for each cell. The descriptor is the result of the concatenation of histograms. The resulting descriptor has a high dimensionality. To handle this, the authors propose the use of an SVM to perform the classification. This approach to building descriptors has been used in several works. The combination of HOG features and SVM has been widely used to detect and classify objects. Malisiewicz et al. [11] proposed the use of a predefined pattern to extract the HOG, taking account of the object size. Their work is focused on learning and classifying several object views. The design of an HOG descriptor is a complex task and it has a strong relationship with the structure of the object that we want to recognize.

In this work, we propose a learning method to deal with the hand posture detection problem. The method is based on performing an efficient search for a set of features. We use a genetic algorithm (GA) to perform this search, instead of manually designing a pattern. We use the standard AdaBoost algorithm as a learning method. We use a perceptron as a weak classifier. To obtain a better performance and reduce the number of false positives, we propose the use of hard negative sampling. We use the Full Image Evaluation framework for the performance evaluation of our approach. We also compare our results with the obtained by the real-time deformable detector [1].

The rest of the paper is organized as follows. In Sect. 2, we describe the proposed methodology to perform the hand detection. The performance of our system is presented in Sect. 3. Finally, some concluding remarks are given in Sect. 4.

2   Methodology

In this section, we describe the proposed methodology to recognize hand postures. We introduce the key elements of the proposed algorithm, and we explain its importance. Our proposed methodology is based on two central concepts: the feature selection step and the design of discriminative descriptors using HOG and variance features.

The feature selection step is the process to find a subset of d useful features from a finite set of features D [12]. The cardinality of the D set is too huge to use all the features at the same time. For this purpose, the GA offers an alternative to search for a subset of features in an efficient way [15].

The manual design of good descriptors involves the analysis of the pattern to classify to select the number of cells, the size, and their positions. The number of cells, the dimension of a cell and its position can be encoded as a candidate solution in the GA. The use of a GA instead of the manual design of a descriptor allows the reduction in the size of the descriptor used to classify the pattern.

We use a perceptron to classify the descriptor. The election of this classifier is based on the learning algorithm. We use AdaBoost as a learning algorithm.

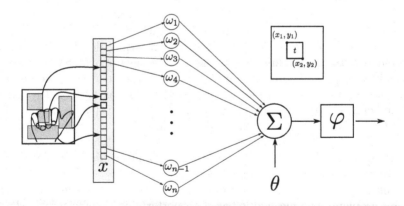

Fig. 1. Description of the proposed weak classifier. The descriptor x is composed of several features. A feature is defined by its type and the support area. The number of features and the types are determined by a genetic algorithm. The descriptor x is the input of the perceptron.

The proposed system is shown in Fig. 1. The result of the feature selection is a descriptor x. Each type of feature is composed of a different number of elements, eight values for the HOG and one value for the variance. The descriptor x is the input of a perceptron. We combine the classification results of several weak classifiers to perform the object detection using AdaBoost algorithm.

2.1   Feature Selection

According to Lillywhite *et al.* [10], feature selection is the process of choosing a subset of features from the original feature space. This selection is based on an optimality criterion. The feature selection step is widely used in learning algorithms. In our proposal, we use two features based on the edge of a region and based on the variance. To select the features, we use a genetic algorithm. The genetic algorithm is useful to perform an efficient search to select the best set of features.

There are several proposed features in the literature: e.g. the Haar-like features proposed by Viola-Jones, the Histogram of oriented gradients (HOG), interest points, etc. All these features have proven their effectiveness as features in methods to detect and classify objects.

The real-time detection restriction imposed to perform a natural and fluent interaction with the robot requires, as a consequence, the need for fast computation of the selected features. To overcome this problem, Viola and Jones proposed the use of features based on integral images. An integral image is a representation that allows us to compute the sum of all elements in a rectangular area of the image in a fast and efficient way. We use two type of features, based on the integral image: the variance [4] and the HOG computed in eight orientations [1].

2.2   Genetic Algorithm

The genetic algorithm (GA) is a useful method to solve optimization problems. The GA performs a heuristic search, inspired by the biological evolution of species. The heuristic search uses a population, where each individual is a possible solution for the problem. A fitness function is used to evaluate the goodness of the individuals and there are three operators to simulate the evolutionary process. The GA operators are the selection, crossover, and mutation. Each of them is similar to the natural processes that appear during the evolution of the species.

We use the GA to find the best combination of features to build a descriptor. A feature is defined by a rectangle inside of the support area. The support area is a square that covers all the object to learn (the hand for our case).

In the GA, each candidate solution (individual) is represented by a string of bits, called a chromosome. The chromosome is divided into sets of bits. Each set is used to represent a variable of the solution. Our chromosome c is composed by quintuples of variables $c = [\mathbf{x}_1, \mathbf{x}_2, \ldots, \mathbf{x}_N]$ where $\mathbf{x} = [t, x_1, y_1, x_2, y_2]$. Each tuple is used to describe a feature, t is used for the type of feature (edge map or variance) and $\{x_1, y_1, x_2, y_2\}$ for the rectangular area of the feature. The rectangular area is represented by two points, upper-left(x_1, y_1) and bottom-right(x_2, y_2).

The fitness function used in our approach is the weighted training error. The weighted training error is used for the AdaBoost algorithm to select the best classifier during the training procedure.

The GA crossover operation is used to obtain a new individual from the combination of two individuals. The point where the individuals are divided and recombined is randomly selected. For our approach, the new individual is the result of the combination of several 5-tuples $\mathbf{x}$. Using this crossover approach, the feature information is preserved across the generations. In the Fig. 2 the crossover operation is represented graphically.

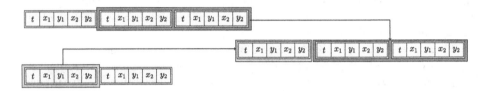

Fig. 2. The crossover operation combines two individuals in a new one. The points where the crossover is performing, are selected randomly. The crossover only combines tuples of values, to keep the feature information during the evolutionary process.

The GA mutation operation is used to modify the individual in a random way. These modifications introduce a new diversity in the individuals. This diversity is useful to explore the solution space. In our proposal, we only mutate the elements $\{x_1, y_1, x_2, y_2\}$. This mutation allows searching for the features in all the pattern.

The GA selection operation is used to retain the best individuals across the generations. There are several methods to perform the selection. We use the roulette method in our approach.

2.3 Learning Method

We use a standard AdaBoost learning algorithm to train our classifier. The AdaBoost is a method that combines the decision of several weak classifiers in a strong classifier. Each weak classifier is weighted according to his discriminative capability. The final decision combines all the results of the weighted classifiers. This weighted sum allows each weak classifiers to focus in different parts of the model to be learned.

The weak classifier used in our approach is a combination of features and the perceptron as a classifier. The perceptron is an Artificial Neural Network with only one layer. In each round, a GA algorithm performs a search for the best combination of features that minimizes the weighted error. The learning procedure is presented in the Algorithm 1. The value of the variable J is the number of individuals in the population.

Algorithm 1. Learning process

---

1 Given training data (x_i, y_i) $x_i \in \mathcal{X}$ and $y_i \in \{-1, 1\}$ Initialize weights
$w_{1,i} = \frac{1}{2a}, \frac{1}{2b}$ where a and b are the total number of positive(hand posture images) and negative examples respectively (images that not contain hand postures). T is the number of weak classifiers. J is the number of individuals in the population

2 **for** $t = 1$ **to** T **do**

3   **for** $j = 1$ **to** J **do**

4   Perform the feature selection process and train the weak classifier h_j. Evaluate weighted classification error:

$$\epsilon_j = \sum_i w_i |h_j(x_i) - y_i|$$

5   **end**

6   Choose the classifier h_j, with the lowest error ϵ_j.

7   Calculate:

$$\alpha_t = \frac{1}{2} \ln \left(\frac{1 - \epsilon_j}{\epsilon_j} \right)$$

8   Update data weights:

$$w_{t+1,i} = w_{t,i} \beta_t^{1-e_i}$$

9   where $e_i = 1$ if the image x_i is classified correctly and $e_i = 0$ otherwise, $\beta_t = \exp(\alpha_t)$.

10   Normalize the weights

$$w_{t+1,i} \leftarrow \frac{w_{t,i}}{\sum_n w_{n,i}}$$

11 **end**

12 The final detector is given by:

$$f(x) = \mathrm{sign} \left(\sum_{k=0}^{N} \omega_k h_k(x) \right)$$

2.4   Hard Negative Mining

The quality of the samples used in the learning setup is crucial to obtain a good classifier. An intuitive approach to get a good classifier is to increase the number of samples used during the training step. The increment of samples used during the training step implies an increase in the computer resources needed and also in the time spent to compute train the classifier.

To avoid these problems, it is preferable to use a small training set composed of useful samples. The process to obtain these samples from a larger dataset is called bootstrapping[1].

---

[1] In the statistical field, the bootstrapping process refers to a re-sampling method.

The bootstrapping methods are an essential component in different object recognition frameworks. To perform the bootstrapping, we need a sampling methodology and a classifier to evaluate the quality of the samples. The training set is constructed actively, during the training process or before the training process, using the hard negative mining (HNM) approach [2]. The selection of the bootstrapping methodology is based on the experimentation.

In our work we use HNM to improve the quality of the final classifier. The HNM allows constructing a small training set with relevant samples from a pool of images. This training set is built in two steps. First, a classifier f_1 is trained with a small set of images from the pool. The images are randomly sampled from the pool. The resulting classifier is then used to obtain higher quality negative samples.

The images in the negative training set, are sequentially evaluated using the classifier f_1. During this procedure, if a sample is misclassified then it is added to the training set. Then, using the new training set a new classifier f_2 is trained. The classifier f_2 is the final classifier.

3   Tests and Results

The results obtained from our proposal are presented in this section. This section is divided into two parts. The first part are the results obtained from the training process. We describe quantitatively the classifier resulting from the training process. In the second part the detection system is evaluated. The protocol used to evaluate our approach is the Full Image Evaluation, used by Dollár et al. [6]. This protocol is useful to obtain a fair comparison among the several approaches. We compare our approach results with those obtained by Hernandez and Ayala [8].

All test were performed in the GNU/Linux operating system using a general-purpose computer using 8 GB RAM and a processor running at 2.7 GHz. No parallel strategy of specific optimization was used in the implementation.

3.1   Training

The training procedure was performed using the AdaBoost learning algorithm and the Hard Negative Mining process. The number of weak learners used for the classifier f_1 was five, and the maximum number of weak classifiers for the second classifier f_2 was twenty. We used the National University of Singapore (NUS) hand posture dataset-II propose by Pisharady et al. [13] in the training and testing process. The NUS-II is divided in images where the hand posture appear alone (NUS dataset A), and where the hand posture appear with people in the scene with human noise(NUS dataset B). For this test we only use one posture.

The NUS dataset A was used for training purposes. The training samples were obtained by cropping the samples from the dataset and resizing them to be rectangles of 50×50 pixel size. We use this size in all our tests.

We use 1000 positive samples and 20000 negative samples in the training step. The parameters used for the GA were: crossover probability 0.8, mutation probability 0.01, two elite members, one hundred individuals in the population and fifty generations. The minimum number of features per individual was one, and the maximum number per individual was 10. These parameters have shown to produce the best results in the training step. Using these parameters, we train 45 classifiers. In each round of the AdaBoost training, the time consumed to find the best descriptor was 11.4175 min., with $\sigma = 0.0677$ min. The elapsed time to obtain f_1 y f_2 was around five hours (25 weak classifiers and HNM procedure). The performance obtained from the best classifier was 0.8900 for true positive classification, 0.1100 for false negative detection and 0.9452 for true negative classification. The mean time to process an image of 340×240 pixels was 96 ms (around 10 frames per second).

Figure 3 depicts the qualitative results for the HNM process using a classifier response heat map. This map is built using the weights of the weak classifiers and the rectangular area of the features. We use the best classifier obtained from the 45 trained classifiers. The object to learn is shown in the Fig. 3a. The first heat map obtained from the classifier f_1 is shown Fig. 3b. In this heat map, the weights of the weak classifiers are concentrated in the center of the pattern. In contrast, the heat map obtained from the classifier f_2 the weights are distributed along the pattern.

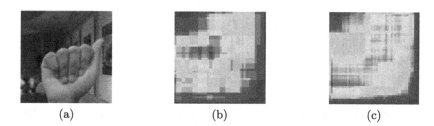

(a)   (b)   (c)

Fig. 3. The number and the type of features vary for each classifier. The heat map is a representation used to determine the areas where the classifier is focused. (a) hand posture (b) is the classifier f_1 before the HNM and (c) is the classifier obtained after applying HNM.

3.2 Full Image Evaluation

The full image evaluation is a methodology to measure the performance of the whole detection system. Using this methodology, the results of several methods can be compared fairly. To use this methodology, we need an annotated dataset. This annotated dataset was used as ground truth to perform the test. We use the ground truth used in [8]. The False Positive Per Image metric is computed $FPPI = FP/NI$, where NI is the number of image in the ground truth. The Miss Rate metric is computed $MR = 1 - TP/NO$, where NO is the number of

object in the ground truth. The results obtained from the full image evaluation
are presented in the Fig. 4. We use the sliding window approach using a scanning
step of 5 pixels in x and y axis and a scale factor of 1.4. Using the NUS Dataset II
A (4(a)), our proposed approach exhibits a better performance. This is because,
our approach has a better detection in all the range and the number of false
positive is less than the RTDD approach. Using the NUS Dataset II A (4(b))
our results are similar to the RTDD approach. The number of false positive
images detected by our approach is moderately greater that the RTDD approach.
Nevertheless, the miss rate for our approach is greater that the RTDD. The full
image evaluation results are promising. We conclude this because, in this test,
we improve the performance of the results obtained by a state of the art method
[8] for one hand posture.

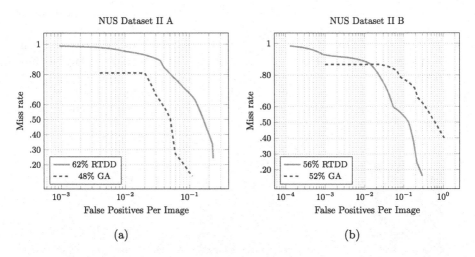

Fig. 4. Results obtained using the full image evaluation. (a) results using the NUS II
dataset A, in this test our approach is better that the RTDD approach. (b) results
using the NUS II dataset B, the performance of our approach is similar to RTDD. The
percentage displayed in the plots is the area under the curve. The lower this value, the
better the performance.

4   Conclusions and Perspectives

In this work, we proposed the use of a feature selection method for hand posture
detection. The feature selection process was performed by using a genetic algorithm and two types of features. The proposed features were implemented using
integral image computations, that are useful for fast computation. The genetic
algorithm combines a different number of features and it varies their configuration. This combination and its variations allow to the genetic algorithm to
find a good solution in a limited time. The results of the genetic algorithm have
proven that this strategy is useful to build image descriptors. We used AdaBoost

algorithm to combine the response of several weak classifiers. A weak classifier is composed by the descriptor and a perceptron. The use of this type of weak classifier proved to be useful in the detection of hand postures. Our proposal was compared with a state of the art algorithm. The results of this comparison were favorables to our approach. From the results, we can say that the generalization ability of the classifier is good enough to detect hand postures. Future work will be to implement a CPU or GPU parallelization approach of the learning algorithm. This parallelization will reduce the time needed to train the classifiers. With a reduced training time, it is possible to perform experimentation about the influence of the training parameters in the resulting classifier.

Acknowledgments. The authors gratefully acknowledge to the Mexico's CONACYT (229784/329356) for the financial support through the scholarship is given by the programs "Convocatoria de Becas Nacionales 2013 Primer Periodo".

References

1. Ali, K., Fleuret, F., Hasler, D., Fua, P.: A real-time deformable detector. IEEE Trans. Pattern Anal. Mach. Intell. **34**(2), 225–239 (2012)
2. Canévet, O., Fleuret, F.: Efficient sample mining for object detection. In: Proceedings of the Asian Conference on Machine Learning (ACML), pp. 48–63 (2014)
3. Chen, Q., Georganas, N.D., Petriu, E.: Hand gesture recognition using Haar-like features and a stochastic context-free grammar. IEEE Trans. Instrum. Meas. **57**(8), 1562–1571 (2008)
4. Correa-Tome, F.E., Sanchez-Yanez, R.E.: Integral split-and-merge methodology for real-time image segmentation. J. Electron. Imaging **24**(1), 013007 (2015). http://dx.org/10.1117/1.JEI.24.1.013007
5. Dalal, N., Triggs, B.: Histograms of oriented gradients for human detection. In: 2005 IEEE Computer Society Conference on Computer Vision and Pattern Recognition CVPR 2005, vol. 1, pp. 886–893 (2005)
6. Dollár, P., Wojek, C., Schiele, B., Perona, P.: Pedestrian detection: an evaluation of the state of the art. IEEE Trans. Pattern Anal. Mach. Intell. **34**, 743–761 (2012)
7. Fürst, L., Fidler, S., Leonardis, A.: Selecting features for object detection using an adaboost-compatible evaluation function. Pattern Recogn. Lett. **29**(11), 1603–1612 (2008)
8. Hernandez-Belmonte, U.H., Ayala-Ramirez, V.: Real-time hand posture recognition for human-robot interaction tasks. Sensors **16**(1), 36 (2016). http://www.mdpi.com/1424-8220/16/1/36
9. Hidaka, A., Kurita, T.: Fast training algorithm by Particle Swarm Optimization and random candidate selection for rectangular feature based boosted detector. In: 2008 IEEE International Joint Conference on Neural Networks (IEEE World Congress on Computational Intelligence), IJCNN 2008, pp. 1163–1169, June 2008
10. Lillywhite, K., Lee, D.J., Tippetts, B., Archibald, J.: A feature construction method for general object recognition. Pattern Recogn. **46**(12), 3300–3314 (2013)
11. Malisiewicz, T., Gupta, A., Efros, A.: Ensemble of exemplar-SVMs for object detection and beyond. In: 2011 IEEE International Conference on Computer Vision (ICCV), pp. 89–96, November 2011

12. Oh, I.S., Lee, J.S., Moon, B.R.: Hybrid genetic algorithms for feature selection. IEEE Trans. Pattern Anal. Mach. Intell. **26**(11), 1424–1437 (2004)
13. Pisharady, P., Vadakkepat, P., Loh, A.: Attention based detection and recognition of hand postures against complex backgrounds. Int. J. Comput. Vision **101**(3), 403–419 (2013)
14. Viola, P., Jones, M.: Rapid object detection using a boosted cascade of simple features. In: 2001 Proceedings of the IEEE Computer Society Conference on Computer Vision and Pattern Recognition, CVPR 2001, vol. 1, pp. 511–518 (2001)
15. Qian, C., Yu, Y., Zhou, Z.H.: Subset selection by Pareto optimization. In: Cortes, C., Lawrence, N.D., Lee, D.D., Sugiyama, M., Garnett, R. (eds.) Advances in Neural Information Processing Systems, vol. 28, pp. 1774–1782. Curran Associates, Inc. (2015)

Activity Recognition in Meetings with One and Two Kinect Sensors

Ramon F. Brena[⊠] and Armando Nava

Tecnologico de Monterrey, Monterrey, Mexico
ramon.brena@itesm.mx, armandnavao@gmail.com

Abstract. Knowing the activities that users perform is an essential part of their context, which become more and more important in modern context-aware applications, but determining these activities could be a daunting task. Many sensors have been used as information source for guessing human activity, such as accelerometers, video cameras, etc., but recently the availability of a sophisticated sensor designed specifically for tracking humans, as is the Microsoft Kinect has opened new opportunities. The aim of this paper is to determine some human activities, such as eating, reading, drinking, etc., while a group of persons are seated, using the Kinect skeleton structure as an input. Further, due to occlusion problems, it could be guessed that a combination of two Kinect sensors could give an advantage in activity recognition tasks, especially in meeting settings. In this paper, we are going to compare the performance of a two Kinect system against a single Kinect in order to determine if there is a significant advantage in using two sensors. Also, we compare several classifiers for the activity recognition task, namely Naive Bayes, Support Vector Machines and K-Nearest Neighbor.

Keywords: Kinect · Naive Bayes · Support Vector Machines · K-Nearest Neighbor · KNN · SVM · NB

1 Introduction

In recent years, there has been an increased interest in recognizing human activities [1], as the user activity is an essential element of her/his context: if the system can recognize and understand the activities performed by a human, they can provide assistance to perform those activities [2]. This is indeed the focus of Ambient Intelligence research (AmI). For example, AmI systems can be used to monitor elder people, perceive their needs and preferences, provide then different services to comfort or helping them applying emergency treatment [3] etc.

Many sensors have been used for activity recognition, including accelerometers [4], microphones video-cameras [5] and other. In recent years, the commercial implementation of the Natal project by Microsoft, that is, the Kinect [6], though it was originally intended as a video-game accessory, started to be used as an experimental human-tracking sensor. The Kinect has a camera with other sensors besides the RGB camera (like the infrared camera and the infrared projector).

© Springer International Publishing Switzerland 2016
J.F. Martínez-Trinidad et al. (Eds.): MCPR 2016, LNCS 9703, pp. 219–228, 2016.
DOI: 10.1007/978-3-319-39393-3_22

The Kinect became popular with researchers because it is useful for AmI research, and because it is cheap.

The capability of tracking the human skeleton in real time with the Kinect sensor makes it possible to guess physical human activities, instead of using computer vision algorithms to the raw video image. So, the Kinect sensor takes part of the processing burden, and delivers a high-level data structure representing the human skeleton, which can be further analyzed by activity-recognition algorithms, allowing these to become the main focus of the research.

We focus on tracking the skeletons of seated persons with one and two Kinect sensors (Fig. 1), and recognize what activities are being performed by them, because this corresponds to meeting situations in enterprises, government, schools, etc. The analysis of activities of participants in meetings could give valuable information concerning the level of focus of participants in a presentation, the level of participation in a discussion, etc.

Tracking seated persons, and recognizing what activity is being performed by each person, is a challenging task, because some several different activities look almost the same. Another challenge is the occlusion that sometimes occurs when a part of the virtual skeleton can not be tracked because something is blocking the Kinect vision. So we proposed to use two Kinect sensors instead of one, as a way to reduce the occlusion, because what is hidden to one Kinect could be visible for the other one.

In this paper we propose specialized algorithms for analyzing the skeleton structure of seated persons, so we can differentiate physical activities such as drinking, eating, paying attention, using laptop, checking watch, checking cellphone, writing, using tablet, attending a call, and participating. Also, we are going to compare the performance of two against one Kinect sensor, to see how effective they are recognizing activities, and if there is actually a significant advantage in using two Kinect sensors, and also we are going to experimentally compare several classification algorithms for this activity recognition task.

The rest of the paper is organized as follows: Sect. 2 outlines the related works, Sect. 3 describes the method process used; experiments and results are in Sect. 4, and finally conclusions are presented in Sect. 5.

Fig. 1. Seated skeleton

2   Related Work

Human activity analysis is made with data coming from a single sensor or a group of sensors. These sensors can be embedded in the environment or can be attached to the person body, so a basic distinction can be made between infrastructure-based approaches, which use fixed devices such as vision cameras, sensitive floor, infrared sensors, etc., and wearable sensors, in which the person about who the activities are going to be recognized is actually wearing the sensor, like a smart watch with accelerometers [4]. Of course wearable approaches favor user privacy, but they also impose requirements on the equipment the user must carry. In this paper we are assuming infrastructure sensors, in particular a set of Kinect sensors. The use of two Kinect sensors was also proposed by Mazurek [7] as well as Sthone et al. [8], who combines data from multiple Kinect sensors to create pose estimates of a human.

2.1 Classifiers for Activity Recognition

Several classification algorithms have been used for activity recognition system, in particular Naive Bayes, Support Vector Machines (SVM) and k-Nearest Neighbors algorithm (k-NN); we review the use of them in the following. The comparison of classification algorithms is one of the goals of the present paper.

From the outset there is no particular reason for using one classifier instead of another when it comes to activity recognition from sensors, but in our experiments (Sect. 4) we found that there could be significant differences in their performance.

Naive Bayes [9] is a simple probabilistic classifier, which uses the Bayes theorem with a naive independent assumptions between the features. It requires a small amount of data to be trained.

Works using the Naive Bayes classifier with infrastructure sensors include Mazurek et al. [10], and others, while Ravi et al. [11] uses Naive Bayes to recognize activities from the data collected from a wearable sensor. A Bayesian classifier is also used by Song et al. [12], who uses two image-related sensors with the depth information from two cameras to track upper body movements.

Support Vector Machines (SVM) can classify linear and nonlinear data. Linear mapping searches for a lineal optimal line to separate "hyperplanes" [9].

Support Vector Machine classifiers are used by Cottone et al. [13], Megavannan et al. [14] and others.

Multiple cameras and a SVM classifier are used by Cohen et al. [15], who uses four cameras to capture 3D human body shapes and infers body postures using SVM to identify the postures and the activity performed by the user.

The specific use of Kinect sensors and SVM is reported by Zhang et al. [16].

K-Nearest Neighbor (k-NN) finds for a given data point the nearest k points in a dataset and assigns it to the most frequent class among those k points, so it is a kind of voting algorithm [9].

The k-NN classifier has been used, together with image sensors, by Gordon et al. [17], Ofli et al. [18] and others.

Wearable sensors with k-NN is used by Abdullah et al. [19], who uses a smartphone attached to a person to recollect data.

The reviewed works fall short in the two aspects that we emphasize in the present work, namely evaluating the advantage of using several Kinect sensors, and comparing several very different classifiers, for assessing their relative strengths.

3   Solution Method

The approach we follow for activity recognition is entirely data-driven: after collecting data from the Kinect sensors during a meeting, we extract features and then train classifiers. Validation experiments use the trained classifiers against test data, and precision assessment is made using ground-truth made by humans who identify the activity of each participant in every frame of a meeting session (Fig. 3).

So the steps of our approach are the following:

1. Collect data from meeting sessions
2. Apply geometric transformation and consolidate skeletons from two Kinect sensors
3. Establish ground truth from video with the activities of each user
4. Extract features
5. Apply data to classifier (Bayes, SVM, k-NN) and obtain predicted activities for each user
6. Check the result against the ground truth and calculate precision

In the following we will present the details of these steps.

3.1   Data Collection

Activities of participants in a meeting are going to be captured by a 60-degree separated arrangement of Kinect sensors connected to a computer and specialized software (see details in Sect. 4).

The activities that we are going to consider are: drinking, paying attention, using laptop (we are going to refer to it as laptop only), checking watch, checking

Fig. 2. Two kinects

Fig. 3. Configuration of the meeting room and kinects

cellphone, writing, using tablet, attending a call, pointing, raising hand, and participating. Each joint coordinate is captured on a frame, a group of frames have a group of joint's coordinates. The skeleton consists on 8 joints, corresponding to the head, shoulders, elbows and hands. The Kinect sensor captures 30 frames each second, so the raw data comprise for each 30th of a second the 3D position of each of the 8 considered joints, that is, 24 numbers for each Kinect, the double for a set of two Kinect sensors.

When we use Two Kinect sensors, each one of them has a different reference for determining the position of body joints, so we need to apply a coordinate transformation to one of them in order to make its measures compatible with the other Kinect. We are going to use one Kinect sensor as a reference and transform the measurements made by the second Kinect using standard methods [20]

Once we have the measurements of two Kinect sensors aligned to the same coordinate system, we can observe that there are small but significant differences in the corresponding position estimations, so we need to apply a *consolidation* operation, in which we take the middle point between the positions reported by each of the two Kinect sensors. This is the position we consider for the two-Kinect experiments.

3.2 Feature Extraction

Once the capture and consolidation process is finished, we have a data matrix and a corresponding synchronized video for establishing the ground truth. The next step in our method is to complement the raw data with derived features that could help the classification process. While most authors use only static (pose, posture) features for activity recognition, we decided to use also dynamic features, such as the body joints speed and acceleration. Also, we included statistical features like Andersson [21] from a 3 s window (that is, 90 frames). Of course, whether or not a feature is useful can only be established once the classification task is performed and precision is established, though there are feature selection techniques such as Principal Component Analysis (we used PCA against our feature set, giving that position is the most important one).

So, the features we considered are:

- Static: Joint Position (x,y,z)
- Dynamic: Absolute Linear Speed
- Dynamic: Absolute Acceleration
- Statistical: Mean position over a 3 s window
- Statistical: Median of position
- Statistical: Standard deviation of position

3.3 Classifiers

In this study we used and compared three different classification algorithms, namely Naive Bayes, Support Vector Machines and k-nearest neighbor.

4   Experiments and Results

The purpose of the validation experiments we are going to present is to compare the performance of activity recognition using one Kinect and two ones, so we test the hypothesis that two Kinect sensors will be able to overcome to a certain degree the occlusion problems. We also want to test the precision of each classifier at recognizing the activities of users. Besides precision, also the specificity and sensitivity were calculated. We use a time window of 3 s to classify the activity performed, 1 s equals 30 frames, so we analyze 90 frames. We use this time window to extract features, the total of extracted features are 11.

For collecting data we need to setup and configure a computer with two Kinect sensors. We used open source tools, in particular OpenNI/NiTE, SimpleOpenNI, and Processing, installed on a Linux machine. OpenNI/NiTE is an open source framework that provides a set of application programming interface (API) that provide support for body motion tracking, hand gestures, and voice command recognition. SimpleOpenNi is a wrapper for processing. It supports all the functions of OpenNI, it provides a simple access to the functionality of OpenNI. Processing is a programming language.

For the experiments with two Kinect sensors these ones were arranged with a 60 degree angle as in Fig. 2, so that users' body parts that were occluded to one Kinect could be visible for the other one. This arrangement has been used in other works, but optimization of the exact arrangement remains to be done (see Sect. 5), and indeed depends on the exact room shape and furniture arrangement.

The experiments consist of three types of meetings. One is an exposition of a class, another type is a discussion meeting, and the third type is a work meeting. The meetings are acted, but the subjects do not follow a script. In each meeting there are two participants seated in front of a table, eventually with a laptop or a block for notes or beverages such as coffee or a snack for eating. Of course, the laptops are going to cause visual occlusion, which is one of the challenges in the activities recognition task.

On average, meetings are 51 min long, and we collected 4 meeting rounds for each type, thus 12 experiments, both for one and for two Kinect sensors, giving a total of 24 individual meetings, each one together with a corresponding video for establishing the ground truth (see below).

Each individual meeting data collection, taken by the Kinect at 30 frames per second, generates a raw data file of around one hundred thousand lines. Overall, the total number of data text lines collected for experiments in the 24 meetings was 2,200,086 text lines.

Then we enrich the tables with the derived features that were presented in Sect. 3.2, including the Mean, Median, Standard Deviation, Variance of the skeleton joint position in a time window of 3 s, as well as the velocity and acceleration of the skeleton's joints.

For all experiments, we are going to use the 80 % of the data collected for training the classifiers, the rest, 20 % are going to be used for testing them.

For the testing data, we use the ground truth as recognized by a human experimenter. Activities are tagged in the frames matrix directly from the video associated to the meeting, which has been synchronized with the Kinect data capture. Needless to say this was an extremely meticulous task. So, the system succeeds in predicting an activity when that prediction matches the tag given by the human experimenter.

The first experiment was an exposition of a class. Users attended the class. Relevant activities performed by the users were *Point, Paying Attention, Drinking Water* and *Raise Hand*.

In the 4 rounds of this experiment we collected 58,235, 110,204, 114,013 and 58,237 lines of skeleton position data from the Kinect sensor (one Kinect) and 114,071, 120,803, 86,990 and 72,113 lines for two Kinect sensors.

In Table 1 we present the precision comparison for experiment 1 using one and two Kinect sensors. Later on we will present the global analysis of experiments.

Table 1. Precision comparison between classifiers, with one Kinect and two Kinect sensors for experiment 1

| | One Kinect | | | Two Kinects | | |
| | Classifiers | | | Classifiers | | |
	NB	SVM	KNN	NB	SVM	KNN
Raise hand	1.70 %	74.26 %	81.21 %	23.37 %	77.19 %	82.96 %
Point	8.19 %	82.34 %	92.01 %	3.85 %	14.47 %	75.00 %
Paying attention	95.44 %	90.78 %	98.17 %	98.56	81.95 %	98.56 %
Drinking water	6.06 %	6.37 %	90.76 %	36.17 %	100 %	100 %
Average	27.84 %	63.43 %	90.53 %	40.48 %	68.40 %	89.13 %

In experiment 2, a discussion meeting is being held. Relevant activities are: *Participate, Paying attention, Drinking Water, Using Tablet, Eat, Raise Hand*.

This experiment also has 4 rounds, both for one and for two Kinect sensors, giving 8 experiment sessions. In this experiment we collected 34,781, 74,996, 102,328 and 114,070 lines of raw data for one Kinect, and 67,123, 110,183, 88,872 and 59,250 lines for two Kinect.

In Table 2 we present the comparison between classifiers and one and two Kinect sensors.

Experiment 3 is about a simulated work meeting, and the experimental settings are very similar to the preceding ones. The relevant activities are *Paying attention, Drinking Water, Check cellphone, Using laptop* and *Participate*; we think they are self-explanatory.

As the preceding ones, this experiment has 4 rounds for one Kinect and 4 for two Kinect. The data collected was composed of 51,251, 94,574, 115,023 and 92,187 lines of data for one Kinect and 98,368, 116,413, 93,345 and 93,406 line for two Kinect sensors.

Table 2. Precision comparison between classifiers, with one Kinect and two Kinect sensors for experiment 2

	One Kinect Classifiers			Two Kinects Classifiers		
	NB	SVM	KNN	NB	SVM	KNN
Participate	17.06	98.56	96.48	43.00	89.01	100.00
Paying attention	61.90	79.17	80.30	89.22	87.30	98.11
Drinking water	9.02	85.57	79.21	13.12	93.11	93.12
Using tablet	95.25	82.23	97.40	84.22	92.11	90.94
Eat	24.55	88.89	96.00	34.22	89.22	94.41
Raise hand	6.23	87.06	88.24	6.15	93.76	95.81
Average	35.67	86.91	89.61	44.99	90.75	95.40

Results of this experiment are presented in Table 3.

Analyzing the data in the tables, we can see that, concerning the classifiers, Naive Bayes fall behind the other two, which alternate being the best depending on the activity (for instance, in experiment 1, raising hand, SVM is better, but in writing KNN is better). In the average of one kinect, we can see that k-NN performs slightly better than two kinects, because maybe an occlusion happened, and it lowered the classifier performance. Overall, the NB precision was under 40 %, which is clearly very low, and on average SVM was below 80 %, while k-NN precision was well above 90 %, establishing a clear advantage above SMV and obviously NB. While the explanation for the low NB performance could be the data independence assumption, which we think is not respected in this task, the advantage of k-NN over SVM could be explained in terms of the high non-linearity of our data; the SVM hyperplanes for data separation imply some degree of linearity that could not be respected in our data.

Table 3. Precision comparison between classifiers, with one Kinect and two Kinect sensors for experiment 3

	One Kinect Classifiers			Two Kinects Classifiers		
	NB	SVM	KNN	NB	SVM	KNN
Paying attention	20.96	94.65	80.00	45.98	81.79	100.00
Drinking water	16.13	70.53	85.73	30.76	77.94	96.88
Check cellphone	16.43	33.44	96.91	26.12	71.22	97.22
Using laptop	89.78	94.43	93.98	95.95	92.96	99.34
Participate	21.01	87.16	90.42	38.56	98.31	100.00
Average	32.86	76.04	89.41	47.47	84.44	98.69

We also see that having two Kinect sensors raises precision noticeably, namely above 7 percent –which was of course something to expect due to occlusion resolution in the case of the two Kinect sensors– but not dramatically. The reasons why having two Kinect sensors does not raise precision 10 points or more is difficult to grasp, as the skeleton detection itself is done inside the Kinect system and could not be analyzed by us.

5   Conclusions

In this paper we have presented a data-driven method for recognizing the activities of users in meeting settings using two Kinect sensors. Indeed, what meeting participants do in organizations get-together is important for assessing their utility and eventual improvements.

We used as input the position of body joints, as given by two Kinect sensors, together with derived features such as velocity and acceleration, as well as statistical features over time windows. We established a reasonably good precision in activity recognition for common activities in meetings such as listening, drinking, taking notes, using a laptop, raising hands for participation, etc.

We compared the performance of three different classifiers for the activity recognition task, namely the Naive Bayes, Support Vector Machine, and the k-Nearest Neighbour, finding that k-NN is the best classifier for this task, followed by SVM and far behind Naive Bayes.

Further, we compared in a rigorous way the performance of the activity recognition using one against two and one sensors, and established exactly how much is the improvements of the two Kinect arrangement. This had never been done to our knowledge.

As future work, we intend to optimize the Kinect arrangement, that is, the angles, the distances and so on, which of course depend of specific meeting rooms, and even tables. We also want to complement the skeleton detection of Kinect with the analysis of sound, as taken from the Kinect microphones, because sound is an additional clue for activities such as participation in a meeting; a two-Kinect arrangements could be particularly suited for detecting the sound source, which could give information about which participant in the meeting was talking.

References

1. Tapia, E.M., Intille, S.S., Larson, K.: Activity recognition in the home using simple and ubiquitous sensors. In: Ferscha, A., Mattern, F. (eds.) PERVASIVE 2004. LNCS, vol. 3001, pp. 158–175. Springer, Heidelberg (2004)
2. Aarts, E., Wichert, R.: Ambient intelligence. In: Bullinger, H.J. (ed.) Technology Guide. Principles - Application - Trends, pp. 244–249. Springer, Heidelberg (2009)
3. Demiris, G., Hensel, B.K., Skubic, M., Rantz, M.: Senior residents' perceived need of and preferences for "smart home" sensor technologies. Int. J. Technol. Assess. Health Care **24**(01), 120–124 (2008)
4. Garcia-Ceja, E., Brena, R.: Long-term activity recognition from accelerometer data. Procedia Technol. **7**, 248–256 (2013)

5. Niu, W., Long, J., Han, D., Wang, Y.-F.: Human activity detection and recognition for video surveillance. In: 2004 IEEE International Conference on Multimedia and Expo, ICME 2004, vol. 1, pp. 719–722. IEEE (2004)
6. Zhang, Z.: Microsoft kinect sensor and its effect. MultiMedia IEEE **19**(2), 4–10 (2012)
7. Mazurek, P., Morawski, R.Z.: Application of naïve bayes classifier in a fall detection system based on infrared depth sensors. In: Proceedings of 8th IEEE International Conference on Intelligent Data Acquisition and Advanced Computing Systems: Technology and Applications (2015)
8. Stohne, V.: Real-time filtering for human pose estimationusing multiple kinects (2014)
9. Han, J., Kamber, M., Pei, J.: Data Mining: Concepts and Techniques: Concepts and Techniques. Elsevier, New York (2011)
10. Mazurek, P., Wagner, J., Morawski, R.Z.: Acquisition and preprocessing of data from infrared depth sensors to be applied for patients monitoring. In: The 8th IEEE International Conference on Intelligent Data Acquisition and Advanced Computing Systems: Technology and Applications (2015)
11. Ravi, N., Dandekar, N., Mysore, P., Littman, M.L.: Activity recognition from accelerometer data. In: AAAI, vol. 5, pp. 1541–1546 (2005)
12. Song, Y., Demirdjian, D., Davis, R.: Multi-signal gesture recognition using temporal smoothing hidden conditional random fields. In: 2011 IEEE International Conference on Automatic Face & Gesture Recognition and Workshops (FG 2011), pp. 388–393. IEEE (2011)
13. Cottone, P., Re, G.L., Maida, G., Morana, M.: Motion sensors for activity recognition in an ambient-intelligence scenario. In: 2013 IEEE International Conference on Pervasive Computing and Communications Workshops (PERCOM Workshops), pp. 646–651. IEEE (2013)
14. Megavannan, V., Bhuvnesh Agarwal, R., Babu, V.: Human action recognition using depth maps. In: 2012 International Conference on Signal Processing and Communications (SPCOM), pp. 1–5. IEEE (2012)
15. Cohen, I., Li, H.: Inference of human postures by classification of 3d human body shape. In: IEEE International Workshop on Analysis and Modeling of Faces and Gestures, AMFG 2003, pp. 74–81. IEEE (2003)
16. Zhang, C., Tian, Y.: Rgb-d camera-based daily living activity recognition. J. Comput. Vis. Image Process. **2**(4), 12 (2012)
17. Gordon, D., Hanne, J.-H., Berchtold, M., Miyaki, T., Beigl, M.: Recognizing group activities using wearable sensors. In: Puiatti, A., Gu, T. (eds.) MobiQuitous 2011. LNICST, vol. 104, pp. 350–361. Springer, Heidelberg (2012)
18. Ofli, F., Chaudhry, R., Kurillo, G., Vidal, R., Bajcsy, R.: Berkeley MHAD: A comprehensive multimodal human action database. In: 2013 IEEE Workshop on Applications of Computer Vision (WACV), pp. 53–60. IEEE (2013)
19. bin Abdullah, M.F.A., Negara, A.F.P., Sayeed, M.S., Choi, D.J., Muthu, K.S.: Classification algorithms in human activity recognition using smartphones. Int. J. Comput. Inf. Eng. **6**, 77–84 (2012)
20. Kaenchan, S., Mongkolnam, P., Watanapa, B., Sathienpong, S.: Automatic multiple kinect cameras setting for simple walking posture analysis. In: 2013 International Computer Science and Engineering Conference (ICSEC), pp. 245–249. IEEE (2013)
21. Andersson, V.O., de Araújo, R.M.: Person identification using anthropometric and gait data from kinect sensor. In: AAAI, pp. 425–431 (2015)

Signal Processing and Analysis

EEG Pattern Recognition: An Efficient Improvement Combination of ERD/ERS/Laterality Features to Create a Self-paced BCI System

Carlos Avilés-Cruz$^{(\boxtimes)}$, Juan Villegas-Cortez,
Andrés Ferreyra-Ramírez, and Arturo Zúñiga López

Azcapotzalco. Departamento de Electrónica, Universidad Autónoma Metropolitana,
San Pablo Xalpa No. 180, Col. Reynosa Tamaulipas, 02200 Mexico, D.F., Mexico
{caviles,juanvc,fra,azl}@azc.uam.mx

Abstract. In this paper, a new method based on an efficient improvement combination of Event-Related Desynchronization (ERD), Event-Related Synchronization (ERS) and lateral activity of sensorimotor cortex features is presented to analyze both left and right hand motor imagery tasks. Our proposal uses delta, theta, alfa and beta rhythms to BCI system. From the spectral power, an efficient combination of ERD/ERS/laterality features was used. Because electroencephalogram signals are non-stationary type and highly vary over time and frequency, a detailed time-frequency analysis is applied. Features coming from time-frequency analysis, where eight frequency bands ranging from 0 to 32 Hz were chosen. Features vectors are classified by Gaussian classifier and the final performance is evaluated in cross-validation scheme. This novel approach was tested using the BCI competition IV data set 1. The detection of the left and right hand motor imagery task was very good, with a result of 96.4 % using BCI-Competition -IV. When comparing results from others competing methods reported in the literature, our approach resulted the best and useful to create a self-paced BCI-system.

Keywords: EEG pattern recognition · EEG · BCI · Motor imagery

1 Introduction

There is a channel for communicating the brain and the external environment based on EEG signals. This channel is the so-called Brain Computer Interface (BCI), which offers an effective help for people with motor disabilities [1] such as amyotrophic lateral sclerosis [2] or spinal cord injury [3]. Several studies have shown that the motor imagery area (contralateral and ipsilateral sensorimotor cortex, respectively) provides information about imagery movements of hands, feet and tongue. In this way, signals are manifested as ERD/ERS (event-related desynchronization/synchronization) [2,4–6]. Due to the fact that the ERD/ERS patterns are opposite, imaginary movements of each hand and foot are suitable

© Springer International Publishing Switzerland 2016
J.F. Martínez-Trinidad et al. (Eds.): MCPR 2016, LNCS 9703, pp. 231–240, 2016.
DOI: 10.1007/978-3-319-39393-3_23

to be classified [3,7] μ and β rhythms registered on the sensorimotor area can be modified by executing either imaginary or observed hand and foot movements. Also, μ and β rhythms may be used to support EEG-based BCI systems (BCIs) [8]. The selection of adequate EEG features in motor imagery movements is crucial, since it determines the accuracy of the classification. Therefore, for attaining good BCI systems, the choice of appropriate and reliable features coming from EEG signals constitutes a very important issue. This is how, in frequency domain, Pregenzer showed the importance of an appropriate preselection of EEG spectral components for accomplishing good classification results [9]. Pregenzer used different frequency bands and resolutions coming from Morlet wavelets, combined with Fisher criteria in order to classify two clases: the left and the right hand motor imagery, obtaining upto 90 % of good classification. In order to set up a realistic comparison between different methods for processing and classifying EEG signals, the BCI Competition was proposed. Specifically, Data Set I for BCI Competition IV [10] was focused on a benchmark to classify two mental tasks: right-hand imaginary movement and left-hand imaginary movement. With this benchmark, signals obtained with EEG were provided and in addition, they were obtained following the same scheme used by the other competition participants. Also, the rules and requirements were laid down. For the present work, the chief functional blocks are shown in Fig. 1. The first module "EEG signal acquisition" reads and transmits EEG signals from an Epoc-Emotiv device to a computer. The second block "signal processing" selects C3 and C4 channel and normalizes amplitudes. The third block develops a time-frequency transformation through a Short-Time Fourier Transform. The fourth block decomposes an EEG signal into 8 frequency bands. The next block "ERS/ERD detection" constitutes the core of our proposal, detecting α, β and μ rhythms over 8 band frequencies. Finally, the "classification motor imagery" block senses 2 mental imagery movements and no-control activity calculating the difference between C3 and C4. The rest of the paper is structured as follows. In Sect. 2, Event-Related Desynchronization/Synchronization is related to alfa,beta and mu rhythms. Materials and Methods are presented in Sect. 3. Section 4 shows and discusses experimental results. Finally, conclusions and future works are given in Sect. 5.

2   Event-Related Desynchronization/Event-Related Synchronization for Brain-Computer Interface

Motor imagery affects the frequency band in the interval of 0.1–32 Hz. Authors typically reference two specific bands: the μ band (8–12 Hz.) and the β band (16–24 Hz.) [3,7,8,11]. The motor imagery process has two leading steps, the first one is related to the Event-Related Desynchronization (ERD), which affects both the μ and the β bands (decrease in activity); the second one event is related to the Event-Related Synchronization (ERS), which particularly affects the β band. In this study, we address for the first time the use of the whole band frequency from 0 to 32 Hz., instead of only the μ and the β bands. Our proposal is based on searching the best representative frequency interval in order to efficiently detect motor imagery self-paced cues.

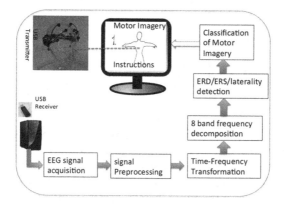

Fig. 1. General functional blocks of the EEG-based BCI system

3 Materials and Methods

3.1 Materials

This study reports results based on data from two sources: the first one constituted by the BCI Competition IV data sets I, provided by the Berlin Group [10]. Those data were selected because they present an asynchronous approach, which is suitable for our main purpose. The second one consists of the EEG recordings, which comes from an EPOC headset provided by Emotiv Systems[1], which is owned by our own laboratory.

BCI Competition IV Data Sets I

The BCI Competition IV Data Sets I contain the EEG signal recorded from 9 healthy people performing motor imagery tasks. The EEG data were recorded from 59 channels, at a rate of 100 samples per second and per subject. The classes of mental task are: (i) imaginary movement of the left hand, (ii) imaginary movement of the right hand and (iii) imaginary movement of any foot. Calibration data were recorded as follows: an arrow was displayed on a computer screen indicating the class of the motor imagery task to be performed, the arrow was presented for a period of 4 s, during which the subject was supposed to imagine the performance of the movement. Periods of time were interleaved with 2 s of blank screen and with 2 s of a cross in the center of the screen. The cross was superimposed to the cue, so it was displayed for 6 s.

3.2 Time-Frequency Representation

General EEG recorded data are represented by the array X (1), where $X \in \Re^{N x M}$, N represents the number of available channels, M represents the number of samples per channel, $x_{c,i}$ represents i channel where $i \in 1...N$. Each signal is processed in one-second window.

---

[1] Emotiv System, http://emotiv.com/.

$$\mathbf{X} = \begin{pmatrix} x_{c1} \\ \vdots \\ x_{ci} \\ \vdots \\ x_{cN} \end{pmatrix} = \begin{pmatrix} x_{c1,1} & \cdots & x_{c1,j} & \cdots & x_{c1,M} \\ \vdots & & \vdots & & \vdots \\ x_{ci,1} & \cdots & x_{ci,j} & \cdots & x_{ci,M} \\ \vdots & & \vdots & & \vdots \\ x_{cN,1} & \cdots & x_{cN,j} & \cdots & cN,M \end{pmatrix} \tag{1}$$

Giving that the EEG signal is non-stationary type, one sample shift is taken from one window to the next one (one-second window each time).

The mean of the signal for each channel $\overline{x_{ci}}$ is subtracted from every x_{ci} row to eliminate the offset and to produce $\widetilde{x_{ci}}$. The spectral power P_{ci} of each channel is calculated using the Short-Fourier Transform (SFT) ($\mathscr{F}$). In order to reduce high frequency artifacts due to windowing process, a blackman window is used (2) to calculate the SFT. In order to get real power spectrum P_{ci}, the spectral power is multiplied by its complex conjugate ($\mathscr{F}^*$) (3).

$$\mathscr{F}\{\widetilde{x_{ci}^k}\} = \sum_{n=-\infty}^{\infty} \widetilde{x_{ci}} W_{Blackman}[n] e^{-j\omega n} \tag{2}$$

$$P_{ci} = \mathscr{F}\{\widetilde{x_{ci}}\} \cdot \mathscr{F}^*\{\widetilde{x_{ci}}\} \tag{3}$$

The whole EEG frequencies (between 0 and 32 Hz.) associated to the delta (δ), theta (θ), alpha (α) and beta (β) rhythms, constitute the most important part of the spectral power P_{ci} since the purpose is to detect ERD/ERS complexes.

3.3   Feature Extraction

From the power representation P_{ci} (3), ranging from 0 to 32 Hz., we compose eight cumulative power values as: $P_{ci}^1 = \sum P_{ci} \in (0-4]\, Hz.$, $P_{ci}^2 = \sum P_{ci} \in (4-8]\, Hz.$, $P_{ci}^3 = \sum P_{ci} \in (8-12]\, Hz.$, $P_{ci}^4 = \sum P_{ci} \in (12-16]\, Hz.$, $P_{ci}^5 = \sum P_{ci} \in (16-20]\, Hz.$, $P_{ci}^6 = \sum P_{ci} \in (20-24]\, Hz.$, $P_{ci}^7 = \sum P_{ci} \in (24-28]\, Hz.$, $P_{ci}^8 = \sum P_{ci} \in (28-32]\, Hz.$ (see Fig. 2), each of them associated to the delta (δ), theta (θ), alpha (α) and beta (β) rhythms.

3.4   Event-Related Desynchronization/Event-Related Synchronization Detection

In this article, any reference to ERD/ERS refers to the frequency combination at a giving time, rather than the sequential-spatial distributed ERD/ERS phenomenon [6]. Motor imagery tasks are detected through out ERD/ERS signals; three states were monitored from eight different frequency bands at the same time: attenuation (related to ERD), enhancement (related to ERS) and laterality (related to $C4 - C3$ difference).

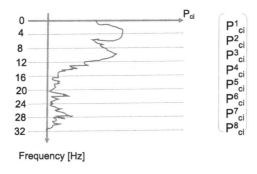

Fig. 2. Interval frequency division of 8 bands

3.5   Imagery Movement Detection

Starting from the spectral power P_{ci}^{j} (for $j = 1, 2, 3...8$), some bands were *attenuated* (ERD) and others were *enhanced* (ERS) (performing imagery movements of hands and feets). Another parameter that was analyzed was the lateral activity (motor imagery movement: left or right hand), called *laterality* (defined as the difference between the C4 and the C3 channel).

For each P_{ci}^{j} signal a low-pass FIR filter was applied in order to eliminate artifacts and to get a flat signal. Filtered P_{ci}^{j} gives us better conditions to determine: (1) attenuated signal (ERD) or enhanced signal (ERS).

The main task was sensing when the frequency band is higher or lower than *upper threshold* or *bottom threshold*; and also sensing if the difference between two channels (C4−C3) was higher or lower than a *lateral threshold*.

One event happens if the signal P_{ci}^{j} combinations were turned *on* (enhancing event) and one of its were turned *off* (attenuating event).

At the end, each $i−$channel results with an attenuating event, an enhancing event, and a lateral event, $\boldsymbol{T}_{ci}^{a}$, $\boldsymbol{T}_{ci}^{e}$ and $\boldsymbol{T}_{ci}^{l}$ respectively. The eight-dimension $\boldsymbol{T}_{ci}^{a,e,l}$ threshold defines a particular motor imagery movement (left or right hand imagery movement).

3.6   Classification

The whole power spectral P_{ci} was used to build a describing feature vector. The feature vector was conformed by slopes (gradient) of P_{ci} for a given time, particularly, around motor imagery event. In order to obtain invariance from one person to other person in the feature vector, whole power espectral matriz was centered-reduced (zero mean value and one standard deviation). The slope was estimated by the difference between left and right power spectral values $\nabla P(k)_{ci}^{CR}$ at a given k time (see Eq. 4).

$$\nabla P(k)_{ci}^{CR} = (P(k+1)_{ci}^{k,CR} - P(k-1)_{ci}^{CR}) \tag{4}$$

From $\nabla P(k)_{ci}^{CR}$, we have conformed four histograms according to four orientations: $h\_1 = \sum slopes \in [0°, 45°)$, $h\_2 == \sum slopes \in [45°, 90°)$, $h\_3 == \sum slopes \in [-45°, 0°)$ and $h\_4 == \sum slopes \in (-90°, -45°)$. The histograms are then weighted with the magnitude of the gradient. Finally, four histograms are obtained for each channel and each band associated to imagery movements.

The final feature descriptor vector $\vec{d}$ was conformed by histograms (8 bands multiplied by 4 histograms). A 32-dimension vector is used for classification. A Gaussian classifier (see Eq. 5) was used in cross-validation mode (50 % − 50 %) to evaluate the classification performance.

$$p(\vec{d}/class_i) = (2\pi)^{-n/2}\|\Sigma_i\|^{-1/2} \exp^{[-\frac{1}{2}(\vec{d}-\overrightarrow{m_{class_i}})^T\Sigma_i^{-1}(\vec{d}-\overrightarrow{m_{class_i}})]} \quad (5)$$

4   Results and Discussions

Proposed pattern recognition methodology for motor imagery detection was applied for offline motor imagery detection with well-known BCI Competition IV data set1. To evaluate motor imagery (intention of movement), a detection rate and an associated noise are calculated from (6) and (7), where TP stands for True Positives, FP for False Positives and FN for False Negatives.

$$\text{Detection Rate} = \frac{TP}{(TP + FN)} . \quad (6)$$

$$\text{Miss Detection Rate} = 1 - \frac{TP}{(TP + FP)} . \quad (7)$$

4.1   Offline Motor Imagery Detection

ERD/ERS/laterality analysis is based on the EEG activity during and after a motor imagery activity. First, the BCI Competition IV Data Sets I was analyzed (five subjects were taken: $BCICIV\_calib\_ds1b$, $BCICIV\_calib\_ds1c$, $BCICIV\_calib\_ds1d$, $BCICIV\_calib\_ds1e$ and $BCICIV\_calib\_ds1g$). From Data set I (59 channels at 100 samples per second; healthy people), only the C3 and the C4 channels were selected, electrodes corresponding to the sensory-motor area (27th and 31st channels from BCI Competition IV data set 1). In order to analyze the EEG recordings and the ERD/ERS/laterality complexes there in contained, the EEG recordings are transformed to a time-frequency representation (spectrogram through Short-Time Fourier Transform). Figure 3 exemplifies the detection of the imagery motor originated from the $BCICIV\_calib\_ds1b$ database. There, the initial points, indicating the beginning of the intention of movement, are shown. The initial points provided by the database are marked as ($\triangleright$ marks) and the the initial points originated from our methodology are marked as ($*$). Also, it is indicated the false positive points and the true positive points.

The signals coming from two different channels are drawn together in two colors: red and blue. Red color indicates C3 channel whereas the blue color indicates C4 channel. In order to illustrate in a clearer manner the detection of the imagery motor, Fig. 3(b) depicts an amplification of a region of the last Fig. (3(a)). There, it is indicated how the two types of initial points, just described, coincide. Figure 3 shows motor imagery cues detected with our proposal ($*$ marks) and with database competition ($\triangleright$ marks). The thresholds T_{ci}^a, T_{ci}^e and T_{ci}^l were estimated experimentally and the values were: $T_{ci}^a = [0.20, 0, 0, 0, 0, 0, 0, 0]$, $T_{ci}^e = [5, 0.02, 0.005, 0.005, 0.0005, 0.0005, 0.0005, 0.0005]$, and $T_{ci}^l = [0.003, 0.005, 0.007, 0.007, 0.0005, 0.0003, 0.0003, 0.0003]$ for an attenuating event, an enhancing event, and a lateral event, respectively. Above thresholds were used for the five people under analysis. Our methodology detects up to 99 % (mean of 96.4 %) of the imagery movements for the five evaluated dataset coming from the BCI Competition IV data. The Miss Classification represents the detection of False Positives and it was around 0.116, as shown in Table 1.

Table 1. Imagery movement detection and miss detection for BCI competition IV dataset I.

Data set	Detection rate	Miss detection rate
BCICIV_calib_ds1b	99 %	0.1
BCICIV_calib_ds1c	97 %	0.2
BCICIV_calib_ds1d	95 %	0.1
BCICIV_calib_ds1e	98 %	0.1
BCICIV_calib_ds1g	93 %	0.08
Mean	96.4 %	0.116

Using the BCI-Competition-IV data [10], we compared the detection performance of our proposal vs. five competitive methods reported in the literature (see Table 2). The five previous methods were: (a) Common Spatial Pattern (CSP), (b) Discriminative Common Spatial Pattern (DCSP) [12], (c) Local Temporal Common Spatial Pattern (LTCSP) [13], (d) Spectrally and Temporally Weighted Classification Method (STWCM) [14] and Time-series discrimination using feature relevance analysis in motor imagery classification (TSDFRAMI) [15]. As you can see in Table 2, our methodology proved a significant improvement on detecting the intention of movement for the right or left arm when it was applied on the data base provided by BCI-Competition IV. Further, when it was compared with the most competitive methods, it provided an improvement by going from 92.86 % (Time-series discrimination using feature relevance analysis in motor imagery classification (TSDFRAMI)) to 96.4 %, i.e. 3.54 % of improvement over the best existing method for detecting imagery movements.

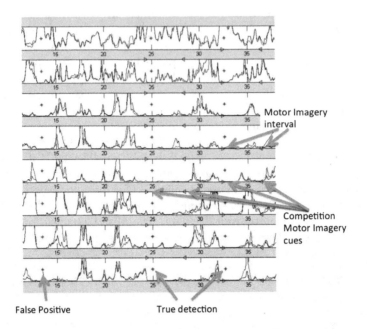

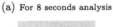

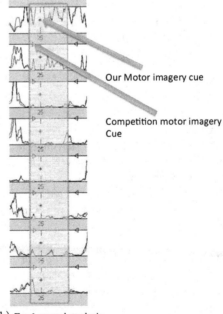

(a) For 8 seconds analysis

(b) For 1 second analysis

Fig. 3. An example of correct cues detection: ∗ corresponds to our methodology detection, and ▷ corresponds to the true competition cues (Color figure online)

Table 2. Detection performance of the most competitive methods in detection of motor imagery for BCI competition IV: Common Spatial Pattern (CSP), Discriminative Common Spatial Pattern (DCSP), Local Temporal Common Spatial Pattern (LTCSP), Spectrally and Temporally Weighted Classification Method (STWCM), Time-series discrimination using feature relevance analysis in motor imagery classification (TSDFRAMI) and the proposed method.

CSP	DCSP	LTCSP	STWCM	TSDFRAMI	Our proposal (BCI-competition)
71%	73%	88%	88%	92.86%	96.4%

5   Conclusion and Future Work

A new methodology to detect and to classify imagery movements was proposed. It is based on an efficient improvement of the combination of attributes originating from the Event-Related Desynchronization (ERD), Event-Related Synchronization (ERS) and the lateral activity of the sensorimotor cortex. These attributes were calculated by analyzing time-frequency spectrograms ranging from 0 to 32 Hz due to the non-stationarity of EEG signals.

Our approach shows its efficiency by using only two channels ($C3$ and $C4$) taken from sensorimotor cortex region, instead of using the whole 59 channels. With these two channels it is possible to detect the activity of the imagery motor (activity of control) or no-activity (activity of no-control) as well as classify left and right motor imagery.

Also, our methodology proved a significant improvement on detecting the intention of movement for the right or left arm when it was applied on the data base provided by BCI-Competition IV. Further, when it was compared with the most competitive methods, it provided an improvement by going from 92.86% (Time-series discrimination using feature relevance analysis in motor imagery classification (TSDFRAMI)) to 96.4%, i.e. 3.54% of improvement over the best existing method for detecting imagery movements.

These results suggest that our approach may be utilize as an efficient switch between activity of control and activity of no-control. Activity of control represents wanting to make a movement with the hands while activity of no-control stands for no intention of any activity, useful to create a asynchronous self-paced BCI-system.

Future work most be focus on the way to automatically determine the following thresholds: $\boldsymbol{T}_{ci}^{a}$, $\boldsymbol{T}_{ci}^{e}$ and $\boldsymbol{T}_{ci}^{l}$, for each person and each kind of test. The authors plan to improve the learning phase (time reduction) by parallel processing, as well as real-time implementation of the whole proposal BCI-asynchronous system. The goal will be to implement control mobile device i.e. tablet, cellphone; by motor imagery.

References

1. Wolpaw, J., Birbaumer, N.: McFarland: brain-computer interfaces for communication and control. Clin. Neurophysiol. **113**, 767–791 (2002)
2. Pfurtscheller, G., Neuper, C., Schlöl, A., Legger, K.: Separability of EEG signals recordered during right and left motor imagery using adaptive autoregressive parameters. IEEE Trans. Rehabil. Eng. **6**(3), 316–325 (1998)
3. Muller-Putz, G., Scherer, R., Pfurtscheller, G., Neuper, C., Rupp, R.: Non-invasive control of neuroprostheses for the upper extremity: temporal coding of brain patterns. In: Engineering in Medicine and Biology Society, EMBC 2009. Annual International Conference of the IEEE, pp. 3353–3356, September 2009
4. Pfurtscheller, G., Brunner, C., Schlogl, A., da Silva, F.L.: Mu rhythm (de) synchronization and eeg single-trial classification of different motor imagery tasks. NeuroImage **31**, 153–159 (2006)
5. Pfurtscheller, G., Neuper, C.: Motor imagery and direct brain-computer communication. Proc. IEEE **89**(7), 1123–1134 (2001)
6. Pfurtscheller, G., da Silva, F.L.: Event-related EEG/MEG synchronization and desynchronization: basic principles. Clin. Neuroph. **110**, 1842–1857 (1999)
7. Pfurtscheller, G., Scherer, R., Muller-Putz, G., da Silva, F.L.: Short-lived brain state after cued motor imagery in naive subjects. Eur. J. Neurosci. **28**(1), 1419–1426 (2008)
8. Pfurtscheller, G., Müller, G.R., Pfurtscheller, J., Gerner, H.J., Rupp, R.: 'Thought' - control of functional electrical stimulation to restore hand grasp in a patient with tetraplegia. Neurosci. Lett. **351**(1), 33–36 (2003)
9. Pregenzer, M., Pfurtscheller, G.: Frequency component selection for an EEG-based brain to computer interface. IEEE Trans. Rehabil. Eng. **7**, 413–419 (1999)
10. Blankertz, B.: Berlin institute of technology, machine learning laboratory, BCI competition iv data set 1 mental imagery, multi-class, June 2004
11. Thomas, E., Fruitet, J., Clerc, M.: Combining ERD and ERS features to create a system-paced BCI. J. Neurosci. Meth. **216**(1), 96–103 (2013)
12. Thomas, K., Guan, C., Lau, C., Vinod, A., Ang, K.: A new discriminative common spatial pattern method for motor imagery brain-computer interfases. IEEE Trans. Biomed. Eng. **56**(11), 2730–2733 (2009)
13. Wang, H., Zheng, W.: Local temporal common spatial pattern for robust single-trial EEG classification. IEEE Trans. Neural Syst. Rehabil. Eng. **16**(2), 131–139 (2008)
14. Tae-Eui, K., Heung-IL, S., Seong-When, L.: Non-homogeneous spatial filter optimization for electroencephalogram (EEG)-based motor imagery classification. Neurocomputing **108**, 58–68 (2013)
15. Alvarez-Meza, A., Velasquez-Martinez, L., Castellanos-Dominguez, G.: Time-series discrimination using feature relevance analysis in motor imagery classification. Neurocomputing **151**(1), 122–129 (2015)

Highly Transparent Steganography Scheme of Speech Signals into Color Images Using Quantization Index Modulation

Diego Renza[✉], Dora M. Ballesteros L., and Jeisson Sanchez

Universidad Militar Nueva Granada, Bogotá, Colombia
{diego.renza,dora.ballesteros,u1400881}@unimilitar.edu.co

Abstract. A highly transparent steganography scheme of speech signals into color images based on Quantization Index Modulation (QIM) is presented in this paper. The proposed method takes advantage of the low distortion in the host image introduced by the scalar quantization of its wavelet coefficients. The stego image is highly similar to the host image, and the secret content is imperceptible. The secret message is recovered at the receiver with high correlation to the original speech signal. For the purpose of increasing the security of the system, an external mask is used to encrypt the secret content before the embedding process. Several tests were carried out in order to quantify the influence of the size of the quantizer (delta) on the quality of the recovered secret content and the transparency of the stego image.

Keywords: Steganography · Discrete wavelet transform · Quantization index modulation · Imperceptibility

1 Introduction

Nowadays, data hiding techniques are very popular due to the need for data privacy or copyright protection. In the first case, steganography consists of embedding a secret message into a host signal with the purpose of concealing the existence of its secret content [7]; in the second case, a mark is embedded in order to protect authorship of the host [11]. This means that in steganography the most important signal is the secret message while in the case of watermarking it is the host signal. Regardless of the data hiding technique, the secret messages and host signals can be text, audio, image or video [1,3].

Some proposals of data hiding embed data in the wavelet transform domain. For example, a speech steganography scheme based on the DWT is proposed in [13]. Authors found that the obtained stego signals are highly correlated (similar) to the host speech signals, and the transparency is guaranteed. These results have been confirmed in other studies [5,6], even in medical applications [8].

On the other hand, Quantization Index Modulation (QIM) has proved to be a good solution for reversible and robust data hiding schemes [4,12]. Nevertheless, some works have demonstrated that security of the systems based on QIM can

© Springer International Publishing Switzerland 2016
J.F. Martínez-Trinidad et al. (Eds.): MCPR 2016, LNCS 9703, pp. 241–250, 2016.
DOI: 10.1007/978-3-319-39393-3_24

be weak and an attacker can reveal the secret content from the analysis of the stego information [9].

In this work we propose a steganography scheme for concealing a speech signal in a color image, by using QIM in the wavelet domain of the host signal. Additionally, we add a level of security by using an XOR operation between an external mask and the speech signal. The result is a stego image with high similarity to the host image, a recovered speech signal with high quality, and an additional level of security to reveal the secret content.

2   Methods and Algorithms

2.1   Discrete Wavelet Transform (DWT)

The DWT is a useful mathematic method for signal decomposition into different resolutions. In the case of images, the input image is passed through two filters to obtain four sub-bands: one approximation sub-band (LL) and three detail sub-bands. The detail coefficients are oriented in one of three directions, horizontal (LH), vertical (HL) or diagonal (HH). The LL sub-band keeps the highest similarity to the image, while HL, LH and HH keep information related to the specific orientation. Figure 1 shows the chart for the decomposition and reconstruction stages.

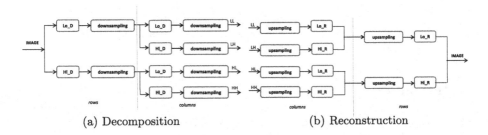

(a) Decomposition   (b) Reconstruction

Fig. 1. Chart of the wavelet transform for images.

Filters used in the decomposition (Lo_D, Hi_D) and reconstruction process (Lo_R, Hi_R) must satisfy the condition of quadrature mirror filters [10].

2.2   Quantization Index Modulation

The QIM is a binary data concealing technique based on quantization rules. The QIM generally refers to data embedding with a specific quantizer or a sequence of quantizers on a host signal. The insertion process is based on the modification of the carrier value (pixel, wavelet coefficient, audio sample, etc.); the carrier value C is approximated to the nearest multiple of the static step value (Δ) obtaining the stego value S.

2.3 Quality Metrics

Different techniques of quality signal analysis such as Structural Similarity Index Metric (SSIM), Gray Value Degree (GVD) and Pearson's Correlation Coefficient (SPCC) were used to evaluate the imperceptibility of the stego image and the quality of the recovered audio signal. Next, the general characteristics of these methods are described.

Structural Similarity Index Metric. SSIM is an efficient method of quality analysis for similarity on image and videos. For two images U and V, the SSIM is given by Eq. 1 [14] and its ideal value is 1.

$$SSIM = \frac{(2\mu_U \mu_V + c_1)(2\sigma_{UV} + c_2)}{(\mu_U^2 + \mu_V^2 + c_1)(\sigma_U^2 + \sigma_V^2 + c_2)} \tag{1}$$

Where μ is the mean, σ is the standard deviation, c_1 and c_2 are stabilization variables.

Gray Value Degree. The Gray Value Degree, (GVD) uses histogram differences between the gray values of adjacent pixels to specify the texture characteristics. Its ideal value is zero. To obtain the GVD of a $U(M \times N)$ image, it is necessary to compute the gray value (GN) and the average neighborhood gray difference (AN), by means of the Eqs. 2–4.

$$GN(x,y) = 4U(x,y) - \sum_{i=0}^{3} U\left(x + \left\lfloor \frac{i}{2} \right\rfloor (-1)^i, y + \left\lfloor \frac{3-i}{2} \right\rfloor (-1)^i\right) \tag{2}$$

Using all the results of GN, the average neighborhood gray difference (AN) is calculated as follows:

$$AG = \sum_{x=2}^{M-1} \sum_{Y=2}^{N-1} GN(x,y) \tag{3}$$

Finally, the gray value degree is calculated as follows:

$$GVD = \frac{AG' - AG}{AG' + AG} \tag{4}$$

Where AG' and AG are the average neighbourhood gray difference of the original image and the modified image, respectively.

Pearson's Correlation Coefficient (SPCC). Audio and image testing were based on correlation coefficient between the original signal and the recovered signal. The correlation function $SPCC$ for two signals U and V is defined by means of Eq. 5, where σ is the deviation and cov the covariance [2].

$$SPCC = \frac{cov(U,V)}{\sigma_U \sigma_V} \tag{5}$$

3   Proposed Scheme

The proposed algorithm conceals an audio signal in an image. Color images are suitable since they can allow one to hide more information than a grayscale image. Moreover, the larger the image size, the greater the concealment capacity. The proposed scheme performs the audio embedding as follows:

3.1   Audio Embedding

The input parameters are the carrier image C and the audio signal A as the secret message. The outputs are the stego image S and the secret key. The audio embedding process is summarized in Fig. 2 and described below.

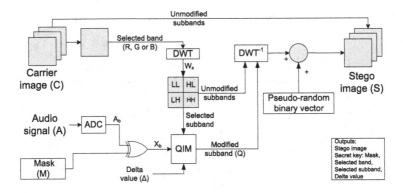

Fig. 2. Illustrative representation of the flow chart for audio embedding.

Step 1. Mask generation: the mask (M) is obtained through the conversion of a text string to a binary representation through the ASCII code. This text string is used to provide security to the embedded message. For example, if the text string is 32-characters length, the total number of bits in M is 256.

Step 2. Binary representation of the speech signal and masking operation: the speech signal (A) is converted to a binary representation (A_b). The total number of bits is equal to the total number of samples multiplied by the number of bits per sample. Then, an XOR operation is performed, $X_b = A_b \oplus M$. If the length of K is lower than the length of A_b, M is repeated as necessary. For example, if M is 256 bits length, and A is a speech signal with $f_s = 8\,KHz$, $2\,\mathrm{s}$ of duration and 16-bits per sample, the length of A_b is 256 K bits. In this case, M is repeated 1000 times, then an XOR operation is made (bit to bit).

Step 3. Extraction of color and wavelet sub-band: the host image is separated in three color bands (R, G, B) and one of them is selected. Then, the DWT is applied to the selected sub-band with one level of decomposition.

Step 4. Data insertion: the algorithm selects one of the wavelet sub-bands; then every bit of the X_b vector is inserted into the wavelet-subband through the

QIM function. Equation 6 represents the way to insert data; where Q is the quantized sub-band:

$$
\begin{cases}
Q(i,j) = \Delta \left\lfloor \frac{W_x(i,j)}{\Delta} \right\rfloor & if \quad X_b = 0 \\[2ex]
Q(i,j) = \Delta \left\lfloor \frac{W_x(i,j)}{\Delta} \right\rfloor + \frac{\Delta}{2} & if \quad X_b = 1
\end{cases}
\tag{6}
$$

For the purpose to recover the secret data at the receiver, delta value, the mask, the selected color band, and the selected wavelet sub-band are transmitted by a private channel. These data form the secret key.

Step 5. Addition of noise to the quantized sub-band: since after the QIM process, only will exist wavelet coefficients multiples of $\Delta/2$, an additive noise is inserted into the quantized sub-band with the purpose to resist attacks based on data distribution.

Step 6. Image reconstruction: the image is reconstructed from the wavelet sub-bands and the color bands. The result is the stego image (S).

3.2 Audio Extraction

The input parameters are the received stego image (Sr) and the parameters used in embedding phase (Δ value in QIM, color band, wavelet sub-band, and the mask). The output is the recovered audio signal (Ar). The audio extraction process is summarized in Fig. 3 and described below.

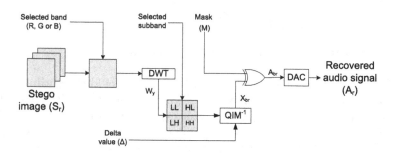

Fig. 3. Illustrative representation of the flow chart for audio extraction.

Step 1. Separation of the three bands of the stego image: the stego image is separated in the colors Red, Green and Blue.

Step 2. Application of the wavelet transform: firstly, one of the color bands is selected according to the information of the secret key. The DWT is applied to this band and then the sub-band specified by the secret key is selected properly.

Step 3. Data extraction: in every pixel of the selected sub-band the inverse QIM function is applied, according to Eq. 7. The result is the recovered masked data, X_{br}.

$$\begin{cases} X_{br} = 1 & if \quad \frac{\Delta}{4} < W_y(i,j) - \Delta \left\lfloor \frac{W_y(i,j)}{\Delta} \right\rfloor \le \frac{3\Delta}{4} \\[2ex] X_{br} = 0 & else \end{cases} \tag{7}$$

Step 4. Unmask the audio: an xor operation between X_{br} and M is performed. M is obtained from the secret key. The result is the recovered bits of the speech signal, $A_{br} = X_{br} \oplus M$.

Step 5. Conversion to analog values: the binary values A_{br} are converted to analog values by using 16-bits data.

3.3 Insertion Capacity

In an information-hiding scheme, the insertion capacity (IC) is a measure of the quantity of information per unit that the method is able to hide. The insertion capacity of the proposed method is given by the Eq. 8,

$$IC = \frac{M \times N}{(2^L)^2} \; (bits/pixel) \tag{8}$$

Where, M and N are the dimensions of the image and L is the number of decomposition levels. The above means that in a 1024×768 image, it is possible to hide 196608 bits. With an audio file sampled at 8 KHz (speech), it is possible to hide a file of 24.576 s of duration. With an audio file sampled at 44.1 KHz (music), it is possible to hide a file of 4.45 s of duration. In any case, it is possible to hide a little phrase or message.

4   Experimental Results and Analysis

To evaluate the effectiveness of the algorithm in terms of imperceptibility of the stego image and quality of the recovered signal we use five RGB images (three standard images and two proprietary images): Baboon ($512 \times 512 \times 3$), Lena ($1600 \times 900 \times 3$), Peppers ($676 \times 470 \times 3$), Penguins ($1024 \times 768 \times 3$) and Flower ($1366 \times 768 \times 3$). Besides, Haar wavelet base was selected and ten Delta values, four wavelet subbands and ten secret messages were used. A total of 2000 tests were done (5 images × 10 Delta values × four subbands × 10 secret messages). For imperceptibility measures (Figs. 4 and 5), we use the following parameters: SSIM, GVD and correlation. To evaluate quality of the recovered audio (Figs. 6 and 7), we use SPCC.

4.1   Imperceptibility of the Stego Image

Imperceptibility aims to quantify the differences between the carrier-image and the stego-image. In this case it is desired that no appreciable distortion exists in the stego image, and therefore that there is no evidence that the image has been modified. To do this, the values of SSIM, GVD and SPCC should be as

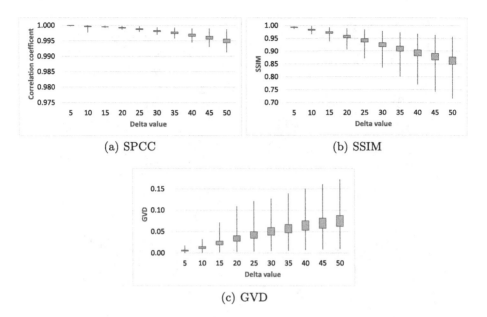

(a) SPCC   (b) SSIM

(c) GVD

Fig. 4. Imperceptibility results for three metrics. Confidence range (95 %) by Delta value.

close to its ideal value, namely, 1, 0 and 1, respectively. To corroborate the imperceptibility of the method, the results for the three metrics above were consolidated by confidence intervals. That is, the ranges in which the 95 % of the values are concentrated were obtained, whereby it is possible to assess the trend of each metric. This evaluation was performed by varying the delta value from 5–50, in increments of 5, as shown in Fig. 4.

Figure 4(a), shows the variation of the correlation coefficient depending on the delta value. Here it is observed that the transparency decreases as the delta value increases, however all values are above 0.99, which means that in terms of correlation coefficient, the transparency is good. It also shows that delta values that present good imperceptibility results in terms of SSIM are lower than 20.

Regarding the SSIM index, Fig. 4(b), shows the trend of this metric at change the delta value. As in the correlation coefficient, the transparency decreases as the delta value increases, but this decrease is more pronounced compared to the previous index. However, for delta values lower than 20, SSIM tends to remain below 0.95.

The GVD meanwhile, gets values remains below 0.15, as shown in Fig. 4(c). As you increase the delta value, the GVD away from its ideal value, so in terms of GVD again small delta values are recommended.

Figure 5 shows the original image and the resulting one after using the proposed scheme. In this case, the HH sub-band, the red band and a delta value of 50 were used as input parameters. The quality metrics are showed in the caption. Although there are differences between the stego band and the original band, at

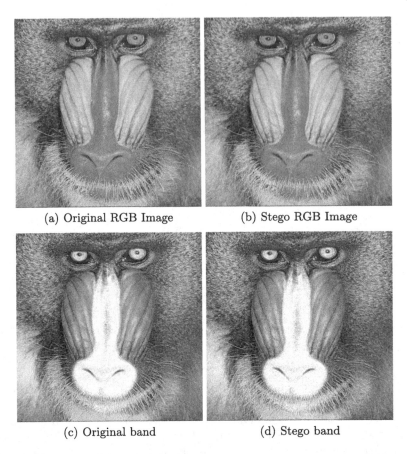

(a) Original RGB Image   (b) Stego RGB Image

(c) Original band   (d) Stego band

Fig. 5. Example of imperceptibility test. Quality metrics: SSIM=0.8240, GVD=0.0147 and SPCC=0.9810.

the end of the hiding process the stego RGB image is very similar to the original RGB image.

4.2   Quality of the Recovered Audio

Quality of the recovered audio signal quantifies the differences between the original audio signal and the recovered signal. In this case, should no exist appreciable distortion in the recovered signal. To do this, similarity between these two signals was calculated by means of the squared pearson correlation coefficient. SPCC values should be as close to its ideal value, namely, 0.

The Fig. 6 shows the confidence intervals for the SPCC between the original audio and the recovered audio signal. Here, unlike the evaluation of imperceptibility, increased delta value means a better quality of the output signal. However, according to the results of Fig. 6, if Delta is higher than 15, SPCC is close to 1 and then the quality of the recovered audio signal is very high.

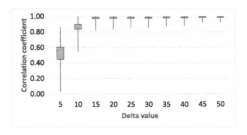

Fig. 6. Correlation coefficient for original and recovered audio signals. Confidence range (95 %) for SPCC by Delta value.

According to the above, the imperceptibility and quality of the recovered signal, have a tradeoff based on the delta value. Note that the higher the delta value, the lower the quality of stego image and lower the quality of the recovered audio. In any case, a delta value that presents a good compromise between imperceptibility and quality of the recovered signal is 15. In order to illustrate the performance of proposed system, Figs. 7(a) and 7(b) show the original audio signal and the recovered audio signal in the time domain. Here it is possible to confirm that the two signals have a high correlation degree.

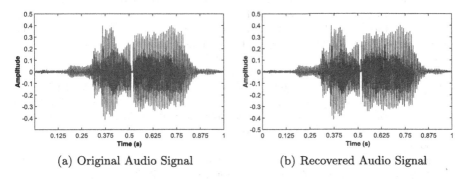

(a) Original Audio Signal   (b) Recovered Audio Signal

Fig. 7. Example of the quality of the recovered audio. Quality metric: SPCC=0.9969.

5 Conclusion

We propose a steganography scheme that allows to hide an audio signal into a RGB image. The process works on the wavelet domain with QIM method. Both imperceptibility and quality of the recovered audio depend on the Delta value. According to the results if Delta value increases, quality increases but imperceptibility decreases. A good tradeoff between them is obtained for a Delta value of 15.

Acknowledgment. This work is supported by the "Universidad Militar Nueva Granada-Vicerrectoría de Investigaciones" under the grant INV-ING-1768 of 2015.

References

1. Ballesteros L, D.M., Renza, D., Rincon, R.: Gray-scale images within color images using similarity histogram-based selection and replacement algorithm. J. Inf. Hiding Multimed. Sig. Process. **6**(6), 1156–1166 (2015)
2. Benesty, J., Chen, J., Huang, Y., Cohen, I.: Pearson correlation coefficient. Noise Reduction in Speech Processing. Springer Topics in Signal Processing, vol. 2, pp. 1–4. Springer, Heidelberg (2009)
3. Bhattacharyya, D., Dutta, P., Balitanas, M.O., Kim, T., Das, P.: Hiding data in audio signal. In: Chang, C.-C., Vasilakos, T., Das, P., Kim, T., Kang, B.-H., Khurram Khan, M. (eds.) ACN 2010. CCIS, vol. 77, pp. 23–29. Springer, Heidelberg (2010)
4. Chen, B., Wornell, G.W.: Quantization index modulation methods for digital watermarking and information embedding of multimedia. J. VLSI Signal Process. Syst. Signal Image video Technol. **27**(1–2), 7–33 (2001)
5. Delforouzi, A., Pooyan, M.: Adaptive digital audio steganography based on integer wavelet transform. Circuits Syst. Signal Process. **27**(2), 247–259 (2008)
6. Hemalatha, S., Acharya, U.D., Renuka, A.: Wavelet transform based steganography technique to hide audio signals in image. Procedia Comput. Sci. **47**, 272–281 (2015)
7. Hopper, N.J., Langford, J., von Ahn, L.: Provably secure steganography. In: Yung, M. (ed.) CRYPTO 2002. LNCS, vol. 2442, pp. 77–92. Springer, Heidelberg (2002)
8. Jero, S.E., Ramu, P., Ramakrishnan, S.: Discrete wavelet transform and singular value decomposition based ECG steganography for secured patient information transmission. J. Med. Syst. **38**(10), 1–11 (2014)
9. Li, S.B., Tao, H.Z., Huang, Y.F.: Detection of quantization index modulation steganography in G.723.1 bit stream based on quantization index sequence analysis. J. Zhejiang Univ. SCIENCE C **13**(8), 624–634 (2012)
10. Mallat, S.G.: A theory for multiresolution signal decomposition: the wavelet representation. IEEE Trans. Pattern Anal. Mach. Intell. **11**(7), 674–693 (1989)
11. Marvel, L.M.: Information hiding: steganography and watermarking. In: Javidi, B. (ed.) Optical and Digital Techniques for Information Security. Advanced Sciences and Technologies for Security Applications, pp. 113–133. Springer, New York (2005)
12. Phadikar, A.: Multibit quantization index modulation: A high-rate robust data-hiding method. J. King Saud Univ. Comput. Inf. Sci. **25**(2), 163–171 (2013)
13. Rekik, S., Guerchi, D., Selouani, S.A., Hamam, H.: Speech steganography using wavelet and fourier transforms. EURASIP J. Audio Speech Music Process. **2012**(1), 1–14 (2012)
14. Wang, Z., Bovik, A.C., Sheikh, H.R., Simoncelli, E.P.: Image quality assessment: from error visibility to structural similarity. IEEE Trans. Image Process. **13**(4), 600–612 (2004)

High Scrambling Degree in Audio Through Imitation of an Unintelligible Signal

Dora M. Ballesteros L., Diego Renza$^{(\boxtimes)}$, and Steven Camacho

Universidad Militar Nueva Granada, Bogotá, Colombia
{dora.ballesteros,diego.renza,u1400943}@unimilitar.edu.co

Abstract. A reversible scheme of audio scrambling based on the imitation of Gaussian unintelligible signals is presented in this paper. It is supported by the similarities it shares with Gaussian unintelligible signals in terms of the Probability Density Function and the entropy. It is feasible for an audio signal to imitate the behavior of a Gaussian noise signal, and then the residual intelligibility is zero. Our proposed scheme, termed ASGI (Audio Scrambling by Gaussian signal Imitation), is tested with four different music genres, and the experimental tests reveal that the scrambled audio signals look like Gaussian noise signals and have high scrambling degrees. Additionally, our scheme preserves the advantages of imitation-based scrambling schemes.

Keywords: Audio scrambling · Imitation-based scheme · Residual intelligibility · Scrambling degree · Entropy value

1 Introduction

Scrambling and encryption are methodologies aimed at protecting the content of information by altering the original content, which can only be recovered with an appropriate key. In image case, some effective methods consist of histogram modification to obtain uniformly distributed behavior [8,13,16]. However, in audio case, methods are focused on the permutation sequences generation without altering data distribution [1,7]. In both types of methodologies, the main conditions to be satisfied are: residual intelligibility, security level of the key, quality of the recovered message, and computational cost [9].

One way to measure residual intelligibility in audio is through the Scrambling Degree (SD). The lower the residual intelligibility, the higher the SD. In [6], the scrambling process obeys iterative displacement and the obtained SD values are lower than 0.8. With Cellular Automata (CA) as key generator, SD can be high, but its value depends greatly on the number of generations (NOG), the neighborhood types and the boundary condition [12]. When CA is mixed with Compressive Sensing (CS), SD depends on the sub-rate and the content type of the audio signal (e.g. voice, instrumental, or a mix of them) [3]. In terms of secure systems, some approaches include progressive scrambling [14,15], high dimensional matrix transformation [11], mixture with watermarking [10], 2D

© Springer International Publishing Switzerland 2016
J.F. Martínez-Trinidad et al. (Eds.): MCPR 2016, LNCS 9703, pp. 251–259, 2016.
DOI: 10.1007/978-3-319-39393-3_25

Arnold transform [2] or bio-inspired process [4], which have demonstrated that their key-spaces are large enough to resist cryptanalysis [5]. However, a good trade-off among the design conditions is still a challenge. In the proposal of Ballesteros and Moreno [4], an auxiliary signal with intelligible content is used to create the key. The original speech signal imitates an auxiliary signal, and the content is transformed in order to resemble the auxiliary signal content. Imitation is feasible if some conditions are satisfied and the system is unconditionally secure. However, a huge database of auxiliary intelligible signals is required to implement the scrambling scheme. To overcome this problem, in [5] a speech scrambling scheme based on imitation of a Gaussian noise signal was proposed. Unlike [4], a database of auxiliary signals is not necessary because the target signal is created in situ. According to several tests, it was validated that a speech signal can imitate a Gaussian noise signal.

In this paper, the work of [5] is continued and an audio scrambling scheme based on imitation of a Gaussian noise signal (Audio Scrambling by Gaussian signal Imitation, ASGI) is proposed. The aim is to verify if in generic audio case (i.e. speech, instrumental or a mixture of both) the imitation process to a Gaussian noise signal is feasible, and if the audio genre influences the results of SD. Since our proposal has the same characteristics of the scrambling schemes based on imitation, it is expected that the ASGI scheme has low computational cost to create the key, complete reversibility, and high security.

The paper is organized as follows. Section 2 presents the proposed ASGI scheme for the scrambling and recovering modules. In Sect. 3, the ASGI scheme is validated in terms of Scrambling Degree and the results are compared with other proposals. Section 4 presents the conclusions of the work.

2   Proposed ASGI Scheme

The scrambling module of the proposed ASGI scheme is carried out by imitating the behavior of a Gaussian noise signal. If the statistical moments of the audio signal are similar to those of the Gaussian noise signal, entropy and probability density distribution (PDF) are similar too. Therefore, if the audio samples are relocated to resemble the Gaussian noise signal, the result (the scrambled audio signal) looks like the Gaussian noise signal, and the original content of the audio is altered in the process. At the receiver, the scrambling process is completely reversed with the appropriate key.

2.1   Scrambling Module

In this module, the places of the audio signal are relocated through the process of imitating a Gaussian noise signal. In our proposal, the key is not an input of the system because it is generated in situ and corresponds with the mapping positions of the signals involved in the imitation. Figure 1 shows a block diagram of the scrambling module. The steps to obtain the scrambled audio signal are as follows:

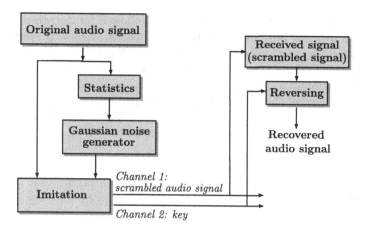

Fig. 1. Block diagram of the proposed ASGI scheme.

1. Calculate the audio signal statistical moments. The first two moments of the audio signal are calculated, namely: mean (Eq. 1) and standard deviation (Eq. 2).

$$\mu = \frac{\sum\limits_{i=1}^{N} x_i}{N-1} \tag{1}$$

$$\sigma = \sqrt{\frac{\sum\limits_{i=1}^{N} (x_i - \mu)^2}{N-1}} \tag{2}$$

Where N is the total number of samples.
2. Generate a Gaussian noise signal from the above statistical moments; the result is an unintelligible signal with similar entropy and PDF values to those of the audio signal.
3. Imitate the Gaussian noise signal; this is the main block of the proposal and it is based on the following hypothesis: "An audio signal can imitate an unintelligible signal if their entropy and PDF values are similar". Suppose that we have two signals (S_a, S_b), S_a with intelligible content and S_b with unintelligible content. Amplitudes of these signals are in the range $x = [x_1, x_2, \ldots, x_n]$, with x_1 as the lowest amplitude and x_n as the highest amplitude. These signals sound different, but they have equal entropy values $(H(S_a), H(S_b))$, and equal PDF. Then, the quantity of samples with amplitude x_1 of the signal S_a is equal to the quantity of samples with amplitude x_1 of the signal S_b, and so on. In other words (Eq. 3),

$$P(S_a = x_i) = P(S_b = x_i) \; for \; i[1, 2, \ldots n] \tag{3}$$

With $P(\cdot)$ as the data probability. Since the signals have the same entropy and PDF, they can look like each other if one of them imitates the other.

Imitation means that the behavior of one is followed by the other, through a relocation process. Suppose S_a wants to imitate S_b. The first step is to find the position of x_n amplitude both in S_a and S_b. These values are kept in the first place of the vectors I_a and I_b, respectively. The second step is to find the position of x_{n-1} amplitude both in S_a and S_b, keeping the results in the second place of the vectors I_a and I_b, respectively. The above procedure is repeated until the positions of x_1 amplitude in S_a and S_b are found and saved. With I_a and I_b, the mapping process follows the Eq. 4:

$$C(I_b) = S_a(I_a) \tag{4}$$

Where C is the scrambled audio signal. It is worth noting that if entropies and PDF of S_a and S_b are equal, then C is equal to S_b. The key is obtained by means of Eq. 5:

$$key(I_a) = I_b \tag{5}$$

The details of the algorithm are summarized in Algorithm 1.

Algorithm 1. Scrambling module

Inputs: Audio signal Sa, Noise signal Sb, amplitudes of these signals ordered from highest to lowest x.
Outputs: Scrambled signal C, key K.
1: **function** SCRAMBLING(Sa,Sb,x)
2:      $audiolength \leftarrow length\ of\ Sa$
3:      **for** $i = 0$ to $audiolength$ **do**
4:          $Ia_i \leftarrow position\ of\ x_{n-i}\ in\ Sa$
5:          $Ib_i \leftarrow position\ of\ x_{n-i}\ in\ Sb$
6:          $C_{Ib_i} \leftarrow Sa_{Ia_i}$
7:          $K_{Ia_i} \leftarrow Ib_i$
8:      **end for**
9: **end function**

In the case of entropy and PDF values of S_a and S_b being similar but not equal, the scrambled audio signal would be similar (but not equal) to the Gaussian noise signal. However, similarity is enough to alter the original content without leaving a trace of it.
4. Transmit the scrambled audio signal by one channel and the key by another channel.

2.2 Recovering Module

With the scrambled audio signal and the appropriate key, the process is completely reversed and the recovered audio signal is equal to the original one. The recovered speech signal, R, is obtained with the reversing Eq. 6.

$$R = C(key) \tag{6}$$

The details of the algorithm are summarized in Algorithm 2.

Algorithm 2. Recovering module

Inputs: Scrambled signal C, key K.
Outputs: Recovered audio signal R.
  1: **function** RECOVERING(C,K)
  2:      $audiolength \leftarrow length\ of\ C$
  3:      **for** $i = 0$ to $audiolength$ **do**
  4:          $R_i \leftarrow C_{k_i}$
  5:      **end for**
  6: **end function**

---

3   Experimental Results and Discussion

To evaluate the performance of our ASGI scheme, we select four musical genres (pop, rock, jazz, and classical), and five songs for each one. Six frames of five seconds are extracted from every song in order to apply the scheme. Since the aim is to alter the original content of the audio, performance is measured through the SD between the original audio signal and the scrambled audio signal. Firstly, the difference of the signal, D, is calculated, as follows (Eq. 7):

$$D(i) = \frac{1}{4} \sum_{i=3}^{m-2} \{4 * S(i) - (S(i-1) + S(i-2) + S(i+1) + S(i+2))\} \quad (7)$$

where $S(i)$ is the ith sample of the audio signal, m is the total number of samples and D is a vector with $m-4$ values.

Next, the sum and the subtraction of the two differences (original signal and scrambled signal) are obtained as follows (Eqs. 8 and 9):

$$B = D_2 - D_1 \quad (8)$$

$$A = D_2 + D_1 \quad (9)$$

where D_2, is the difference of the scrambled signal and D_1 is the difference of the original signal, calculated by means of Eq. 7.

Finally, the Scrambling Degree, SD, is calculated according to Eq. 10:

$$SD = B/A \quad (10)$$

where B/A solves the system of linear equation $SD * A = B$ for SD. The highest value of SD is 1 and the lowest is 0. If the result is 1, it means that the original content of the audio signal has been completely altered in the scrambling process, or in other words, that the residual intelligibility is zero.

In order to illustrate the performance of the proposed scheme, we show the results of two audio file frames. Then, the results of one hundred and twenty simulations are summarized. Figure 2 shows an example with the rock genre. It contains voices and sounds of musical instruments. In this case, entropy of the Gaussian noise signal is 4.3202 and entropy of the scrambled audio signal is

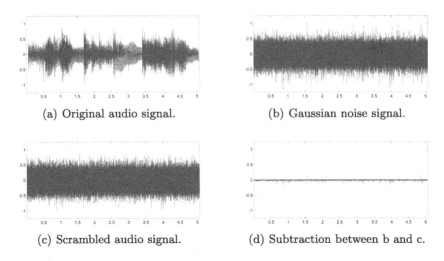

(a) Original audio signal.

(b) Gaussian noise signal.

(c) Scrambled audio signal.

(d) Subtraction between b and c.

Fig. 2. Example of the proposed ASGI scheme for the rock genre.

4.3186. SD between the original audio signal and the scrambled audio signal is 0.9818.

Figure 3 shows an example with the classical genre. It contains only sounds of musical instruments. In this case, entropy of the Gaussian noise signal is 4.3232 and entropy of the scrambled audio signal is 4.6199. SD between the original audio signal and the scrambled audio signal is 0.9999.

Figure 4 shows the summary of the tests. Every genre contains the results of thirty simulations (five songs by six frames by song) with a confidence range of 95 %. These results are higher than 0.87, which is high enough to guarantee very low residual intelligibility. On the other hand, the proposed scheme works with different kinds of audio content, such as speech, instrumental or a mixture of both.

Finally, our proposed scheme is compared with some of the existing techniques (Fig. 5). We use the results reported in the works of Madain et al. [12] and Augustine et al. [3]. In the first one, the results of NOG (number of generations) equal to 1 and 15 were taken into account. In the second one, we use the results of subrate (SR) equal to 0.1 and 0.5. In the work of Madain et al., the value of SD depends very much on the number of generations (NOG). In the proposal of Augustine et al., the subrate (SR) influences the quality of the scrambled signal. In all cases, our proposal has better performance in terms of SD and this value does not depend on adjustable parameters.

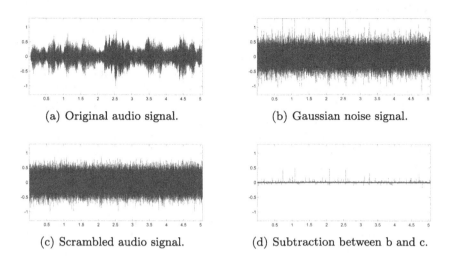

(a) Original audio signal.   (b) Gaussian noise signal.

(c) Scrambled audio signal.   (d) Subtraction between b and c.

Fig. 3. Example of the proposed ASGI scheme for the classic genre.

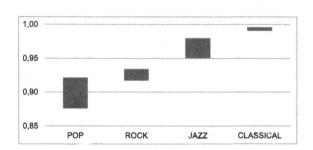

Fig. 4. SD results for the proposed Scheme. Confidence range (95 %) for SD by genre (120 simulations).

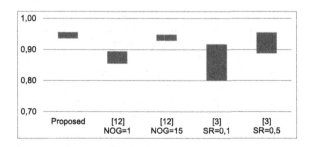

Fig. 5. Comparative results for SD. SD Confidence range (95 %) for the proposed scheme and Madain et al. [12] and Augustine et al. [3] works.

4 Conclusion

In this paper, a scheme of audio scrambling that uses a generalization of the proposal in [5] was presented. Our approach, ASGI, exploits the imitation ability of audio signals and their Gaussian behaviour. The scrambled audio signals look and sound like Gaussian noise signal and there is no trace of the original content. According to our tests, the values of SD are higher than other schemes based on cellular automata [3,12] with the advantage that the behaviour does not depend on initial conditions. On the other hand, we tested the ASGI scheme with different audio genres and it was found that classical music gives a better performance that other genres. However, ASGI works with different audio contents, such as speech, instrumental or a mixture of both.

Acknowledgment. This work is supported by the "Universidad Militar Nueva Granada - Vicerrectoría de Investigaciones" under the grant INV-ING-1910 of 2015.

References

1. Alwahbani, S.M., Bashier, E.: Speech scrambling based on chaotic maps and one time pad. In: 2013 International Conference on Computing, Electrical and Electronics Engineering (ICCEEE), pp. 128–133. IEEE (2013)
2. Augustine, N., George, S.N., Deepthi, P.P.: Compressive sensing based audio scrambling using arnold transform. In: Martínez Pérez, G., Thampi, S.M., Ko, R., Shu, L. (eds.) SNDS 2014. CCIS, vol. 420, pp. 172–183. Springer, Heidelberg (2014)
3. Augustine, N., George, S.N., Deepthi, P.: Sparse representation based audio scrambling using cellular automata. In: 2014 IEEE International Conference on Electronics, Computing and Communication Technologies (IEEE CONECCT), pp. 1–5. IEEE (2014)
4. Ballesteros L, D.M., Moreno A, J.M.: Speech scrambling based on imitation of a target speech signal with non-confidential content. Circ. Syst. Sig. Process. **33**, 3475–3498 (2014)
5. Ballesteros L, D.M., Renza, D., Camacho, S.: An unconditionally secure speech scrambling scheme based on an imitation process to a gaussian noise signal. J. Inf. Hiding Multimedia Sig. Process. **7**(2), 233–242 (2016)
6. Chen, G., Han, B.: An audio scrambling degree measure based on information criteria. In: 2010 2nd International Conference on Signal Processing Systems (ICSPS), vol. 1, pp. V1–181. IEEE (2010)
7. Da-hui, H., Zhi-guo, D.: An audio watermarking based on logistic map and m-sequence. Int. J. Digital Content Technol. Appl. **6**(1), 1–10 (2012)
8. Ghebleh, M., Kanso, A., Noura, H.: An image encryption scheme based on irregularly decimated chaotic maps. Signal Process. Image Commun. **29**(5), 618–627 (2014)
9. Kulkarni, N.S., Raman, B., Gupta, I.: Multimedia encryption: a brief overview. Recent advances in multimedia signal processing and communications. Springer, Heidelberg (2009)

10. Kwon, G.R., Wang, C., Lian, S., Hwang, S.S.: Advanced partial encryption using watermarking and scrambling in mp3. Multimedia Tools Appl. **59**(3), 885–895 (2012)
11. Li, H., Qin, Z., Shao, L., Zhang, S., Wang, B.: Variable dimension space audio scrambling algorithm against MP3 compression. In: Hua, A., Chang, S.-L. (eds.) ICA3PP 2009. LNCS, vol. 5574, pp. 866–876. Springer, Heidelberg (2009)
12. Madain, A., Dalhoum, A.L.A., Hiary, H., Ortega, A., Alfonseca, M.: Audio scrambling technique based on cellular automata. Multimedia Tools Appl. **71**(3), 1803–1822 (2014)
13. Murillo-Escobar, M., Cruz-Hernández, C., Abundiz-Pérez, F., López-Gutiérrez, R., Del Campo, O.A.: A rgb image encryption algorithm based on total plain image characteristics and chaos. Signal Process. **109**, 119–131 (2015)
14. Oo, T.T., Onoye, T.: Progressive audio scrambling via complete binary tree's traversal and wavelet transform. In: Asia-Pacific Signal and Information Processing Association, 2014 Annual Summit and Conference (APSIPA), pp. 1–7. IEEE (2014)
15. Oo, T.T., Onoye, T.: Progressive audio scrambling via wavelet transform. In: 2014 IEEE Asia Pacific Conference on Circuits and Systems (APCCAS), pp. 97–100. IEEE (2014)
16. Zhou, Y., Bao, L., Chen, C.P.: Image encryption using a new parametric switching chaotic system. Signal Process. **93**(11), 3039–3052 (2013)

A Dynamic Indoor Location Model for Smartphones Based on Magnetic Field: A Preliminary Approach

Carlos E. Galván-Tejada[1]([⊠]), Jorge I. Galván-Tejada[1],
José M. Celaya-Padilla[2], J. Rubén Delgado-Contreras[3],
Vanessa Alcalá-Ramírez[1], and Luis Octavio Solís-Sánchez[2]

[1] Unidad Académica de Ingeniería Eléctrica, Universidad Autónoma de Zacatecas,
Ave. Ramón López Velarde 801, 98064 Zacatecas, Zacatecas, Mexico
ericgalvan@uaz.edu.mx
[2] Laboratorio de Innovación y Desarrollo Tecnológico en Inteligencia
Artificial (LIDTIA), Universidad Autónoma de Zacatecas,
Ave. Ramón López Velarde 801, 98064 Zacatecas, Zacatecas, Mexico
[3] Instituto Tecnológico Superior Zacatecas Sur (ITZaS),
Tlaltenango, Zacatecas, Mexico

Abstract. Due to an increase interest for providing services based on user location, several indoor location approaches based on mobile devices have been proposed recently. This paper focuses on the use of a novel crowdsourcing approach for indoor location of a mobile device that uses social collaboration to improve the accuracy and magnetic field signal as information source using feature extraction and a deterministic method that allows us to include information from new users that improves the fitness of the model. Four phases were included in the methodology: Raw data collection, Data pre-process, Feature extraction and Social collaboration. An experiment was succesfully carried out to test the proposed methodology. On the whole, good results were obtained on computational cost, recalculation time and accuracy improvement.

Keywords: Indoor positioning · Crowdsourcing · Social collaboration · ILS

1 Introduction

Recently there has been increased interest for providing services based on user location. These services are characterized by integrating the location or position of a mobile device with other value-added information to a user [5]. However, to achieve this, there are several challenges, such as the proposal of a model of accurate Indoor Location System (ILS) based on mobile devices, which does not require a dedicated infrastructure and uses a minimal amount of computer resources.

© Springer International Publishing Switzerland 2016
J.F. Martínez-Trinidad et al. (Eds.): MCPR 2016, LNCS 9703, pp. 260–269, 2016.
DOI: 10.1007/978-3-319-39393-3_26

To tackle the mentioned challenges, several indoor location approaches based on mobile devices have been proposed, for instance, (i) *inertial based*: This approach is characterized by using the inertial sensors such as accelerometer and gyroscope that are used to estimate the location of the user. The accelerometer can be used to determine the changes in the user's position produced when an acceleration is detected in one or more axes. While the gyroscope can be used to detect the changes in the direction to improve the location estimation [10,13], (ii) *camera based systems*: This approach uses the camera of the mobile device to capture information from the user's location (e.g. an image, video, markers or codes). The captured information is then compared with the reference information that was previously collected [11,17]. The main disadvantage of this approach is that it requires high processing capacity and the accuracy of the system is reduced when the quality of captured information presents low resolution or motion blur, (iii) *mobile signal processing*: This approach uses the sensors embedded in a mobile device to detect, measure and capture signals that are emitted by other devices and then conveyed inside the indoor environment (e.g. Wi-fi, Bluetooth), as well as the natural signals that are commonly found in those indoor environments (e.g. magnetic-field, ambient sound). In these systems the position estimation is commonly performed through methods such as "fingerprinting". This latter is composed of two phases: training and position determination. Firstly, a map of the observed signal strength values measured at different locations is recorded during a training phase. Secondly, the signal strength values observed at a user's device are compared with the map values by using proximity matching algorithms, including [2,15], but not limited to k-Nearest Neighbors (k-NN) [12]. The main disadvantage of this approach is the calibration phase. The calibration phase consist in point-by-point mapping of a given indoor environment, measuring the magnitude and/or direction of a specific signal at each point, and then, using this signal map for location purposes, finding the most similar place in the signal map to the one detected at a given point.

In order to avoid the calibration phase at Indoor Location Systems (ILS) we proposed a novel approach called crowdsourcing, described as a new online distributed problem solving and production model in which networked people collaborate to complete a task [1,8,18]. When exploring this approach, several works has been proposed to tackle the indoor location and mapping issue, for instance CrowdInside [1] proposes a data collection using inertial sensor data that is generated by the user's motion, this raw data is then used to construct indoor maps. In other hand, CrowdMap [4] uses a combination of computational vision and other embedded sensors to reconstruct maps, they divide their proposal in 4 steps, including one that process video to find key-frames. Nevertheless, these approaches requires an expensive computational cost phase to integrate the new information that comes from new collaborative users. This paper focuses on the use of a crowdsourcing approach for indoor location of a mobile device that uses social collaboration to improve the accuracy and magnetic field signal as information source using feature extraction and a deterministic method that allows us to include information from new users that improves the fitness of the ILS.

2   Methodology

We modified the methodology proposed by Galvan-Tejada et al. [6] with the aim of using a crowdsourcing approach for indoor location of a mobile device. Although in essence the methodology is similar, feature selection methods based on filters were used and a new phase named "social collaboration" was added as is shown in Fig. 1. This modified methodology is described as follows:

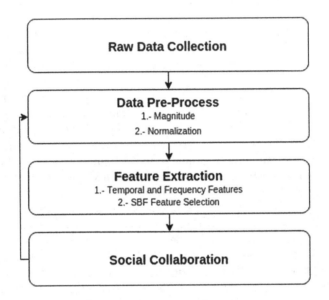

Fig. 1. Four phases proposed methodology.

2.1   Phase 1: Raw Data Collection

At this stage the magnetic field intensity of an indoor environment is captured and stored through a mobile application that uses the magnetometer of a smartphone. This sensor returns three float values, indicating three magnetic field axes in the environment when the lecture is done.

To capture magnetic field measures the user should walk in the indoor environment with the active application on the smartphone. The mobile device must be on hand at waist level and the walking must be done at average walk speed during 10 seconds. The data lectures were collected at a frequency of 100 lectures per second. We named *signatures* to this basic set of data that represent a room.

2.2   Phase 2: Data Pre-process

Once the raw data was collected, a pre-process is needed to get usable data without outliers and anomalies in the final data set. The pre-process is done following the next two steps:

Magnitude. Raw data from the magnetic field is composed by three components. For the sake of simplicity, a single data value of the magnitude was calculated using the Eq. 1.

$$|M| = \sqrt{M_x^2 + M_y^2 + M_z^2} \tag{1}$$

Magnitude allows us to avoid the constraint of smartphone position (i.e. screen position respect to the hand).

Normalization. To avoid the spatial scaling a Z norm normalization was applied to each light lecture signature using the Eq. 2, $z_{i,d}$ where $z_{i,d}$ is the normalized reading, $r_{i,d}$ refers to the ith observation of the signature in dimension d μ_d is the mean value of the signature for dimension d and σ_d is the standard deviation of the signature for dimension d.

$$\forall i \in m : z_{i,d} = \frac{r_{i,d} - \mu_d}{\sigma_d} \tag{2}$$

Equation 2 was applied for all dimensions in R^d.

Once initial magnetic signatures were collected in the rooms and stored in a data set created during the signature collection, feature extraction and selection were carried out to get the initial classification model that allows us to estimate the location of users.

2.3 Phase 3: Feature Extraction

The feature extraction is a process that consists in carrying out an efficient data reduction while preserving the appropriate amount of signal information. In this sense, the feature extraction step allow us to reduce the amount of data to develop the indoor location estimation model. To apply this step we must select the features that will be extracted from the signal, these activities are described below.

Signal Features. To select the signal features that describe the behavior of the magnetic field signal for ILS problem, we used the set features proposed by Galvan-Tejada et al. [6] because it includes first and second order statistical descriptors and Digital Signal Processing (DSP) which are well-known features. This set of features are appreciated in Table 1.

Temporal features were computed from the waveform of magnetic field signal that is generated from the raw data previously recorded, and *Spectral features* extracted from the signal after performing a P-point Fast Fourier Transform to each signature [16].

Feature Selection. To improve the model fitness and getting an indoor location model with less signal features a feature selection process was applied.

To perform a feature selection several techniques have been proposed, for instance filter, wrapper and embedded methods [14]. From these methods,

Table 1. Features extracted.

Features	Temporal Domain	Frequency Domain
Kurtosis	*	*
Mean	*	*
Median	*	*
Standard Deviation	*	*
Variance	*	*
Coefficient of Variation (CV)	*	*
Inverse CV	*	*
1,2,3 Quartile	*	*
1,5,95,99 Percentile		*
Trimmed Mean	*	*
Shannon Entropy		*
Slope		*
Spectral Flatness		*
1-10 Spectrum Components		*

we identified that filter methods have two advantages over other methods: (1) Returns features that can be used with different classifiers and (2) Less computationally expensive. From filter methods we selected the Selection By Filter (SBF) method that was used to the feature selection process. This is a cross validated feature selection strategy [9]. The SBF is set up to fit with a Random Forest function. Once the features are selected, these can be used with several classifiers to get a model to estimate the location, which is the core of an ILS.

2.4   Phase 4: Social Collaboration

To improve the accuracy of the ILS, signatures from different users and sensors must be added to the data set of signatures. Data from single users must be procesed with the first two phases and added to the data set of signatures before the phase 3 is carried on. This procedure has the aim to perform a feature extraction and selection from a bigger signatures data set, that includes more variability given the new signatures of different users with different devices.

3   Experiment

The experiment was carried out in a school building shown in Fig. 2. Experimental data is available in the research group website[1]. Thirteen (13) rooms were selected considering their distribution. Some spaces share magnetic field data in some parts of rooms (i.e. classrooms that were divided in two).

---

[1] http://ingsoftware.reduaz.mx/.

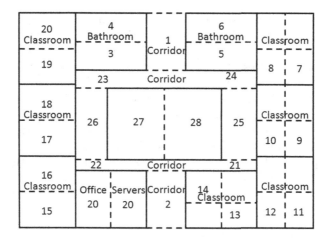

Fig. 2. Distribution plane of school building used in our experiment.

Even when our proposal is a methodology that can be improved adding new signatures, to begin the experimentation and get the initial model to the ILS, we use the Eq. 3 as is proposed by Galvan-Tejada et al., were x is the minimal number of signatures, and N is the number of variables used in the experiment. In our experiment N is equal 598 (13 rooms multiplied by 46 features each). We got a total of 10.22 magnetic signatures, finally uppered to 11.

$$x = log_2(N) + 1 \tag{3}$$

We choose a Motorola G2, with BOSH BMC150 magnetic sensor to record the initial magnetic data, using an Android application, developed with the aim to get magnetic sensor data. For the second user, we select a Sony Xperia 7 with MPC Magnetic Field sensors, and finally a third user with a Nexus 7. All sensors were selected to be different with the purpose of test the ILS in a common situation in where we have a diverse type of sensors and users.

After the raw data was collected, the pre-process step was applied using a script in R, which is a freely available language and environment for statistical computing and graphics[2], because it is multi-plataform and an environment for statistical computing.

After feature matrix is normalized, the SBF is applied using the CARET package in R [7]. The frame work used consist in five phases:

1. Data equalization and normalization.
2. Use the SBF to obtain their meaningful features.
3. Obtain a classification rate using the filtered model with Random Forest.
4. Obtain a classification rate using 46 selected features with Random Forest.
5. Calculate a Confusion Matrix for comparison.

---

[2] http://www-r-project.org/.

Once the SBF features are selected, a Random Forest classifier composed of 5,000 trees is used to evaluate indoor location model conformed by features selected by filter method. This classifier was proposed by Breiman et al. [3], and selected because it provides tree ensembles that depend on the values of a random feature vector and provides the same distribution to all the trees included in the forest. The decision of this classifier depends on the decision of several trees.

All classification process was performed using Random Forest R package.

In order to improve the fitness of the ILS we proceed to add signatures from the second and third user. The signatures following the pre-process and feature extraction steps. The data from the second user is added first to the data set of features generated from the first user, then, the model is recalculated. Finally, the features from the last user were added to the data set, and the model is recalculated.

4   Results

The SBF fitted model was a 23 feature model. In Table 2 those features included on the model are described.

Table 2. SBF Model Features.

Features	Temporal Domain	Frequency Domain
Kurtosis		
Mean	*	*
Median	*	
Standard Deviation	*	*
Variance	*	*
Coefficient of Variation (CV)	*	*
Inverse CV	*	*
1,2,3 Quartile		
95,99 Percentile	*	
1,5,95,99 Percentile		*
Trimmed Mean		*
Shannon Entropy		*
Slope		*
Spectral Flatness		*
2,8 Spectrum Components		*

After experimental test, the final filtered model from the first user data set, was a 23 feature model of classification, the obtained model can classify correctly 41.11 % of the samples. To improve the accuracy of the model, we added data from a second user and recalculated the model, this phase was done in less than a second given the deterministic approach to calculate a new model. After adding the data from the second user, the SBF model was the same 23 features model that was acquired with the data from a single user, however the accuracy was improved dramatically to 98.89 %. Finally, a third user was added to the data set; again the model was recalculated in less than a second even when the raw data was increased to 1,404,000 data points. After the SBF was applied, the same 23 features model was acquired and the accuracy was improved reaching a non error classification (100 %).

To verify the improvement using our proposal we tested a Random Forest (RF) using the 46 features extracted from the magnetic signal. This test was carried on in the 3 cases presented with the SBF approach. In Table 3 is shown the accuracy of a random forest model using 46 features against the 23 model acquired with the SBF selection. For all the cases, RF uses twice the information than the cases of the model acquired with SBF given the number of features needed to develop the model.

Table 3. Accuracy with different number of users.

Users	Random Forest Accuracy	SBF model Accuracy
One User	39.40	41.11
Two User	100	98.11
Three User	100	100

5   Concluding Remarks

In this paper we presented a novel approach for indoor location that uses social collaboration to improve the accuracy and magnetic field signal as information source.

The following important aspects regarding the proposal were identified:

- *Computational cost.* Compared against other ILS proposals, as the case of Genetic Algorithms (GA), the computational cost is lower given that wrapped methods avoid the stochasticity of other methods (as those based in a GA approach).
- *Recalculation Time.* Time became the most important characteristic to take into account when developing a crowdsourcing ILS because the recalibration step is very time consuming. However using a deterministic feature selection method (SBF) allows us to do this task in less than a second.
- *Accuracy Improvement:* Because a diversity applications of an ILS, is a must to consider that will be used by several users with different devices, walking behaviors and morphological characteristics and this implies that ILS must be able to estimate location even with those differences. In this paper we presented how adding new users considering the aforementioned differences, not just can be predicted; moreover, the model increase the accuracy of location estimation of the users.

Future work should take note the possibility of sampling a larger space and try to use different types of feature selection and ways to integrate new information (social collaboration) to the ILS. Additionally other tools should be used to improve the performance of mobile phones and tablets.

References

1. Alzantot, M., Youssef, M.: Crowdinside: automatic construction of indoor floorplans. In: Proceedings of the 20th International Conference on Advances in Geographic Information Systems, pp. 99–108. ACM (2012)
2. Bilke, A., Sieck, J.: Using the magnetic field for indoor localisation on a mobile phone. Progress in Location-Based Services. Springer, Heidelberg (2013)
3. Breiman, L.: Random forests. Mach. Learn. **45**(1), 5–32 (2001)
4. Chen, S., Li, M., Ren, K., Qiao, C.: Crowd map: Accurate reconstruction of indoor floor plans from crowdsourced sensor-rich videos. In: 2015 IEEE 35th International Conference on Distributed Computing Systems (ICDCS), pp. 1–10. IEEE (2015)
5. D'Roza, T., Bilchev, G.: An overview of location-based services. BT Technol. J. **21**(1), 20–27 (2003)
6. Galván-Tejada, C.E., García-Vázquez, J.P., Brena, R.F.: Magnetic field feature extraction and selection for indoor location estimation. Sensors **14**(6), 11001–11015 (2014)
7. Jed Wing, M.K.C., Weston, S., Williams, A., Keefer, C., Engelhardt, A., Cooper, T., Mayer, Z.: The R Core Team: caret: Classification and Regression Training (2014), r package version 6.0-24. http://CRAN.R-project.org/package=caret
8. Jiang, Y., Xiang, Y., Pan, X., Li, K., Lv, Q., Dick, R.P., Shang, L., Hannigan, M.: Hallway based automatic indoor floorplan construction using room fingerprints. In: Proceedings of the 2013 ACM International Joint Conference on Pervasive and Ubiquitous Computing, pp. 315–324. ACM (2013)
9. Kuhn, M.: Variable selection using the caret package (2012). URL: http://cran.cermin.lipi.go.id/web/packages/caret/vignettes/caretSelection.pdf
10. Li, F., Zhao, C., Ding, G., Gong, J., Liu, C., Zhao, F.: A reliable and accurate indoor localization method using phone inertial sensors. In: Proceedings of the 2012 ACM Conference on Ubiquitous Computing, pp. 421–430. ACM (2012)
11. Mulloni, A., Wagner, D., Barakonyi, I., Schmalstieg, D.: Indoor positioning and navigation with camera phones. Pervasive Comput. IEEE **8**(2), 22–31 (2009)
12. Ni, L.M., Liu, Y., Lau, Y.C., Patil, A.P.: Landmarc: indoor location sensing using active rfid. Wirel. Netw. **10**(6), 701–710 (2004)
13. Pratama, A., Widyawan, H.R.: Smartphone-based pedestrian dead reckoning as an indoor positioning system. In: 2012 International Conference on System Engineering and Technology (ICSET), pp. 1–6, Sept 2012
14. Saeys, Y., Inza, I., Larrañaga, P.: A review of feature selection techniques in bioinformatics. Bioinformatics **23**(19), 2507–2517 (2007)
15. Storms, W., Shockley, J., Raquet, J.: Magnetic field navigation in an indoor environment. In: Ubiquitous Positioning Indoor Navigation and Location Based Service (UPINLBS), pp. 1–10. IEEE (2010)
16. Tsai, W.H., Tu, Y.M., Ma, C.H.: An FFT-based fast melody comparison method for query-by-singing/humming systems. Pattern Recogn. Lett. **33**(16), 2285–2291 (2012). http://www.sciencedirect.com/science/article/pii/S016786551200284X

17. Werner, M., Kessel, M., Marouane, C.: Indoor positioning using smartphone camera. In: 2011 International Conference on Indoor Positioning and Indoor Navigation (IPIN), pp. 1–6. IEEE (2011)
18. Zhang, X., Yang, Z., Wu, C., Sun, W., Liu, Y., Liu, K.: Robust trajectory estimation for crowdsourcing-based mobile applications. Parallel Distrib. Syst. IEEE Trans. **25**(7), 1876–1885 (2014)

Using N-Grams of Quantized EEG Values for Happiness Detection

David Pinto[(⊠)], Darnes Vilariño, Illiana Morales, Cristina Aguilar, and Mireya Tovar

Faculty of Computer Science Language and Knowledge Engineering Lab, Benemérita Universidad Autonóma de Puebla, Puebla, Mexico
{dpinto,darnes,mtovar}@cs.buap.mx, illiana.mrls.t@gmail.com, crisaguilarc@hotmail.com
http://www.lke.buap.mx

Abstract. When applying classification methods for the automatic detection of happiness in human beings using electroencephalographic signals, the major research works in literature report the employment of power spectral density as the main feature. However, the aim of this paper is to explore wheter or not the use of N-grams of quantized EEG values as new features may help to improve the classification process. N-grams is a standard method of data representation in the area of natural language processing which usually reports good results. In this type of input data make sense to employ this kind of representation because the happiness signal is made up of a sequence of values which naturally matches the N-grams paradigm. The results obtained show that this kind of representation obtains better results than others reported in literature.

Keywords: EEG · N-grams · Happiness detection · Classification

1 Introduction

Among different emotional states the human being has, there is one defined by positive or pleasant emotion which ranges from content to intense joy named "happiness". The automatic detection of emotional states, in particular happiness, is the aim of this paper. Constructing a computational method able to recognize this emotional state in people surrounding us is an important part of human interaction, and also human-machine interaction. There have been different studies in literature for approaching the automatic detection of emotional states which can generally be categorized into three approaches, when the input signal is taken into account: (a) based on facial expression analysis, (b) based on electroencephalographic signals (EEG) and, (c) based on peripheral physiological signals [1]. The experiments carried out in this paper use EEG signals for

M. Tovar–This research work has been partially supported by PRODEP-SEP project (EXB-792) DSA/103.5/15/10854.

© Springer International Publishing Switzerland 2016
J.F. Martínez-Trinidad et al. (Eds.): MCPR 2016, LNCS 9703, pp. 270–279, 2016.
DOI: 10.1007/978-3-319-39393-3_27

detecting whether or not a human being experiments an emotion of happiness, therefore, this paper can be categorized into the second approach.

There is, however, other types of categories we may use for those papers reporting emotional states detection in literature. Some papers differ from others because of the type of features employed in the signal representation. Some authors, for example, use EEG signals as input data considering Power Spectral Density (PSD) as the main feature with accuracies around 75 % and 65 % [2].

In [3], the authors propose a novel feature named "functional connectivity" which can be used together with other features in the task of automatic identification of emotional states. The authors report an accuracy between 0.4 and 0.65 for the experiments carried out. Even if there are other works in literature reporting results for the detection of emotional states by employing EEG signals, to the best of our knowledge works employing N-grams for the representation of data nearly have been reported in literature.

The protocol employed for acquisition is an issue that some research papers emphasize, for example, in [4], the authors have designed an acquisition protocol based on the recall of past emotional life episodes to acquire data from both peripheral and EEG signals. They report the performance of several classifiers for distinguishing between the three areas of the valence-arousal space, corresponding to negatively excited, positively excited, and calm-neutral states. The same authors propose an approach for affective representation of movie scenes based on the emotions that are actually felt by spectators [5].

The remaining of this paper is structured as follows. In Sect. 2 we present the concepts related with electroencephalography, in particular, we describe the wave patterns employed as input signals. The methodology proposed in this paper is given in Sect. 3. Here we describe the dataset, the quantization process, the data representation model and the algorithms for automatic identification of happiness. The obtained results are discussed in Sect. 4. Finally, in Sect. 5 the conclusions are given.

2   Electroencephalograpy

EEG refers to the recording of the brain's spontaneous electrical activity over a period of time [6]. Usually, this recording is done by using multiple electrodes which are placed on the scalp of a human being. The EEG is typically described in terms of rhythmic activity, which is divided into bands by frequency. It has been noted that these frequency bands have certain biological significance and distribution over the scalp.

In general, waveforms are subdivided into bandwidths known as *alpha*, *beta*, *theta*, and *delta* to describe the majority of the EEG signals used in clinical practice [7]. In the following subsection we briefly describe each one of these waveforms.

2.1 Wave Patterns

Delta is the frequency range up to 4 Hz. It tends to be the highest in amplitude and the slowest waves. This waveform is usually most prominent frontally in adults, whereas it is most prominent posteriorly in children. It is said that this waveform is normally seen in adults when they are in slow wave sleep. It is also seen in babies.

Theta is the frequency range from 4 Hz to 7 Hz. It is associated with some reactions as drowsiness or arousal in older children and adults, but it could be also seen when the person is in meditation [8]. When this waveform is presented in excess, it may represent abnormal activity. However, this range can be also be associated with reports of relaxed, meditative and creative states.

Alpha is the frequency range from 7 Hz to 14 Hz. Hans Berger named the first rhythmic EEG activity he saw as the "alpha wave" [9]. This was the "posterior basic rhythm" seen in the posterior regions of the head on both sides, higher in amplitude on the dominant side. It emerges with closing of the eyes and with relaxation, and attenuates with eye opening or mental exertion. The alpha frequency range in young children is slower than 8 Hz and, therefore, it is technically in the theta range.

Beta is the frequency range from 15 Hz to about 30 Hz. It is associated to motor behavior and it is normally attenuated during active movements [10]. When this signal presents low amplitude with multiple and varying frequencies, it is frequently associated with active, busy or anxious thinking and active concentration. Rhythmic beta with a dominant set of frequencies is associated with various pathologies and drug effects. This is the dominant waveform in persons who are anxious or alert.

In Table 1 we show the frequency ranges for each one of the waveforms aforementioned.

Table 1. EEG Frequency bands

Band	Frequency (Hz)
Delta	< 4
Theta	= 4 and < 8
Alpha	= 8 and < 14
Beta	= 14

3 Methodology Proposed

In this section we describe the design cycle we have proposed for the analysis of EEG signals with the purpose of identifying whether a given sequence of waveforms expresses an emotion of happiness.

3.1   Dataset Construction

In order to collect a dataset for the experiments, we have created an ad-hoc software which is able to record the brain signal of a given person while he or she is observing a video containing scenes which trigger an emotion of happiness. The video has previously been annotated in order to determine which parts are associated to happiness and the parts of the video that are not. In this paper we are not interested in annotating other type of emotions, because we are only determining those sequences associated to happiness, whereas the remaining sequences are just annotated as unhappiness.

We employed three different videos with a length time of 4.2, 4.1 and 6.3 min, respectively. The three videos were shown to 20 different persons in order to construct the dataset. The brain signals of each person were parallelly recorded while they were observing the video. The adquisition software developed was able to automatically annotate the output signal with the corresponding tag associated to happiness emotion (H) or another tag associated to unhappiness emotion (N) for each one of the four waveforms: Delta, Theta, Alpha and Beta.

3.2   Quantization

Quantization is the process of mapping a large set of input values to a (countable) smaller set. This process may help to improve the performance of different tasks because all those similar values are grouped together in a single value which represent to all of them. The quantization process we have applied to the EEG signals are based on the mean and standard deviation values obtained from the EEG signals. In particular, we have obtained these two values (μ and σ) for each one of the four channels Delta, Theta, Alpha and Beta, and, therefore, the original signal may be grouped together into three different clusters, LOW, MEDIUM and HIGH, which may be better as discrete values used instead of the continuous ones. The limits of the clusters were found considered that the distribution of the waveforms for the person may be modeled as a normal distribution. In Table 2 we may see the ranges used for the quantization process.

Table 2. Quantization ranges

Discrete value	Symbol	Range
LOW	L	$\text{EEG\_Signal} < \mu - \sigma$
MEDIUM	M	$\mu - \sigma \leq \text{EEG\_Signal} < \mu + \sigma$
HIGH	H	$\text{EEG\_Signal} \geq \mu + \sigma$

Although, these quantization thresholds have performed well in the experiments carried out, we consider that more investigation needs to be done with respect to the task of determining the optimal thresholds, so as to analyze whether or not, another probabilistic distribution should be employed instead of the normal one.

In order to distinguish the four waveforms in the discretization process, we have prefixed each quantized EEG signal (represented with the symbols: L, M and H) with the first letter of the waveforms, i.e., we used the letter "A" for Alpha, "D" for Delta and so on. In Table 3 we may observe a sample of the quantized sequences obtained. Each value of the sequence represents one second of measure of the person human brain. Each sequence has been annotated with a tag indicating whether or not the sequence is associated with an emotion of happiness.

Table 3. Sample of quantized sequence of EEG signals. The H tag means "Happiness", whereas the U tag means "Unhappiness"

Sequence of quantized EEG signals	Tag
AL AL AL AL AL AL AL AL	U
DL DL DL DL DM DL DL DL	U
TL TL TL TL TL TL TL TL	U
BL BL BL BL BL BL BL BL	U
AL AL AM	H
DL DM DM	H
TL TM TM	H
BL BL BM	H
AL AL AL AL AL AL AL AL	U
DL DL DL DL DL DL DL DL	U
TL TL TL TL TL TL TL TL	U
BL BL BL BL BL BL BL BL	U
AM AH AL AM AL AL AL	H
DL DM DL DL DL DL DL	H
TM TM TL TL TL TL TL	H
BM BL BL BL BL BL BL	H
AL AL AL AL AH AM	U
DL DL DL DL DM DM	U
TL TL TL TL TM TM	U
BL BL BL BL BM BM	U

3.3 Data Representation

The quantization process previously applied allow us to have sequences of quantized values associated to a given tag. In the particular area of natural language processing, we describe this sequence of values as a string. Each string needs to be correctly represented in order to apply machine learning methods. The feature extraction process is then applied by using the N-grams representation technique. Each string is split out into sequences of $n = 2, 3$ values, calculating the frequency of each N-gram. Since, the frequency is not sufficient for determining the degree of discrimination each N-gram has, we have employed a technique of term weighting known as TF-IDF in which each N-gram is weighted in terms of its frequency in the string and also proportional to the inverse document frequency (the number of strings containing the N-gram). The complete dataset is made up of documents manually annotated and represented by N-grams weighted with the TF-IDF schema.

3.4 Classification Model

The supervised machine learning techniques are able to learn the human process of identifying emotions based on features fed in the classifier by means of the manually annotated corpus.

We have selected one learning algorithm from four different types of classifiers: Bayes, Lazy, Functions and Trees in order to investigate the one that performs better in the particular task of automatic identification of happiness. The following four learning algorithms were chosen:

NaïveBayes: This is the standard probabilistic Naïve Bayes classifier.

K-Star: This is the k-nearest neighbor classifier with a generalized distance function.

SMO: This is a sequential minimal optimization algorithm for support vector classification.

J48: This is the C4.5 decision tree learner which implements the revision 8 of C4.5.

In the following subsection we describe the measures we employed for evaluating the performance of the experiments carried out.

3.5 Evaluation Measures

In order to evaluate the quality of the results obtained, we have used the following standard measures for the evaluation: Precision, Recall and F-Measure [11].

The measures employed make use of a set of values calculated when the classification process is carried out. The terms "true positive - (TP)", "true negative - (TN)", "false positive - (FP)", and "false negative - (FN)" compare the results obtained by the classifier under test set with a given gold standard wich is usually obtained by external judgments (manual annotated data). The terms "positive" and "negative" refer to the classifier's prediction, and the terms "true" and "false" refer to whether that prediction corresponds to the external judgment.

Thus, the Precision and Recall is calculated as follows:

$$Precision = \frac{TP}{TP + FP}$$

$$Recall = \frac{TP}{TP + FN}$$

The F-Measure combines Precision and Recall as the harmonic mean of these two values. The traditional F-measure or balanced F-score is calculated as follows:

$$F - Measure = \frac{Precision * Recall}{Precision + Recall}$$

In this paper we also use the Accuracy measure for reporting the results obtained. This value is calculated as follows:

$$Accuracy = \frac{TP + TN}{TP + TN + FP + FN}$$

4   Experimental Results

In this section we are presenting the accuracy obtained by each classifier when attempting to identify whether or not a sequence of EEG signals correspond to a signal of human happiness.

Tables 4, 5, 6 and 7, show the detailed accuracy by class using the KStar, Naïve Bayes, SMO and J48 supervised classifiers, respectively. As can be seen, in all the cases the identification of Class 1 (when the set of N-grams of quantized EEG signals is associated to a human emotion of happiness) obtained a better performance than the identification of Class 2 (when the set of features does not correspond to a human emotion of happiness). Even if the difference is not so significant, this issue is important. As future work, we need to provide better features for improving the results obtained. On the one hand, the KStar classifier was the one that obtained the worst results with a weighted average F-Measure of 0.745. On the other hand, the other three classifiers obtained similar results with a weighted average F-Measure of around 0.79.

Table 4. Detailed accuracy by class using the KStar classifier

Class	Precision	Recall	F-Measure
Class 1 (Happiness)	0.667	0.786	0.721
Class 2 (¬Happiness)	0.760	0.633	0.691
Weighted Avg	0.715	0.707	0.745

Table 5. Detailed accuracy by class using the Naïve Bayes classifier

Class	Precision	Recall	F-Measure
Class 1 (Happiness)	0.742	0.875	0.803
Class 2 (¬Happiness)	0.860	0.717	0.782
Weighted Avg	0.803	0.793	0.792

Table 6. Detailed accuracy by class using the SMO classifier

Class	Precision	Recall	F-Measure
Class 1 (Happiness)	0.735	0.893	0.806
Class 2 (¬Happiness)	0.875	0.700	0.778
Weighted Avg	0.808	0.793	0.792

Table 7. Detailed accuracy by class using the J48 classifier

Class	Precision	Recall	F-Measure
Class 1 (Happiness)	0.732	0.929	0.819
Class 2 (¬Happiness)	0.911	0.683	0.781
Weighted Avg	0.825	0.802	0.799

Table 8. Percentage of correctly vs. incorrectly instances classified

Classifier	Type	Correct (%)	Incorrect (%)
K-Star	Lazy	70.69	29.31
Naïve Bayes	Bayes	79.31	20.68
SMO	Functions	79.31	20.68
J48	Trees	80.17	19.82

In Table 8 we show the percentage of instances classified correctly and incorrectly. Basically, this table summarize the weighted average results of the previously shown result tables. Actually, we have included a graph (see Fig. 1) with the aim of showing the summarized average results.

As can be seen, the J48 classifier performed similar to SMO and Naïve Bayes, with an accuracy of 80.17 %. Also, it is noticeable that these three classifiers outperformed the K-Star classifier.

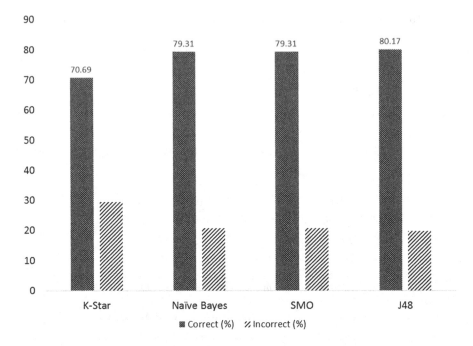

Fig. 1. Comparison among the different classifiers employed in the experiments

5   Conclusions

In this paper we have presented a novel representation based on N-grams of quantized EEG values which obtained good results for the particular task of automatic happiness detection. The obtained accuracy results are up to 80 %, and the F-measure is about 0.8. Unfortunately, the obtained results can not be directly compared with those previously reported in the literature because we are not using the same datasets, nor the same type of sensors. A fair comparison among different methods should take into account that the experiments should be executed in very similar conditions.

The employment of N-grams is capturing information about a sequence of signals with acceptable performance, however, we consider we could improve the results obtained up to now, by proposing a different quantization method or even better, by increasing the number of samples used in the training phase.

In this paper we have considered that the data are distributed according to the normal distribution, a very common continuous probability distribution. However, we need still to investigate whether or not this assumption is correct for the quantization process. Instead, we could employ another probability distribution such as the gamma or the poison one.

References

1. Wang, X.-W., Dan Nie, B.L.L.: Emotional state classification from EEG data using machine learning approach. Neurocomputing **129**(1), 94–106 (2014)
2. Jatupaiboon, N., Pan-ngum, S., Israsena, P.: Real-time EEG-based happiness detection system. Sci. World J. **2013**, 52–61 (2013)
3. Lee, Y.-Y., Hsieh, S.: Classifying different emotional states by means of EEG-basedfunctional connectivity patterns. PLoS ONE **9**(4), e95415 (2014)
4. Chanel, G., Kierkels, J.J., Soleymani, M., Pun, T.: Short-term emotion assessment in a recall paradigm. Int. J. Hum. Comput. Stud. **67**(8), 607–627 (2009)
5. Soleymani, M., Chanel, G., Kierkels, J.J.M., Pun, T.: Affective characterization of movie scenes based on content analysis and physiological changes. Int. J. Semant. Comput. **3**(2), 235–254 (2009)
6. Niedermeyer, E., da Silva, F.L.: Electroencephalography: Basic Principles, Clinical Applications, and Related Fields, 5th edn. Lippincott Williams & Wilkins, Baltimore (2004)
7. Tatum, W.O.: Ellen r. grass lecture: Extraordinary EEG. Neurodiagnostic J. **54**, 3–21 (2014)
8. Cahn, B.R., Polich, J.: Meditation states and traits: EEG, ERP, and neuroimaging studies. Psychol. Bull. **132**(2), 180–211 (2006)
9. Millet, D.: The origins of EEG. In: Seventh Annual Meeting of the International Society for the History of the Neurosciences (ISHN), Los Angeles, California, USA Department of Neurology, UCLA Medical Center (2004)
10. Pfurtscheller, G., de Lopes, F.H.: Event-related EEG/MEG synchronization and desynchronization:basic principles. Clin. Neurophysiol. **110**(11), 1842–1857 (1999)
11. Fawcett, T.: ROC graphs: Notes and practical considerations for researchers. Technical report, HP Labs (2004)

LSTM Deep Neural Networks Postfiltering for Improving the Quality of Synthetic Voices

Marvin Coto-Jiménez[1,2(✉)] and John Goddard-Close[2]

[1] University of Costa Rica, San José, Costa Rica
marvin.coto@ucr.ac.cr
[2] Autonomous Metropolitan University, Mexico, DF, Mexico
jgc@xanum.uam.mx

Abstract. Recent developments in speech synthesis have produced systems capable of providing intelligible speech, and researchers now strive to create models that more accurately mimic human voices. One such development is the incorporation of multiple linguistic styles in various languages and accents. HMM-based speech synthesis is of great interest to researchers, due to its ability to produce sophisticated features with a small footprint. Despite such progress, its quality has not yet reached the level of the current predominant unit-selection approaches, that select and concatenate recordings of real speech. Recent efforts have been made in the direction of improving HMM-based systems. In this paper, we present the application of long short-term memory deep neural networks as a postfiltering step in HMM-based speech synthesis. Our motivation stems from a desire to obtain spectral characteristics closer to those of natural speech. The results described in the paper indicate that HMM-voices can be improved using this approach.

Keywords: LSTM · HMM · Speech synthesis · Statistical parametric speech synthesis · Postfiltering · Deep learning

1 Introduction

Text-to-speech synthesis (TTS) is the technique of generating intelligible speech from a given text. Applications of TTS have grown from early systems which aid the visually impaired, to in-car navigation systems, e-book readers, spoken dialog systems, communicative robots, singing speech synthesizers, and speech-to-speech translation systems [1].

More recently, TTS systems have moved from the task of producing intelligible voices, to the more difficult challenge of generating voices in multiple languages, with different styles and emotions [2]. Despite these trends, there are unresolved obstacles, such as improving the overall quality of the voices. Some researchers are striving to create TTS systems which try to mimic natural human voices more closely.

The statistical methods for TTS, which arose in the late 1990s, have grown in popularity [3], particularly those based on Hidden Markov Models (HMMs).

© Springer International Publishing Switzerland 2016
J.F. Martínez-Trinidad et al. (Eds.): MCPR 2016, LNCS 9703, pp. 280–289, 2016.
DOI: 10.1007/978-3-319-39393-3_28

HMMs are known for their flexibility in changing speaker characteristics, having a low footprint, and their capacity to produce average voices. Previously, HMMs were utilized extensively in the inverse task to TTS of speech recognition. Here they have proved to be successful at providing a robust representation of the main events into which speech can be segmented [4], using efficient parameter estimation algorithms.

More than twenty statistical speech synthesis implementations have been developed for several different languages from around the world. For example [5–16], are a few of the recent publications. Every implementation of a new language, or one of it's dialects, requires the adaptation of HMM-related algorithms by incorporating their own linguistic specifications, and making a series of decisions regarding the type of HMM, decision trees, and training conditions.

In this paper, we present our implementation of a statistical parametric speech synthesis system based on HMM, together with the use of long short-term memory postfilter neural networks for improving its spectral quality.

The rest of this paper is organized as follows: Sect. 2 provides some details of an HMM-based speech synthesis system and in Sect. 3, long short-term memory neural networks are briefly described. Section 4 gives the proposed system and the experiments carried out in order to test the postfilter. Section 5 presents and discusses the results and objective evaluations conducted, and finally, some conclusions are given in Sect. 6.

2   Speech Synthesis Based on HMM

An HMM is a Markov process with unobserved or hidden states. The states themselves emit observations according to certain probability distributions.

In Fig. 1, a representation of a left-to-right HMM is shown, where there is a first state to the left from which transitions can occur to the same state or to the next one on the right, but not in the reverse direction. In this p_{ij} represents the probability of transition from state i to state j, and O_k represents the observation emitted in state k.

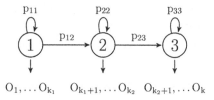

Fig. 1. Left to right example of an HMM with three states

In HMM-based speech synthesis, the speech waveforms can be reasonably reconstructed from a sequence of acoustic parameters learnt and emitted as vectors from the HMM states [1]. Typical implementation of this model includes vectors of observations comprising of the pitch, $f0$, the mel frequency cepstral coefficients, MFCC and their delta and delta features, for an adequate modeling of the dynamic features of speech. A common tool used to build these HMM-based speech systems is known as HTS [17], which we also use in this paper.

In order to improve the quality of the results, some researchers have recently experimented with postfiltering stages, in which the parameters obtained from HTS voices have been enhanced using deep generative architectures [18–21], for example restricted boltzmann machines, deep belief networks, bidirectional associative memories, and recurrent neural networks (RNN).

In the next section, we present our proposal to incorporate long short-term memory recurrent neural networks in order to improve the quality of HMM-based speech synthesis.

3   Long Short-Term Memory Recurrent Neural Networks

Among the many new algorithms developed to improve some tasks related to speech, such as speech recognition, several groups of researchers have experimented with the use of Deep Neural Networks (DNN), giving encouraging results. Deep learning, based on several kinds of neural networks with many hidden layers, have achieved interesting results in many machine learning and pattern recognition problems. The disadvantage of using such networks is they cannot directly model the dependent nature of each sequence of parameters with the former, something which is desirable in order to imitate human speech production. It has been suggested that one way to solve this problem is to include RNN [22, 23] in which there is feedback from some of the neurons in the network, backwards or to themselves, forming a kind of memory that retains information about previous states.

An extended kind of RNN, which can store information over long or short time intervals, has been presented in [24], and is called long short-term memory (LSTM). LSTM was recently successfully used in speech recognition, giving the lowest recorded error rates on the TIMIT database [25], as well as in other applications to speech recognition [26]. The storage and use of long-term and short-term information is potentially significant for many applications, including speech processing, non-Markovian control, and music composition [24].

In a RNN, output vector sequences $\mathbf{y} = (y_1, y_2, \ldots, y_T)$ are computed from input vector sequences $\mathbf{x} = (x_1, x_2, \ldots, x_T)$ and hidden vector sequences $\mathbf{h} = (h_1, h_2, \ldots, h_T)$ iterating Eqs. 1 and 2 from 1 to T [22]:

$$h_t = \mathcal{H}\left(\mathbf{W}_{xh}x_t + \mathbf{W}_{hh}h_{t-1} + b_h\right) \tag{1}$$

$$y_t = \mathbf{W}_{hy}h_t + b_y \tag{2}$$

where $\mathbf{W}_{ij}$ is the weight matrix between layer i and j, b_k is the bias vector for layer k and $\mathcal{H}$ is the activation function for hidden nodes, usually a sigmoid function $f : \mathbb{R} \to \mathbb{R}, f(t) = \frac{1}{1+e^{-t}}$.

Each cell in the hidden layers of a LSTM, has some extra gates to store values: an input gate, forget gate, output gate and cell activation, so values can be stored in the long or short term. These gates are implemented following the equations:

$$i_t = \sigma \left(\mathbf{W}_{xi} x_t + \mathbf{W}_{hi} h_{t-1} + \mathbf{W}_{ci} c_{t-1} + b_i \right) \tag{3}$$

$$f_t = \sigma \left(\mathbf{W}_{xf} x_t + \mathbf{W}_{hf} h_{t-1} + \mathbf{W}_{cf} c_{t-1} + b_f \right) \tag{4}$$

$$c_t = f_t c_{t-1} + i_t \tanh \left(\mathbf{W}_{xc} x_t + \mathbf{W}_{hc} h_{t-1} + b_c \right) \tag{5}$$

$$o_t = \sigma \left(\mathbf{W}_{xo} x_t + \mathbf{W}_{ho} h_{t-1} + \mathbf{W}_{co} c_t + b_o \right) \tag{6}$$

$$h_t = i_t \tanh \left(c_t \right) \tag{7}$$

where σ is the sigmoid function, i is the input gate activation vector, f the forget gate activation function, o is the output gate activation function, and c the cell activation function. $\mathbf{W}_{mn}$ are the weight matrices from each cell to gate vector.

4 Description of the System

Often, the resulting voices from the HTS system have notable differences with the original voices used in their creation. It is possible to reduce the gap between natural and artificial voices by additional learning directly applied to the data [18]. In our proposal, we use aligned utterances from natural and synthetic voices produced by the HTS system to establish a correspondence between each frame.

Given a sentence spoken using natural speech and also with the voice produced by the HTS, we extract a representation consisting of one coefficient for f0, one coefficient for energy, and 39 MFCC coefficients, using the system Ahocoder [27]. The inputs to the LSTM network correspond to the MFCC parameters of each frame for the sentences spoken using the HTS voice, while the output corresponds to the MFCC parameters given by the natural voice for the same sentence. In this way, we have an exact correspondence given by the alignment between the vectors from each utterance using the HTS voice and the natural voice.

Hence, each LSTM network attempts to solve the regression problem of transforming the values of the speech produced by the artificial and natural voices. This allows a further improvement to the quality of newly synthesized utterances with HTS, and uses the network as a way of refining these synthetic parameters to more closely resemble those of a natural voice. Figure 2 outlines the proposed system.

4.1 Corpus Description

The CMU_Arctic databases were constructed at the Language Technologies Institute at Carnegie Mellon University. They are phonetically balanced, with several US English speakers. It was designed for unit selection speech synthesis research.

The databases consist of around 1150 utterances selected from out-of-copyright texts from Project Gutenberg. The databases include US English male and female speakers. A detailed report on the structure and content of the database and the recording conditions is available in the Language Technologies Institute Tech Report CMU-LTI-03-177 [28]. Four of the available voices were selected: BDL (male), CLB (female), RMS (male) and SLT (female).

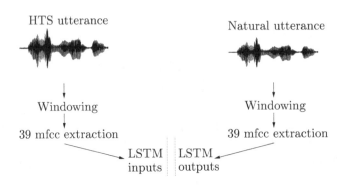

Fig. 2. Proposed system. HTS and Natural utterances are aligned frame by frame

4.2 Experiments

Each voice was parameterized, and the resulting set of vectors was divided into training, validation, and testing sets. The amount of data available for each voice is shown in Table 1. Despite all voices uttering the same phrases, the length differences are due to variations in the speech rate of each speaker.

Table 1. Amount of data (vectors) available for each voice in the databases

Database	Total	Train	Validation	Test
BDL	676554	473588	135311	67655
SLT	677970	474579	135594	67797
CLB	769161	538413	153832	76916
RMS	793067	555147	158613	79307

The LSTM networks for each voice had three hidden layers, with 200, 160 and 200 units in each one respectively.

To determine the improvement in the quality of the synthetic voices, several objective measures were used. These measures have been applied in recent speech synthesis experiments and were found to be reliable in measuring the quality of synthesized voices [29,30]:

- Mel Cepstral Distortion (MCD): Excluding silent phonemes, between two waveforms v^{targ} and v^{ref} it can be measured following Eq. 8 [31]

$$\text{MCD}\left(v^{\text{targ}}, v^{\text{ref}}\right) = \frac{\alpha}{T} \sum_{t=0}^{T-1} \sqrt{\sum_{d=s}^{D} \left(v_d^{\text{targ}}(t) - v_d^{\text{ref}}(t)\right)^2} \qquad (8)$$

where $\alpha = \frac{10\sqrt{2}}{\ln 10}$, T is the number of frames of each utterance, and D the total number of parameters of each vector.

– mfcc trajectory and spectrogram visualization: Observation of these figures allow a simple visual comparison between the similitude of the synthesized and natural voices.

These measures were applied to the test set after being processed with the LSTM networks, and the results were compared with those of the HTS voices. The results and analysis are shown in the following section.

5   Results and Analysis

For each synthesized voice produced with HTS and processed with LSTM networks, MCD results are shown in Table 2. It can be seen how this parameter improved when all voices were processed with LSTM networks.

This shows the ability of these networks to learn the particular regression problem of each voice.

Table 2. MCD between HTS and natural voices, and between LSTM postfiltering and natural voices

Database	HTS to natural	LSTM-pf to natural
BDL	8.46	7.98
CLB	7.46	6.87
SLT	7.03	6.65
RMS	7.66	7.60

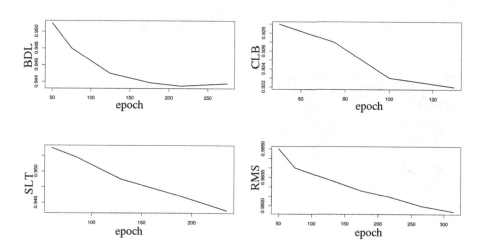

Fig. 3. Evolution of MCD improvement in LSTM postfiltering during training epochs

The best result of MCD improvement with the LSTM postfiltering is CLB (11.2 %) and the least best was RMS (1 %). Figure 3 shows how MCD evolves with the training epochs for each voice. All HTS voices, except one, were improved by the LSTM neural network postfilter for MCD after the first 50 epochs of training.

The differences in the number of epochs required to reach convergence in each case are notable. This can be explained by the difference in MCD between HTS

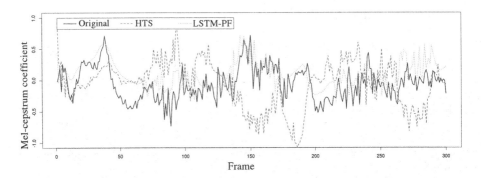

Fig. 4. Illustration of enhancing the 5th mel-cepstral coefficient trajectory by LSTM postfiltering

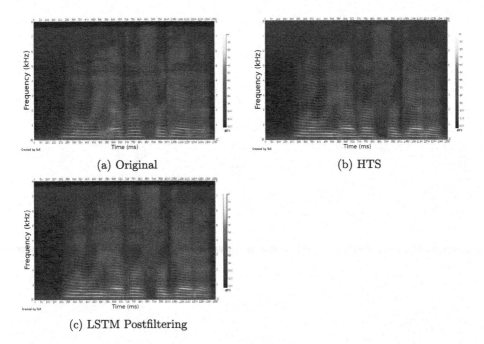

(a) Original   (b) HTS

(c) LSTM Postfiltering

Fig. 5. Comparison of spectrograms

and natural voices. The gap between them is variable and the LSTM network requires more epochs to model the regression function between them.

An example of the parameters generated by the HTS and the enhancement pursuit by the LSTM postfilter is shown in Fig. 4. It can be seen how the LSTM postfilter fits the trajectory of the mfcc better than the HTS base system.

In Fig. 5 a comparison of three spectrograms of the utterance "Will we ever forget it?" for the voices of: (a) Original (b) HTS and (c) LSTM postfilter enhanced, is shown. The HTS spectrogram usually shows bands in higher frequencies not present in the natural voice, and the LSTM postfilter helps to smooth it, making it closer to the spectrogram of the original voice.

6   Conclusions

We have presented a new proposal to improve the quality of synthetic voices based on HMM with LSTM networks. The method shows how to improve an artificial voice and make it mimic more closely the original natural voice in terms of its spectral characteristics.

We evaluated the proposed LSTM postfilter using four voices, two masculine and two feminine, and the results show that all of them were improved for spectral features, such as MCD measurement, spectrograms, and mfcc trajectory generation.

The improvement of the HTS voices in MCD to the original voices were observed from the first training epochs of the LSTM neural network, but the convergence to a minimum distance took many more epochs. Due to the extensive amount of time required to train each epoch, further exploration should determine new network configurations or training conditions to reduce training time.

Future work will include the exploration of new representations of speech signals, hybrid neural networks, and fundamental frequency enhancement with LSTM postfilters.

Acknowledgements. This work was supported by the SEP and CONACyT under the Program SEP-CONACyT, CB-2012-01, No.182432, in Mexico, as well as the University of Costa Rica in Costa Rica.

References

1. Tokuda, K., Nankaku, Y., Toda, T., Zen, H., Yamagishi, J., Oura, K.: Speech synthesis based on hidden markov models. Proc. IEEE **101**(5), 1234–1252 (2013)
2. Black, A.W.: Unit selection and emotional speech. In: Interspeech (2003)
3. Yoshimura, T., Tokuda, T., Masuko, T., Kobayashi, T., Kitamura, T.: Simultaneous modeling of spectrum, pitch and duration in HMM-based speech synthesis. In: Proceedings of the Eurospeech, pp. 2347–2350 (1999)
4. Falaschi, A., Giustiniani, M., Verola, M.: A hidden markov model approach to speech synthesis. In: Proceedings of the Eurospeech, pp. 2187–2190 (1989)

5. Karabetsos, S., Tsiakoulis, P., Chalamandaris, A., Raptis, S.: HMM-based speech synthesis for the greek language. In: Sojka, P., Horák, A., Kopeček, I., Pala, K. (eds.) TSD 2008. LNCS (LNAI), vol. 5246, pp. 349–356. Springer, Heidelberg (2008)
6. Pucher, M., Schabus, D., Yamagishi, Y., Neubarth, F., Strom, V.: Modeling and interpolation of austrian german and viennese dialect in HMM-based speech synthesis. Speech Commun. 52(2), 164–179 (2010)
7. Erro, D., Sainz, I., Luengo, I., Odriozola, I., Sánchez, J., Saratxaga, I., Navas, E., Hernáez, I.: HMM-based speech synthesis in basque language using HTS. In: Proceedings of the FALA (2010)
8. Stan, A., Yamagishi, Y., King, S., Aylett, M.: The romanian speech synthesis (RSS) corpus: building a high quality HMM-based speech synthesis system using a high sampling rate. Speech Commun. 53(3), 442–450 (2011)
9. Kuczmarski, T.: HMM-based speech synthesis applied to polish. Speech Lang. Technol. 12, 13 (2010)
10. Hanzlíček, Z.: Czech HMM-based speech synthesis. In: Sojka, P., Horák, A., Kopeček, I., Pala, K. (eds.) TSD 2010. LNCS, vol. 6231, pp. 291–298. Springer, Heidelberg (2010)
11. Li, Y., Pan, S., Tao, J.: HMM-based speech synthesis with a flexible mandarin stress adaptation model. In: Proceedings of the 10th ICSP2010 Proceedings, Beijing, pp. 625–628 (2010)
12. Phan, S.T., Vu, T.T., Duong, C.T., Luong, M.C.: A study in vietnamese statistical parametric speech synthesis based on HMM. Int. J. 2(1), 1–6 (2013)
13. Boothalingam, R., Sherlin, S.V., Gladston, A.R., Christina, S.L., Vijayalakshmi, P., Thangavelu, N., Murthy, H.A.: Development and evaluation of unit selection and HMM-based speech synthesis systems for Tamil. In: National Conference on Communications (NCC), pp. 1–5. IEEE (2013)
14. Khalil, K.M., Adnan, C.: Implementation of speech synthesis based on HMM using PADAS database. In: 12th International Multi-Conference on Systems, Signals & Devices (SSD), pp. 1–6. IEEE (2015)
15. Nakamura, K., Oura, K., Nankaku, Y., Tokuda, K.: HMM-based singing voice synthesis and its application to japanese and english. In: IEEE International Conference on Acoustics, Speech and Signal Processing (ICASSP), pp. 265–269 (2014)
16. Roekhaut, S., Brognaux, S., Beaufort, R., Dutoit, T.: Elite-HTS: a NLP tool for French HMM-based speech synthesis. In: Interspeech, pp. 2136–2137 (2014)
17. HMM-based Speech Synthesis System (HTS). http://hts.sp.nitech.ac.jp/
18. Chen, L.H., Raitio, T., Valentini-Botinhao, C., Ling, Z.H., Yamagishi, J.: A deep generative architecture for postfiltering in statistical parametric speech synthesis. IEEE/ACM Trans. Audio, Speech Lang. Process. (TASLP) 23(11), 2003–2014 (2015)
19. Takamichi, S., Toda, T., Neubig, G., Sakti, S., Nakamura, S.: A postfilter to modify the modulation spectrum in HMM-based speech synthesis. In: IEEE International Conference on Acoustics, Speech and Signal Processing (ICASSP), pp. 290–294 (2014)
20. Takamichi, S., Toda, T., Black, A.W., Nakamura, S.: Modified post-filter to recover modulation spectrum for HMM-based speech synthesis. In: IEEE Global Conference on Signal and Information Processing (GlobalSIP), pp. 547–551 (2014)
21. Prasanna, K.M., Black, A.W.: Recurrent Neural Network Postfilters for Statistical Parametric Speech Synthesis. arXiv preprint (2016). arXiv:1601.07215
22. Fan, Y., Qian, Y., Xie, F.L., Soong, F.K.: TTS synthesis with bidirectional LSTM based recurrent neural networks. In: Interspeech, pp. 1964–1968 (2014)

23. Zen, H., Sak, H.: Unidirectional long short-term memory recurrent neural network with recurrent output layer for low-latency speech synthesis. In: IEEE International Conference on Acoustics, Speech and Signal Processing (ICASSP), pp. 4470–4474 (2015)

24. Hochreiter, S., Schmidhuber, J.: Long short-term memory. Neural Comput. 9(8), 1735–1780 (1997)

25. Graves, A., Jaitly, N., Mohamed, A.: Hybrid speech recognition with deep bidirectional LSTM. In: IEEE Workshop on Automatic Speech Recognition and Understanding (ASRU) (2013)

26. Graves, A., Fernández, S., Schmidhuber, J.: Bidirectional LSTM networks for improved phoneme classification and recognition. In: Duch, W., Kacprzyk, J., Oja, E., Zadrożny, S. (eds.) ICANN 2005. LNCS, vol. 3697, pp. 799–804. Springer, Heidelberg (2005)

27. Erro, D., Sainz, I., Navas, E., Hernaez, I.: Improved HNM-based vocoder for statistical synthesizers. In: InterSpeech, pp. 1809–1812 (2011)

28. Kominek, J., Black, A.W.: The CMU Arctic speech databases. In: Fifth ISCA Workshop on Speech Synthesis (2004)

29. Zen, H., Senior, A., Schuster, M.: Statistical parametric speech synthesis using deep neural networks. In: IEEE International Conference on Acoustics, Speech and Signal Processing (ICASSP) (2013)

30. Zen, H., Senior, A.: Deep mixture density networks for acoustic modeling in statistical parametric speech synthesis. In: IEEE International Conference on Acoustics, Speech and Signal Processing (ICASSP) (2014)

31. Kominek, J., Schultz, T., Black, A.W.: Synthesizer voice quality of new languages calibrated with mean mel cepstral distortion. In: SLTU (2008)

Applications of Pattern Recognition

Detecting Pneumatic Failures on Temporary Immersion Bioreactors

Octavio Loyola-González[1,2(✉)], José Fco. Martínez-Trinidad[1],
Jesús A. Carrasco-Ochoa[1], Dayton Hernández-Tamayo[2],
and Milton García-Borroto[3]

[1] Instituto Nacional de Astrofísica, Óptica y Electrónica,
Luis Enrique Erro No. 1, Sta. María Tonanzintla, 72840 Puebla, Mexico
{octavioloyola,fmartine,ariel}@inaoep.mx
[2] Centro de Bioplantas, Universidad de Ciego de Ávila,
Carretera a Morón km 9, 69450 Ciego de Ávila, Cuba
{octavioloyola,dayton}@bioplantas.cu
[3] Instituto Superior Politécnico José Antonio Echeverría,
Calle 114 No. 11901, 19390 Marianao, La Habana, Cuba
mgarciab@ceis.cujae.edu.cu

Abstract. Temporary immersion bioreactors are an effective procedure
to increase plant multiplication rates. The pneumatic system is an important part of a bioreactor, which should be controlled to guarantee both
the efficiency and efficacy in the system. Therefore, bioreactors have
been automated using a pneumatic drive to execute the immersion time.
Sometimes, the pneumatic system presents failures which can affect the
plant quality; therefore, pneumatic failure detection is an important task.
Since failures are a few compared with the normal behavior, it is a class
imbalance problem. In this paper, we study the use of contrast pattern-
based classifiers, designed for class imbalance problems, for creating an
understandable and accurate model for detecting pneumatic failures on
temporary immersion bioreactors. Our experiments over eight real-world
databases show that a decision tree ensemble obtains significantly better
AUC results than other tested classifiers.

Keywords: Supervised classification · Contrast patterns · Imbalanced
databases · Failure detection

1 Introduction

Temporary immersion systems have been widely used as an effective technology
for increasing the plant multiplication rates and plant quality [2,10,12,13,24].
A temporary immersion bioreactor (TIB) is a type of temporary immersion system which allows immerse periodically the plants into nutrient medium. A TIB
has several variables which should be controlled, as duration and frequency of
nutrient medium through air pressure, for guaranteeing both the efficiency and

© Springer International Publishing Switzerland 2016
J.F. Martínez-Trinidad et al. (Eds.): MCPR 2016, LNCS 9703, pp. 293–302, 2016.
DOI: 10.1007/978-3-319-39393-3_29

the efficacy[1] in TIBs. Therefore, temporary immersion bioreactors have been automated using a pneumatic drive to execute the immersion time in specific time intervals [10–13]. An important task for temporary immersion bioreactors is detecting failures on the pneumatic system, since these failures could affect the multiplication rate, the plant quality, or even discontinue the plant micro-propagation into the temporary immersion bioreactor [13]. Failure detection on the pneumatic system of temporary immersion bioreactors could be considered as a class imbalance problem, because failures usually are a few compared with the normal process of immersion.

In real-world imbalanced problems, the objects are not equally distributed among classes, which produces a bias of classification results to the majority class (the class with more objects). Frequently, the most interesting class, for an expert of the application domain, is the one that contains significantly less objects (minority class), because it is commonly associated to rare cases [28]. This type of problems is known as the class imbalance problem. Currently, there are several classifiers designed to deal with the class imbalance problem [20–22]. However, not all classifiers make a model that can be understood by experts in the application domain [5,9,16,22].

Contrast pattern-based classifiers are an important family of both under-standable and accurate classifiers. These classifiers have been used to solve real-world problems in fields like bioinformatics [9], intruder detection [5], anomaly detection in network connection data [29], rare event forecasting [27], and pri-vacy preserving data mining [1], which are well-known class imbalance problems [20,22].

In this paper we present a study of the use of contrast pattern-based classi-fiers for pneumatic failures detection in temporary immersion bioreactors, using eight real-world databases of pineapple plants. Through this study, we show that the Area Under the receiver operating characteristic Curve (AUC) [17] is signif-icantly improved when the HeDex classifier [19] is used. The main contribution of this paper is the use of contrast pattern-based classifiers for pneumatic fail-ure detection in temporary immersion bioreactors (a class imbalance problem), which allows creating an understandable and accurate model that forewarns the experts in practice about failures in the temporary immersion bioreactor.

The rest of the paper has the following structure. Section 2 provides a brief introduction to temporary immersion bioreactors and their pneumatic system. Section 3 reviews some of the most popular contrast pattern-based classifiers designed to deal with class imbalance problems. Section 4 presents our study about pneumatic system failure detection using the classifiers presented in Sect. 3 over eight real-world databases, including the experimental setup and a brief interpretation, issued by the pneumatic system expert, of the most frequent patterns belonging to the minority class (failure). Finally, Sect. 5 provides con-clusions and future work.

---

[1] By efficiency and efficacy we mean high plant growing rate, high quality plants, and low production cost.

2 Pneumatic Systems in Temporary Immersion Bioreactors

Temporary immersion bioreactors are a type of temporary immersion system, which are commonly categorized into mechanically agitated and pneumatically agitated bioreactors [24].

The temporary immersion bioreactor at the *Centro de Bioplantas*[2] is classified as a pneumatically agitated bioreactor, which is constituted by two transparent glass (or plastic) containers, autoclavable silicone tubes, hydrophobic air filters, electric valves, and an air compressor (see Fig. 1) [10]. One container is for growing plants and the other container is for liquid medium. These are connected by silicone tubes. In each case, the air flow is sterilized by passage through hydrophobic filters. Then, air pressure from an air compressor pushes the medium from one container to the other to immerse the plants completely. The air flow is reversed to withdraw the medium from the culture container. Also, three-way solenoid valves provided on/off operation; where the frequency and length of the immersion period is controlled by using a programmable logic controller (PLC) [3] connected with a supervisory control and data acquisition (SCADA) system through the *mobus* protocol [4].

The main failures that arise in the pneumatic system of this temporary immersion bioreactor are:

(i) Bad air pressure into the central distribution line.
(ii) Bad air pressure into the plant container.

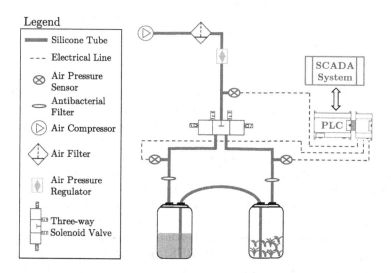

Fig. 1. Temporary immersion bioreactor diagram.

---

[2] www.bioplantas.cu.

(iii) Bad air pressure into the liquid medium container.

(iv) There is an obstruction[3] into the silicone tube connecting the containers.

If these failures are not detected as soon as they take place in the pneumatic system then the plant micropropagation into the temporary immersion bioreactor is discontinued.

In this paper, we propose to use contrast pattern-based classifiers, which allow creating understandable and accurate models. In this way, the pneumatic system experts could introduce this model into the PLC to forewarn failures in the temporary immersion bioreactor.

Despite there are several contrast pattern-based classifiers reported in the literature, not all can deal with class imbalance problems.

3   Contrast Pattern-Based Classifiers in Class Imbalance Problems

There are three main approaches to deal with the class imbalance problem [21, 22]; data level, algorithm level, and cost-sensitive. The cost-sensitive approach has as drawback that the cost matrix is commonly unknown. On the other hand, the data level approach has well-known drawbacks like representative object exclusion or promoting of classifier overfitting. However, ensemble methods (algorithm level) have reported good classification results in class imbalance problems [18, 26]. Therefore, we will be focusing on the application of ensembles of contrast pattern-based classifiers to detect pneumatic failures on temporary immersion bioreactors.

The following classifiers are among the most popular contrast pattern-based classifiers designed to deal with the class imbalance problem.

Coverage creates balanced subsamples without oversampling the minority class in order to create a decision tree ensemble [18]. The main goal is to establish robustness to the *consolidated tree construction* algorithm (CTC) [25]. CTC is an algorithm based on decision trees that requires the use of several subsamples of the training sample (as a lot of multiple classifier systems), but it returns a single decision tree. Coverage uses internally the CTC algorithm.

HeDex is a decision trees ensemble [19] using the *Hellinger Distance* [6] as a decision tree splitting criterion. It builds extremely randomized ensemble trees through the randomization on both attribute selection and split-point selection, which allows to achieve high level of variety of decision trees. HeDex has shown good results over several imbalanced databases [19].

RUSBoost is a decision tree ensemble that uses a boosting algorithm [26]. RUSBoost applies a resampling method (RUS) that randomly removes examples from the majority class. RUSBoost, also is based on the SMOTEBoost algorithm (which use the AdaBoost.M2 algorithm [14]) but RUSBoost uses RUS

---

[3] The most frequent obstruction is produced by waste of plant material.

rather than SMOTE. The RUSBoost classifier presents a simpler, faster, and less complex alternative to SMOTEBoost for learning from imbalanced databases [26].

iCAEP performs **i**nformation-based **C**lassification by **A**ggregating **E**merging **P**atterns [30]. It uses the minimum encoding inference approach to classify an object, instead of the aggregation of support. iCAEP selects a smaller but more representative subset of contrast patterns from the object to be classified.

4 Detecting Pneumatic Failures on Temporary Immersion Bioreactors Through Contrast Pattern-Based Classifiers

This section presents an empirical study on using contrast pattern-based classifiers for detecting pneumatic failures on temporary immersion bioreactors, which is a class imbalance problem. The experimental setup is presented in Sect. 4.1 and the experimental results are presented in Sect. 4.2.

All dataset partitions used in this paper as well as the experimental results are available for downloading from our supplementary material website[4].

4.1 Experimental Setup

For our experiments, we use eight pineapple databases, which were collected from the temporary immersion bioreactor at *Centro de Bioplantas* [11]. All databases contain 210 numerical attributes from the sensors of the pneumatic system. Each object represent an immersion time (of 70 seconds) in the temporary immersion bioreactor. Each database contains 70 attributes corresponding to measures of air pressure into the central distribution line, 70 attributes corresponding to measures of air pressure into the liquid medium container, and the last 70 attributes corresponding to measures of air pressure into the plant container (the air pressure sensors can be visualized in Fig. 1). Each object labeled as *failure* represents a problem during the immersion time, more specifically it means that the liquid medium was not transferred from a container to another.

Table 1 shows for each database: the name, the number of objects belonging to the minority (or failure) class (#Objects_Min), the number of objects belonging to the majority class (#Objects_Maj), and the class imbalance ratio (IR).

All databases were partitioned using 5-fold and distribution optimally balanced stratified cross validation (DOB-SCV) [23] with the goal of avoiding problems into data distribution on highly imbalanced databases [21].

The iCAEP classifier takes advantage of the patterns extracted through a contrast pattern miner; therefore, we used the *bagging miner* algorithm [15] to extract the contrast patterns to be used by the iCAEP classifier. The main reasons is that this miner has reported good classification results over databases

---

[4] http://sites.google.com/site/octavioloyola/papers/DPFonTIBs.

Table 1. Summary of the imbalanced databases used in our study

Name	#Objects_Min	#Objects_Maj	IR
Pineapple1	17	236	13.88
Pineapple2	11	236	21.45
Pineapple3	25	237	9.48
Pineapple4	14	237	16.93
Pineapple5	19	236	12.42
Pineapple6	16	237	14.81
Pineapple7	15	237	15.80
Pineapple8	12	234	19.50

with numerical attributes [15], and given that our databases contain only numerical attributes we chose bagging miner. Additionally, in [6] the authors suggested to use the bagging algorithm with the Hellinger distance to obtain good classification accuracy; therefore, we used the Hellinger distance as a decision tree splitting criterion into the bagging miner.

For the contrast pattern methods presented in Sect. 3, we used the parameter values recommended by their authors. For Coverage, we used the WEKA[5] implementation provided by its authors. For RUSBoost, we used the implementation into the KEEL[6] Data-Mining tool. For HeDex, iCAEP, and Bagging Miner we implemented our own versions.

We used the Friedman test [7] and the Bergmann-Hommel dynamic post-hoc procedure [8] to statistically compare our results. Moreover, post-hoc results are presented using *critical distance* (CD) diagrams [7]. Usually, in a CD diagram, the position of the classifier within the segment represents its rank value, where the rightmost classifier is the best one. If two or more classifiers share a thick line it means they have statistically similar behavior.

We used the AUC measure [17] to evaluate the classification performance because it is the most used measure for class imbalance problems [18,19,21,22,25].

4.2   Experimental Results

Figure 2 shows a CD diagram with the classification ranking for each contrast pattern-based classifier using all imbalanced databases described in Table 1.

From our results, the best contrast pattern-based classifier for detecting pneumatic failures on a temporary immersion bioreactor is HeDex, which statistically outperforms the remainder other classifiers used in our experiments. A possible explanation for this behavior is that HeDex uses randomization on both attribute selection and split-point selection, which can achieve high level of variety of decision trees. Furthermore, HeDex selects more than one split-point to

---

[5] http://www.sc.ehu.es/aldapa/2014/coverage-ctc.
[6] http://www.keel.es.

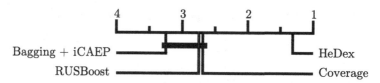

Fig. 2. CD diagram with a statistical comparison (using $\alpha = 0.10$) of the AUC results for contrast pattern-based classifiers over all the tested databases.

find a sub-optimal split-point that attains better results [19]. Also, HeDex uses the Hellinger distance as a decision tree splitting criterion, which has been widely used to deal with the class imbalance problem [6].

On the other hand, Coverage is based on a resampling strategy using multiple subsamples, which could exclude some representative objects to train the classifier. Additionally, Coverege uses a decision tree induction algorithm (CTC) which does not uses a skew-insensitive splitting criteria. In the case of RUSBoost, it uses an undersampling method (RUS), which could exclude some important objects. Finally, bagging miner jointly with iCAEP obtained the worst AUC results, which could be attributed to the lower support of the patterns belonging to the minority class (failure).

We analyzed the patterns extracted from all databases using the HeDex algorithm where the AUC results were equal to 1.0. Among them, we can see, several times, the following patterns:

(i) Air pressure into the central distribution line is lower than or equal to 0.104
(ii) Air pressure into the plant container is lower than or equal to 0.136 and air pressure into the central distribution line is greater than 0.127

These patterns indicate to the expert that the temporary immersion bioreactor has a failure in the pneumatic system. These contrast patterns were analyzed and confirmed, by the pneumatic expert in the temporary immersion bioreactor, as useful patterns to classify failures in this temporary immersion system. The explanation issued by the expert for the first pattern is that the air compressor does not operate correctly or the central distribution line is broken. For the second pattern the explanation is that the plant container has an air escape. Accordingly, the PLC was reprogrammed including these patterns, which will forewarn failures in the temporary immersion bioreactor.

Finally, it is important to highlight that the use of contrast pattern-based classifiers for detecting pneumatic failures on temporary immersion bioreactors was qualified by the experts as meaningful and very useful, since it is very difficult for them to find these patterns manually.

5   Conclusions and Future Work

The main contribution of this paper is an empirical study of the use of contrast pattern-based classifiers for detecting pneumatic failures on temporary immersion bioreactors.

From our study, we can conclude that HeDex obtains the best AUC results for detecting pneumatic failures on temporary immersion bioreactors. Statistical tests prove that the differences among HeDex and the other tested contrast pattern-based classifiers are statistically significant. Additionally, through our study, we find useful patterns which were analyzed by a pneumatic system expert and these patterns were introduced into the PLC to forewarn failures in the temporary immersion bioreactor.

Finally, as future work, following the same approach presented in this paper, we will extend our study to other types of failures that can occur in temporary immersion bioreactors, e.g. if the liquid medium container is empty or if the plant grow rate is inappropriate. These studies would help to improve the plant quality in temporary immersion bioreactors.

Acknowledgment. This work was partly supported by National Council of Science and Technology of Mexico under the scholarship grant 370272. Also, the authors want to thank the laboratory members for cell culture and tissues at *Centro de Bioplantas* for their valuable suggestions related to the plant micropropagation and the pneumatic system in the temporary immersion bioreactor, which significantly improved the quality of our work.

References

1. Andruszkiewicz, P.: Lazy approach to privacy preserving classification with emerging patterns. In: Ryżko, D., Rybiński, H., Gawrysiak, P., Kryszkiewicz, M. (eds.) Emerging Intelligent Technologies in Industry. SCI, vol. 369, pp. 253–268. Springer, Heidelberg (2011)
2. Barretto, S., Michoux, F., Nixon, P.J.: Temporary immersion bioreactors for the contained production of recombinant proteins in transplastomic plants. In: MacDonald, J., Kolotilin, I., Menassa, R. (eds.) Recombinant Proteins from Plants. MMB, vol. 1385, pp. 149–160. Springer, Heidelberg (2016)
3. Bolton, W.: Programmable logic controllers, 6th edn. Newnes, Bolton (2015)
4. Buchanan, W.J.: Modbus. In: The Handbook of Data Communications and Networks, vol. 1, pp. 677–687. Springer, Heidelberg (2004)
5. Chen, L., Dong, G.: Using emerging patterns in outlier and rare-class prediction. In: Contrast Data Mining: Concepts, Algorithms, and Applications, chap. 12, pp. 171–186. Data Mining and Knowledge Discovery Series, Chapman & Hall/CRC (2012)
6. Cieslak, D., Hoens, T., Chawla, N., Kegelmeyer, W.: Hellinger distance decision trees are robust and skew-insensitive. Data Min. Knowl. Disc. **24**(1), 136–158 (2012)
7. Demšar, J.: Statistical comparisons of classifiers over multiple data sets. J. Mach. Learn. Res. **7**, 1–30 (2006)
8. Derrac, J., García, S., Molina, D., Herrera, F.: A practical tutorial on the use of nonparametric statistical tests as a methodology for comparing evolutionary and swarm intelligence algorithms. Swarm Evol. Comput. **1**(1), 3–18 (2011)

9. Dong, G.: Overview of results on contrast mining and applications. In: Contrast Data Mining: Concepts, Algorithms, and Applications, chap. 25, pp. 353–362. Data Mining and Knowledge Discovery Series, Chapman & Hall/CRC, United States of America (2012)

10. Escalona, M., Lorenzo, C.J., González, B., Daquinta, M., González, J., Desjardins, Y., Borroto, G.C.: Pineapple (ananas comosus l. merr) micropropagation in temporary immersion systems. Plant Cell Rep. **18**(9), 743–748 (1999)

11. Escalona, M., Lorenzo, J., González, B., Daquinta, M., Fundora, Z., Borroto, C., Espinosa, P., Espinosa, D., Arias, E., Aspiolea, M.: New system for in-vitro propagation of pineapple (ananas comosus (l.) merr). Trop. Fruits Newsl. **29**, 3–5 (1998)

12. Escalona, M., Samson, G., Borroto, C., Desjardins, Y.: Physiology of effects of temporary immersion bioreactors on micropropagated pineapple plantlets. Vitro Cell. Dev. Biol. Plant **39**(6), 651–656 (2003)

13. Etienne, H., Berthouly, M.: Temporary immersion systems in plant micropropagation. Plant Cell Tissue Organ Culture **69**(3), 215–231 (2002)

14. Freund, Y., Schapire, R.E., et al.: Experiments with a new boosting algorithm. In: 13th International Conference on Machine Learning (ICML 1996). vol. 96, pp. 148–156 (1996)

15. García-Borroto, M., Martínez-Trinidad, J.F., Carrasco-Ochoa, J.A.: Finding the best diversity generation procedures for mining contrast patterns. Expert Syst. Appl. **42**(11), 4859–4866 (2015)

16. García-Borroto, M., Martínez-Trinidad, J., Carrasco-Ochoa, J.: A survey of emerging patterns for supervised classification. Artif. Intell. Rev. **42**(4), 705–721 (2014)

17. Huang, J., Ling, C.X.: Using AUC and accuracy in evaluating learning algorithms. Knowl. Data Eng. IEEE Trans. **17**(3), 299–310 (2005)

18. Ibarguren, I., Pérez, J.M., Muguerza, J., Gurrutxaga, I., Arbelaitz, O.: Coverage-based resampling: Building robust consolidated decision trees. Knowl. Based Syst. **79**, 51–67 (2015)

19. Kang, S., Ramamohanarao, K.: A robust classifier for imbalanced datasets. In: Tseng, V.S., Ho, T.B., Zhou, Z.-H., Chen, A.L.P., Kao, H.-Y. (eds.) PAKDD 2014, Part I. LNCS, vol. 8443, pp. 212–223. Springer, Heidelberg (2014)

20. López, V., Fernández, A., García, S., Palade, V., Herrera, F.: An insight into classification with imbalanced data: Empirical results and current trends on using data intrinsic characteristics. Inf. Sci. **250**, 113–141 (2013)

21. López, V., Fernández, A., Herrera, F.: On the importance of the validation technique for classification with imbalanced datasets: Addressing covariate shift when data is skewed. Inf. Sci. **257**, 1–13 (2014)

22. Loyola-González, O., Martínez-Trinidad, J.F., Carrasco-Ochoa, J.A., García-Borroto, M.: Study of the impact of resampling methods for contrast pattern based classifiers in imbalanced databases. Neurocomputing **175**, 935–947 (2016). Part

23. Moreno-Torres, J.G., Saez, J.A., Herrera, F.: Study on the impact of partition-induced dataset shift on k-fold cross-validation. Neural Netw. Learn. Syst. IEEE Trans. **23**(8), 1304–1312 (2012)

24. Paek, K.Y., Chakrabarty, D., Hahn, E.J.: Application of bioreactor systems for large scale production of horticultural and medicinal plants. In: Hvoslef-Eide, A.K., Preil, W. (eds.) Liquid Culture Systems for in vitro Plant Propagation, pp. 95–116. Springer, Heidelberg (2005)

25. Pérez, J.M., Muguerza, J., Arbelaitz, O., Gurrutxaga, I., Martín, J.I.: Combining multiple class distribution modified subsamples in a single tree. Pattern Recogn. Lett. **28**(4), 414–422 (2007)

26. Seiffert, C., Khoshgoftaar, T.M., Van Hulse, J., Napolitano, A.: RUSBoost: A hybrid approach to alleviating class imbalance. Syst. Man Cybern. Part A: Syst. Hum. IEEE Trans. **40**(1), 185–197 (2010)

27. Tsai, C.H., Chang, L.C., Chiang, H.C.: Forecasting of ozone episode days by cost-sensitive neural network methods. Sci Total Environ. **407**(6), 2124–2135 (2009)

28. Weiss, G.M., Tian, Y.: Maximizing classifier utility when there are data acquisition and modeling costs. Data Min. Knowl. Discovery **17**(2), 253–282 (2008)

29. Xue, J., Hu, C., Wang, K., Ma, R., Zou, J.: Metamorphic malware detection technology based on aggregating emerging patterns. In: Proceedings of the 2nd International Conference on Interaction Sciences: Information Technology, Culture and Human, pp. 1293–1296 (2009)

30. Zhang, X., Dong, G.: Information-based classification by aggregating emerging patterns. In: Leung, K.-S., Chan, L., Meng, H. (eds.) IDEAL 2000. LNCS, vol. 1983, pp. 48–53. Springer, Heidelberg (2000)

Classification of Motor States from Brain Rhythms Using Lattice Neural Networks

Berenice Gudiño-Mendoza[1]([✉]), Humberto Sossa[2], Gildardo Sanchez-Ante[1], and Javier M. Antelis[1]

[1] Tecnologico de Monterrey, Campus Guadalajara Av. Gral. Ramón Corona 2514, 45201 Zapopan, Jalisco, Mexico
{bereniceg,gildardo,mauricio.antelis}@itesm.mx
[2] Instituto Politécnico Nacional-CIC, Av. Juan de Dios Batiz S/N, Gustavo A. Madero, 07738 Mexico, D.F., Mexico
hsossa@cic.ipn.mx

Abstract. The identification of each phase in the process of movement arms from brain waves has been studied using classical classification approaches. Identify precisely each movement phase from relaxation to movement execution itself, is still an open challenging task. In the context of Brain-Computer Interfaces (BCI) this identification could accurately activate devices, giving more natural control systems. This work presents the use of a novel classification technique Lattice Neural Networks with Dendritic Processing (*LNNDP*), to identify motor states using electroencephalographic signals recorded from healthy subjects, performing self-paced reaching movements. To evaluate the performance of this technique 3 bi-classification scenarios were followed: (i) relax vs. intention, (ii) relax vs. execution, and (iii) intention vs. execution. The results showed that *LNNDP* provided an accuracy of (i) 65.26%, (ii) 69.07%, and (iii) 76.71% in each scenario respectively, which were higher than the chance level.

Keywords: Lattice Neural Network · Brain-Computer Interface · Electroencephalogram · Motor states

1 Introduction

Brain-Computer Interfaces (BCI) are a novel emergent technology that allows an alternative interaction with the environment using only brain signals [1]. BCI systems can have several uses in the context of rehabilitation or assistance for motor disabled people, and also they can be used in ludic activities. Non-invasive methods can be used to record input signals in the form of EEG (Electroencephalography). BCI technology is of great interest, since it allows the real time characterization for motor activity to obtain information related to actions.

Recognizable changes on the EEG activity allows the description of cortical processes associated with motor actions [2]. In a moving arm, three phases can be identified: (i) at the beginning the person is resting, in a relaxation state (*relax*), (ii) then, there is a mental process of motor planning where internally

© Springer International Publishing Switzerland 2016
J.F. Martínez-Trinidad et al. (Eds.): MCPR 2016, LNCS 9703, pp. 303–312, 2016.
DOI: 10.1007/978-3-319-39393-3_30

and unconsciously the person generates electrical brain signals to prepare the body for the movement (*intention*), (iii) finally the execution of the movement by the arm (*execution*). Recognition or discrimination of these motor states is an important key in the design of BCI. Particularly, anticipate any movement execution leads towards a more natural BCI technology with more high temporal precision for the control of neuroprostetic devices or virtual applications, which allows developing new ways of inducing neural rehabilitation [3–7].

It is well established the correlation between EEG activity and motor states. The decrease/increase of oscillatory activity in specific frequency bands (ERD/S) allows to identify every motor state [8]. There are some works related to the identification of movement intention following a static decoding [9–12]; all of them try to identify movement intention before a movement execution; they make use of classical algorithms for the classification part like naive Bayesian (BSC) and linear discriminant analysis (LDA); as feature vector some of them use spectral power and others common spatial patterns. The accuracy of motor states detection is still an open issue that can be explored by the use of promising advance techniques in the area of pattern recognition. This work presents the application of a novel classification technique named Lattice Neural Networks with Dendritic Processing (*LNNDP*) [13]. This method has been successfully applied in other topics like [14] and recently, with brain signals [15]. However, to our knowledge, this is the first time in which this classification problem is tackled with a *LNNDP*.

The goal of this work is to apply and evaluate the *LNNDP* classification technique for the identification of 3 motor states: relax, movement intention, and movement execution. To this end, real EEG signals were recorded from healthy subjects performing a self-paced arm movement, this innovative experiment is inspired in a daily life activity. At the end, an offline signal analysis was performed, which includes signal processing, feature selection through a r^2 analysis and classification. For the classification part a two-class discrimination strategy was set for training and test: (i) *relax* versus *intention*, (ii) *relax* versus *execution*, and (iii) *intention* versus *execution*. A multi-class tactic could be considered, but the results in these two-class strategies are enough for giving us an idea in how close we are to build an accurate feedback for a BCI system.

The paper is organized as follows: Sect. 2 describes the details in the methodology that supports the experiment, signal processing, and results; Sect. 3 describes the obtained results; and finally in Sect. 4 the conclusions are included.

2   Methods and Materials

2.1   Description of the EEG Dataset

Data Recording: Eighteen healthy right handed students (6 males and 12 females) voluntarily participated in this study (average age 20.3 years). None of them presented neurological or motor disease. All participants were informed about the experiment and all signed informed consent forms. EEG and EMG (Electromyographic) activity were acquired at a sampling frequency of 2048Hz and no filtering was applied. EEG signals were recorded according to the $10-10$ international system from 21 scalp positions ($Fp1$, $Fp2$, $F7$, $F3$, Fz, $F4$, $F8$, $T3$, $C3$, Cz, $C4$, $T4$, $T5$, $P3$, Pz, $P4$, $T6$, $O1$, $O2$, $A1$, $A2$). The impedance for all electrodes was kept below $5\,k\Omega$. The ground and reference electrodes were placed on Fpz and on the left earlobe, respectively. Bipolar EMG signals from the biceps were also recorded from both arms. The impedance for these electrodes were kept below $20\,k\Omega$.

During the execution of the experiments, the participants were comfortably seated with both arms resting on the chair's arm. The execution of the experiment was controlled by visual cues presented in a computer screen located in front of the participants. The experimental task consisted in the execution of many trials of natural reaching movements performed individually with the left arm or the right arm. The execution of each trial was controlled by three visual cues (see Fig. 1a). The first cue showed for three seconds the text "relax" and instructed to relax the body without performing any movement. The second cue showed for twelve seconds an image of a "left" or "right" arrow and instructed to stay relaxed for about five seconds and then to naturally move the corresponding arm towards the center of the screen. Notice that the actual movement initiation is different across trials. The last cue showed for three seconds the text "rest" and indicated to rest, move or blink while adopting the initial relaxed position. The experiment was performed in blocks of 24 trials each. Therefore, a total of 96 trials were recorded per participant.

Data Pre-processing: Firstly, EEG and EMG data were re-sampled to $256\,Hz$ and EEG signals were band-pass filtered from $0.1\,Hz$ to $100\,Hz$ using a zero-phase four-order Butterworth filter and re-referenced using the common average reference (CAR) filter. Subsequently, EEG and EMG data were segmented in trials starting from the first visual cue and up to the third visual cue. In consequence the length of the resulted trials was fifteen seconds. Then, the EMG signal of the moved arm of each trial was used to estimate the time instant of the movement onset. This EMG-based movement onset estimation was performed using the Hilbert transform. Subsequently, trials in which the time of the movement onset was lower than $3s$ and greater that $11s$ relative to the second visual cue were excluded and not used in the study. The time axis of all trials were then referenced to the EMG-based movement onset, thus $t = 0$ represents the start of movement execution. Finally, all trials were trimmed from the trial's initiation up to $1s$ relative to the EMG-based movement onset. Therefore, all trials have the same movement initiation ($t = 0$) but different trial's initiation (t_{ini}) and trial's length. An illustration of the time axis of the trials is presented in Fig. 1b. As a result of these pre-processing steps, the total number of trials across all subjects was on average 93.8 ± 2.0 (minimum of 89 and maximum of 96), while the time of the trial initiation across all subjects was on average -10.1 ± 1.5 (minimum of -13.9 and maximum of -6.0).

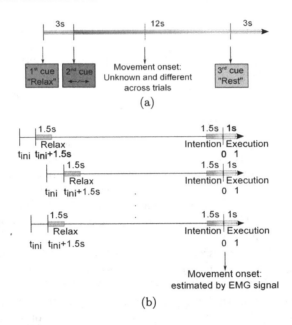

Fig. 1. (a) Graphical description of the timeline of a trial during the execution of the experiments. The first cue instructed to relax, the second cue instructed to self-initiate a reaching movement with the corresponding arm, and the third cue instructed to rest. (b) Illustration of the time axis across trials. All trials are referenced to the movement onset at $t = 0$ (estimated with the EMG activity) while all of them have different initiation time at t_{ini}. The figure also shows the segments used to evaluate the classification of motor states: *relax* segment at $[t_{ini} + 1.5, t_{ini} + 3]$ s; *intention* segment at $[-1.5, 0]$ s; *execution* segment at $(0, 1]$s.

2.2   Classification of Motor States Using *LNNDP*

The aim of this study was to employ *LNNDP* to asses the recognition of motor states from brain signals in three bi-classification scenarios: (i) *relax* versus *intention*, (ii) *relax* versus *execution*, and (iii) *intention* versus *execution*.

Features Extraction and Selection: The Power Spectral Density (PSD) of the EEG signals were used as features to recognize between motor states. It is well known, the existing changes in the spectral power for alpha and beta rhythms during movement execution, intention and relax phases, especially over sensory-motor cortex areas; PSD was chosen to be one of the most robust methods and the standard approach for feature extraction. The PSD was computed based on the Welch's averaged modified periodogram method in the frequency range between $2\,Hz$ and $40\,Hz$ at a resolution of $1\,Hz$ using Hanning-windowed epochs of length 500ms. For each trial, the PSD was computed in three different segments (see Fig. 1b): (i) *relax* segment $[t_{ini} + 1.5, t_{ini} + 3]$s; (ii) *intention* segment $[-1.5, 0]$s, and (iii) *execution* segment $(0, 1]$s. Note that the *relax* segment is the second half of the total relaxation phase, this was to avoid artifacts in the EEG induced by the rest phase of the previous trail, while the *intention* segment includes EEG activity exclusively previous to the movement execution. For each classification scenario, a r^2 analysis was performed to examine significant differences in the PSD between the two conditions, and subsequently to select the features (channel-frequency pairs) with the higher discriminative power. The r^2 was computed for each electrode and frequency as the square of the Pearson's correlation between the PSD values and their corresponding labels. Then, the eight channel-frequency pairs (from channels located above the motor cortex and from frequencies bins within the motor-related α $[8, 12]Hz$ and β $[13 - 30Hz]$ frequency bands) with the maximum values of r^2 were selected and used as features. This resulted in a feature vector of dimension $m = 8$. Therefore, the feature vector is $\boldsymbol{x} \in \mathbb{R}^8$ with an associated class label $\boldsymbol{y} \in \{relax, intention, execution\}$.

Lattice Neural Network with Dendritic Processing (*LNNDP*): Lattice Neural Networks with Dendritic Processing (*LNNDP*) is a recent classification method that considers computation in the dendritic structure as well as in the body of the neuron [13]. It requires no hidden layers, is capable of multiclass discrimination, presents no convergence problems and produces close separation surfaces between classes. The diagram of this model is presented in Fig. 2. The *LNNDP* has n input and m output neurons, where n is the number of features in the input vector, and m is the number of classes of the problem. A finite number of dendrites $(D_1, ..., D_k)$ establish the connection between input and output neurons. The input neuron N_i has at most two connections on a given dendrite. The weight w_{ijk}^l associated between neuron N_i and dendrite D_k, where $l \in \{0,1\}$ distinguishes between excitatory ($l = 1$, black dot in Fig. 2) and inhibitory ($l = 0$, empty dot in Fig. 2). In order to increase the tolerance to noise, it is possible to add a margin M, this margin is a number greater or equal to zero. For this work, the optimal M was selected following an strategy of exhaustive search, among 1000 values uniformly distributed between 0 and 1. This M value was computed exclusively from training data.

2.3 Evaluation Experiments

Evaluation Procedure: Classification performance was assessed for each subject independently using a ten-fold cross-validation procedure. The full set of trials were randomly partitioned into ten subsets which were used to construct mutually exclusive training and test sets. Nine of the subsets were used to train the classifier while the remaining was used to measure performance. This process was repeated until all the ten combinations of train and test sets were selected. The parameter M for the *LNNDP* classification method and feature extraction and selection were performed individually for each fold. For each combination, classification accuracy or CA (percentage of correct classifications) is computed, which is defined as:

$$CA = \frac{TP + TN}{TP + TN + FP + FN} \ . \tag{1}$$

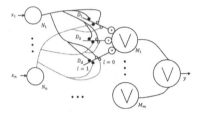

Fig. 2. Representation of a lattice neural network with dendritic computing [15]

where TP(True Positives), TN(True Negatives), FP(False Positives), FN(False Negatives).

The metric F1-score or $f1$ [16] (weighted average of the precision and recall, it reaches its best value at 1 and worst at 0) was also calculated, it can be defined as follows:

$$f1 = \frac{2 \times (precision \times recall)}{precision + recall} \, . \tag{2}$$

The significant chance level of the classification accuracy or CA_{chance} was computed using the cumulative binomial distribution [17] at the confidence level of $\alpha = 0.05$. This can be done by assuming that the classification errors obey a binomial cumulative distribution, where for a total of N_{trials} and c classes, the probability to predict the correct class at least z times by chance is given by:

$$p(z) = \sum_{i=z}^{N_{trials}} \binom{N_{trials}}{i} \times \left(\frac{1}{c}\right)^i \times \left(\frac{c-1}{c}\right)^{N_{trials}-1} \, . \tag{3}$$

For this work $c = 2$ is the number of classes and $N_{trials} = 89$ is the minimum number of trials across all participants. Hence, $CA_{chance} = 62\%$ is the bound above which classification is significant. Finally, distributions CA where constructed for each subject and for all of them, and significant differences between the median of the distributions and CA_{chance} were assessed using the Wilcoxon signed rank test at the significant chance level of $\alpha = 0.05$.

3   Results

3.1   PSD Features Analysis

Figure 3a shows PSD averaged across all participants for electrode Cz. For classification scenarios *relax* vs. *intention* and *relax* vs. *execution*, the PSD in the α and β frequency bands is greater in *relax* than in *intention* and *execution*, respectively. For scenario *intention* vs. *execution*, differences in the PSD are observed from the β, up to higher frequency bands. To examine the differences between the two conditions, the r^2 analysis across all participants is presented in Fig. 3b. For scenarios *relax* vs. *intention* and *relax* vs. *execution*, the largest

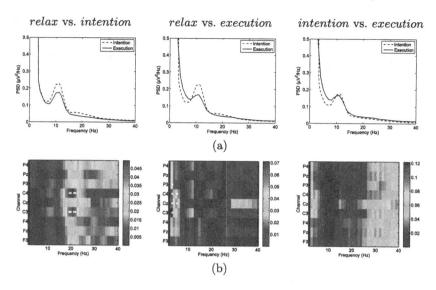

Fig. 3. (a) PSD and (b) r^2 analysis across all participants for the three studied scenarios: *relax* vs. *intention*; *relax* vs. *execution*; *intention* vs. *execution*. White crosses is the r^2 analysis represents the eight channel-frequency pairs with the maximum r^2 that were selected as attributes.

differences between the two conditions are observed in electrodes located above the motor cortex ($C3$, Cz and $C4$) and in the motor-related frequency bands (α and β), while for scenario *intention* vs. *execution*, the largest differences between the two conditions are observed across all electrodes and frequencies higher than 15 Hz. As illustration, these plots also display as white crosses the eight channel-frequency pairs with the maximum r^2 which were selected as attributes to discriminate between the two conditions.

3.2 Classification Results

Figure 4 shows the distributions of CA computed for each participant and under every scenario. In scenario *relax* vs. *intention*, seven subjects (1, 6–8, 10, 13, and 17) showed that the median of their distributions of CA are higher and significantly different than the CA_{chance} ($p < 0.05$). In scenario *relax* vs. *execution*, the median of the distributions of CA were significantly different and above chance level CA_{chance} ($p < 0.05$) in ten of the eighteen participants (3–7, 9–12 and 18). For the last scenario *intention* vs. *execution*, the median of the distributions of CA were significantly different ($p < 0.05$) and greater than the CA_{chance} for all the participants except for subject 13.

Table 1 presents a summary of CA metric in each scenario. In scenario 1, the mean across all subjects is $65.26 \pm 5.73\%$, and TPR/TNR are 60.17% and 70.95%, respectively. For scenario 2, in the average CA is $69.07 \pm 6.48\%$ across all subjects, and 67.85% and 70.75% values can be observed for TPR and TNR.

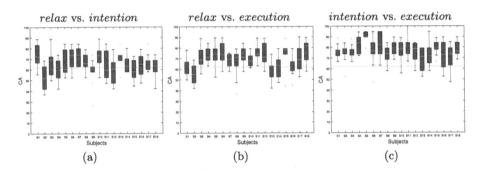

Fig. 4. Distribution of the classification accuracy for (a) *relax* vs. *intention*; (b) *relax* vs. *execution*; and (c) *intention* vs. *execution* across all subjects. Horizontal line over every bloxplot represents the median, and the black dot represents the mean. The red dotted line over the plot represents the chance level (Color figure online).

In scenario 3, the average of CA is $76.71 \pm 5.24\%$, and the specific values for TPR and TNR are 76.40% and 77.26%. Note that the mean for all scenarios are greater than CA_{chance}, and particularly the last bi-class classification scenario shows the better results, even the minimal value 68.42% is higher than the chance level.

The $f1$ metric considers both precision and recall measures to compute the score. These values are presented in Table 2. For the three different classification scenarios the values 0.64, 0.69, and 0.77 were obtained.

Table 1. Summary of CA, TPR and TNR computed across all subjects

	Metrics (%)					
	CA	Min	Max	TPR	TNR	CA_{chance}
relax vs. intention	65.26 ± 5.73	53.68	73.33	60.17	70.95	
relax vs. execution	69.07 ± 6.48	56.32	78.42	67.85	70.75	62 %
intention vs. execution	76.71 ± 5.24	68.42	90.47	76.40	77.26	

Table 2. Summary for *Precision*, *Recall*, and $f1$ computed across all subjects

	Metrics		
	Precision	Recall	$f1$
relax vs. intention	0.68 ± 0.06	0.60 ± 0.07	0.64 ± 0.06
relax vs. execution	0.70 ± 0.07	0.68 ± 0.08	0.69 ± 0.07
intention vs. execution	0.77 ± 0.05	0.76 ± 0.06	0.77 ± 0.05

4   Conclusions

In this work, the performance of the novel classification algorithm Lattice Neural Networks with Dendritic Processing ($LNNDP$) was evaluated in three motor states under arm movement using EEG signals. A daily life experiment of an auto-initiated movement arm over healthy participants was conducted. In every experiment EEG and EMG activity was recorded. Power spectral density was used as feature vector. First of all, through a r^2 analysis, significant power spectral differences were founded between three scenarios: (i) *relax* vs. *intention*, (ii) *relax* vs. *execution*, (iii) *intention* vs. *execution*. This observation was used as feature selector, and it allowed to choose the best feature vector. The most discriminative power was observed in the electrodes around the motor cortex ($C3$, Cz and $C4$) and in the motor-related frequency bands (α and β). For the classification and test evaluation, a two-scheme was followed to classify each condition under every scenario. The results showed that the $LNNDP$ classification technique provides a classification accuracy significantly different and superior to chance level ($p < 0.05$, Wilcoxon signed-rank test) for seven, ten and seventeen from eighteen participants in the scenarios: (i) *relax* vs. *intention*, (ii) *relax* vs. *execution*, and (iii) *intention* vs. *execution*, respectively. The average of CA is (i) 65.26%, (ii) 69.07%, and (iii) 76.71%; in each case. The better results can be observed for the third classification scenario. However, these results cannot be compared against the related state of the art. Several important differences exist, here some of them: (i) the experiments are quite different (finger extension, curl toes, tongue movement, etc.), (ii) some classification schemes try to identify more than two motor states, (iii) and others include participants with some disable motor condition. This work shows how the classification technique Lattice Neural Network with Dendritic Processing allows to obtain a confident classification accuracy in each movement arm phase, particularly the accurate identification for movement intention can be used in the context of Brain-Computer Interfaces to trigger neurorehabilitation devices. Further schemes for the testing part can be performed; like a continuous sliding window, to detect other interesting measure: the time of movement intention initiation. It could be also interesting as a next step, to perform an on-line experiment to test the applicability over a real neurorehabilitation BCI system.

Acknowledgments. The first author acknowledge the support from CONACYT through a postdoctoral fellowship. M. Antelis thanks to COECYTJAL for the partial financial support, project 3232-2015. H. Sossa would like to thank IPN-CIC under project SIP 20161126, and CONACYT under projects 155014 and 65 within the framework of call: Frontiers of Science 2015, for the economic support to carry out this research.

References

1. Lebedev, M.A., Nicolelis, M.A.: Brain-machine interfaces: past, present and future. Trends Neurosci. **29**, 536–546 (2006)
2. Serruya, M.D., Hatsopoulos, N.G., Paninski, L., Fellows, M.R., Donoghue, J.P.: Brain-machine interface: Instant neural control of a movement signal. Nature **416**, 141–142 (2002)
3. Ang, K., Guan, C., Chua, K., Ang, B., Kuah, C., Wang, C.: Clinical study of neurorehabilitation in stroke using EEG-based motor imagery brain-computer interface with robotic feedback. In: Annual International Conference of the IEEE 2010 on Engineering in Medicine and Biology Society (EMBC), pp. 5549–5552 (2010)
4. Broetz, D., Braun, C., Weber, C., Soekadar, S.R., Caria, A., Birbaumer, N.: Combination of brain-computer interface training and goal-directed physical therapy in chronic stroke: a case report. Neurorehabilitation Neural Repair **24**, 674–679 (2010)
5. Daly, J., Wolpaw, J.: Brain-computer interfaces in neurological rehabilitation. Lancet Neurol. **7**, 1032–1043 (2008)
6. Dobkin, B.H.: Brain-computer interface technology as a tool to augment plasticity and outcomes for neurological rehabilitation. J. Physiol. **579**, 637–642 (2007)
7. Fukuda, O., Tsuji, T., Ohtsuka, A., Kaneko, M.: EMG based human-robot interface for rehabilitation aid. In: Proceedings of the IEEE International Conference on Robotics and Automation, vol. 4, pp. 3492–3497 (1998)
8. Pfurtscheller, G., da Silva, F.H.L.: Event-related EEG/MEG synchronization and desynchronization: basic principles. Clin. Neurophysiol. **110**, 1842–1857 (1999)
9. Morash, V., Bai, O., Furlani, S., Lin, P., Hallett, M.: Classifying EEG signals preceding right hand, left hand, tongue, and right foot movements and motor imageries. Clin. Neurophysiol. **119**, 2570–2578 (2008)
10. Muralidharan, A., Chae, J., Taylor, D.: Extracting attempted hand movements from EEGs in people with complete hand paralysis following stroke. Front. Neurosci. **5**, 1–7 (2011)
11. Ibánez, J., Serrano, J., del Castillo, M., Minguez, J., Pons, J.: Predictive classification of self-paced upper-limb analytical movements with EEG. Med. Biol. Eng. Comput. **53**, 1201–1210 (2015)
12. Salvaris, M., Haggard, P.: Decoding intention at sensorimotor timescales. PLoS ONE **9**, e85100 (2014)
13. Sossa, H., Guevara, E.: Efficient training for dendrite morphological neural networks. Neurocomput. **131**, 132–142 (2014)
14. Vega, R., Guevara, E., Falcon, L.E., Sanchez-Ante, G., Sossa, H.: Blood vessel segmentation in retinal images using lattice neural networks. Adv. Artif. Intell. Its Appl. **8265**, 532–544 (2013)
15. Ojeda, L., Vega, R., Falcon, L.E., Sanchez-Ante, G., Sossa, H., Antelis, J.M.: Classification of hand movements from non-invasive brain signals using lattice neural networks with dendritic processing. In: Carrasco-Ochoa, J.A., Martí-nez-Trinidad, J.F., Sossa-Azuela, J.H., Olvera López, J.A., Famili, F. (eds.) MCPR 2015. LNCS, vol. 9116, pp. 23–32. Springer, Heidelberg (2015)
16. Goutte, C., Gaussier, É.: A probabilistic interpretation of precision, recall and F-score, with implication for evaluation. In: Losada, D.E., Fernández-Luna, J.M. (eds.) ECIR 2005. LNCS, vol. 3408, pp. 345–359. Springer, Heidelberg (2005)
17. Combrisson, E., Jerbi, K.: Exceeding chance level by chance: The caveat of theoretical chance levels in brain signal classification and statistical assessment of decoding accuracy. J. Neurosci. Method **250**, 126–136 (2015)

Crime Detection via Crowdsourcing

Daniel L. Pimentel-Alarcón$^{(\boxtimes)}$ and Claudia R. Solís-Lemus

University of Wisconsin-Madison, Madison, USA
pimentelalar@wisc.edu, claudia@stat.wisc.edu

Abstract. In this paper we propose a novel yet simple scheme for criminal detection. Rather than tracking the criminal that committed a particular crime, the police will rank the houses suspected to host criminals according to patterns on citizens' tips. We show that this strategy will provably identify the desired houses under reasonable assumptions. This will aid detectives decide where to focus their efforts. We also give related problems of great interest to the community where the same ideas may be applied with similar results. We complement our theoretical findings with experiments that illustrate the effectiveness of this approach.

Keywords: Statistical signal processing · Confidence bounds · Detection of criminal patterns · Crowdsourcing · Anonymous tips

1 Introduction

When there is a major criminal in a neighborhood (drug dealer, kidnapper, serial killer), the police work can be compared to finding a needle in a haystack. The community wants to help, but the number of calls can be overwhelming and the citizens' noblest intentions to contribute can be translated to countless unsubstantiated clues. More importantly, the police cannot follow up all the tips from the community because of limited resources. But what if instead of treating tips as unrelated data, we group them and analyze them to identify patterns?

Recent years have shown us that the active collaboration of a large community, also known as crowdsourcing, can play a decisive role at solving challenging tasks [1,2]. Examples include finding a lost boat in thousands of satellite images [3], studying migration patterns of birds [4], searching for anomalous archaeological patterns to locate the lost tomb of Genghis Khan [5], propagating information to bring relief in natural disasters [6], tracking stolen vehicles using social media [7], and aiding the transparency and accountability of the justice system [8].

In this paper we formalize the idea of crowdsourcing criminal detection: using the citizens' tips to rank the houses in a community according to the likelihood that they accommodate a criminal. We show that if reasonable assumptions are met, the strategy will provably succeed at locating houses hosting criminals. We extend the model to incorporate major drawbacks like geographic proximity, personal resentment or prejudice, and we will present other settings where

© Springer International Publishing Switzerland 2016
J.F. Martínez-Trinidad et al. (Eds.): MCPR 2016, LNCS 9703, pp. 313–323, 2016.
DOI: 10.1007/978-3-319-39393-3_31

similar strategies may be applied with very promising results. We complement our theoretical findings with experiments that illustrate our approach and show its effectiveness.

Organization of the Paper. In Sect. 2 we introduce our model and our main results, which we prove in Sect. 3. In Sect. 4 we present experiments that support our theory. In Sect. 5 we give a brief discussion of our findings, along with simple generalizations and other settings where our ideas may be applied.

2   Model and Main Results

Suppose there is a criminal that lives in one of the $n + 1$ houses of a city. The goal is to identify $h_\star$, the house that hosts the criminal. The police receives m tips from the citizens, and each tip suggests one house suspected to be $h_\star$. In the end we will select the most suggested house, $\hat{h}$.

Let $\mathcal{H} = \{h_1, h_2, \ldots, h_n, h_\star\}$ denote the set of all houses. Suppose that if a citizen provides a tip, he will independently suggest $h_\star$ with probability $p_\star$, and h_j with probability p_j. We will assume without loss of generality that $p_1 \geq p_2 \geq \cdots \geq p_n$. This way, h_1 is the most suspicious (with highest probability of being suggested) among the innocent houses. Intuitively, $p_\star$ models the accuracy of the citizens' perception and p_1 models their level of prejudice or other sources of inaccuracy.

Our main result is presented in the following theorem. It essentially states that as long as $p_\star$ (the citizens' accuracy) is slightly larger than p_1 (the level of prejudice), then with high probability the most suggested house will indeed be the one hosting the criminal.

Theorem 1. *Let $\epsilon > 0$ be given and suppose*

$$p_\star \; \geq \; p_1 + \sqrt{\tfrac{2}{m} \log\left(\tfrac{n}{\epsilon}\right)}. \tag{1}$$

Then $\hat{h} = h_\star$ with probability at least $1 - \epsilon$.

The proof of Theorem 1 is given in Sect. 3. Equivalently, Theorem 1 states that as long as we have enough tips to overcome the gap between $p_\star$ and p_1, we will identify $h_\star$ with high probability. This result is related to survey sampling. For a fixed n, the gap between $p_\star$ and p_1 is $O(1/\sqrt{m})$. A conservative two-sample test for difference in proportions q_1 and q_2 states that one is able to distinguish between the two proportions if their confidence intervals do not overlap. The width of each confidence interval is $O(1/\sqrt{m})$ for m the sample size.

2.1   Geographic Dependency

In practice, it is more likely that citizens perceive suspicious activities on houses that they frequently see, e.g., neighboring houses or houses on their way to work. We can model this by weighting the *inherent* probabilities $\{p_1, p_2, \ldots, p_n, p_\star\}$ by the exposure that citizens have to the houses, e.g., by the distances between citizens and houses.

To this end we introduce the matrix $\mathbf{G}$ that encodes the information of the geographic dependency. Essentially, $\mathbf{G}$ will specify the proximity of each citizen to each house, and this will determine the probability that each citizen perceives suspicious activities in each house. More precisely, let $\mathbf{G} \in \mathbb{R}^{m \times (n+1)}$. If citizen i lives in house j, then $\mathbf{G}_{ij} := 0$. Otherwise, $\mathbf{G}_{ij}$ denotes how close citizen i is from house j. Intuitively, if citizen i does not live in house j, then the closest citizen i is to house j, the larger $\mathbf{G}_{ij}$. The setup in the previous section is the particular case where all entries in $\mathbf{G}$ are equal.

Example 1. Consider a street with $n + 1$ houses. Suppose there is one citizen living in each house, and that each reported one tip to the police, such that $m = n + 1$. Suppose we measure the geographic dependency between citizen i and house h_j using the number of houses between h_i and h_j, such that

$$\mathbf{G} = n + 1 - \begin{bmatrix} n+1 & 1 & 2 & 3 & \\ 1 & n+1 & 1 & 2 & \cdots \\ 2 & 1 & n+1 & 1 & \\ 3 & 2 & 1 & n+1 & \\ & \vdots & & & \ddots \end{bmatrix}.$$

In this case, the closer h_i is to h_j, the more likely it is that citizen i notices suspicious activities in house h_j.

Let $\mathbf{p}$ be the diagonal matrix with diagonal elements taking the values in $\{p_1, p_2, \ldots, p_n, p_\star\}$. Let the $(i, j)^{th}$ entry of $\mathbf{P} := \mathbf{Gp}$ (with normalized rows) denote the probability that citizen i suggests h_j (given that citizen i provided a tip). In particular, we use $\mathbf{P}_{i\star}$ to denote the probability that citizen i suggests $h_\star$. In order to present our next result, let us introduce the set of γ-perceptive citizens, defined as

$$\mathcal{C}_\gamma := \{ i : \mathbf{P}_{i\star} - \mathbf{P}_{ij} \geq \gamma \;\; \forall j \}.$$

Intuitively, $\mathcal{C}_\gamma$ is the set of citizens that are at least γ more likely to suggest $h_\star$ than any other house.

The next theorem is a generalization of Theorem 1. It states that if there are enough tips from sufficiently perceptive citizens, then with high probability the most suggested house will indeed be the one hosting the criminal.

Theorem 2. *Let $\epsilon > 0$ be given. For any $k \in \mathbb{N}$ define γ_k as:*

$$\gamma_k := \frac{m-k}{k} + \sqrt{\frac{2}{k} \log\left(\frac{n}{\epsilon}\right)}. \qquad (2)$$

Assume without loss of generality that each citizen provided one tip. If there is a $k \in \mathbb{N}$ such that the set $\mathcal{C}_{\gamma_k}$ has at least k elements, then $\hat{h} = h_\star$ with probability at least $1 - \epsilon$.

The proof of Theorem 2 is given in Sect. 3. Notice the double dependency on k in Theorem 2. First, k determines the number of perceptive citizens required, that is, the number of citizens in $\mathcal{C}_{\gamma_k}$. And second, it determines how perceptive each of them must be, which is given γ_k. The larger k, the more perceptive citizens are required, but the less perceptive each needs to be.

Also notice that $r := \frac{m-k}{k}$ represents the ratio of non-perceptive citizens versus perceptive citizens (that provided tips). So in words, Theorem 2 states that as long as there is a group $\mathcal{C}_{\gamma_k}$ of k perceptive citizens that are more likely to suggest $h_\star$ over any other house by *a little more* than r, then with high probability we will identify $h_\star$. This *little more* is given by $\sqrt{2/k \log(n/\epsilon)}$. In a nutshell, Theorem 2 requires to have enough citizens that provide tips (at least k) with sufficient accuracy (at least γ_k).

Finally, observe that γ_k is monotonically decreasing. This implies that $\mathcal{C}_{\gamma_{k+1}}$ allows citizens with less perception than $\mathcal{C}_{\gamma_k}$, which in turn implies

$$\mathcal{C}_{\gamma_1} \subset \mathcal{C}_{\gamma_2} \subset \mathcal{C}_{\gamma_3} \subset \cdots.$$

So the question is: as k grows and γ_k shrinks, will $\mathcal{C}_{\gamma_k}$ grow enough to contain at least k citizens? This will depend on $\mathbf{P}$, which in turn depends on $\mathbf{p}$ and $\mathbf{G}$. Fortunately, given $\mathbf{p}$ and $\mathbf{G}$, we can iteratively test whether $\mathcal{C}_{\gamma_k}$ has at least k elements. If so, by Theorem 2 we will identify $h_\star$ with high probability. See Fig. 1 to build some intuition.

We point out that Theorem 2 considers the worst-case scenario in which all non-perceptive citizens may even be providing tips collaboratively and adversarially to confuse the police. More about this is discussed in Sect. 5.

2.2 Tipping Prior

The matrix $\mathbf{P}$ determines how the vote of each citizen *would* be distributed *if* he provided a tip. In this section we add one simple layer to our model to account for the distribution of citizens that provide tips. To this end, observe that

$$\mathbf{P}_{ij} = \mathsf{P}(\text{citizen } i \text{ suggests } h_j \mid \text{citizen } i \text{ provides a tip})$$

by definition. Letting π_i denote the probability that citizen i provides a tip, it follows that:

$$\mathsf{P}(\text{citizen } i \text{ suggests } h_j) = \mathbf{P}_{ij}\, \pi_i.$$

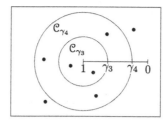

Fig. 1. Theorem 2 asks for a set $\mathcal{C}_{\gamma_k}$ with at least k citizens, such that each of these citizens has a gap between $\mathbf{P}_{i\star}$ and any $\mathbf{P}_{ij}$ at least as large as γ_k. If such set exists, then with high probability we will identify $h_\star$. Notice that γ_k is monotonically decreasing. This implies that $\mathcal{C}_{\gamma_1} \subset \mathcal{C}_{\gamma_2} \subset \cdots$. So the question is: as k grows and γ_k shrinks, will $\mathcal{C}_{\gamma_k}$ grow enough to contain at least k citizens? In this figure, $\mathcal{C}_{\gamma_3}$ only contains 2 citizens (represented with points). It follows that $|\mathcal{C}_{\gamma_3}| = 2 < 3 = k$, and so $\mathcal{C}_{\gamma_3}$ is not large enough to satisfy the conditions of Theorem 2. On the other hand, $\mathcal{C}_{\gamma_4}$ contains 5 citizens. This time $|\mathcal{C}_{\gamma_4}| = 5 > 4 = k$, and so $\mathcal{C}_{\gamma_4}$ satisfies the conditions of Theorem 2. Since there is a set that satisfies these conditions, namely $\mathcal{C}_{\gamma_4}$, we conclude that with high probability we will identify $h_\star$.

It is then clear that the number of citizens that suggest h_j, and hence the outcome of our procedure, will depend on π_i. This probability can be modeled in different ways. For instance, it is reasonable to assume that citizens are more likely to provide a tip if they live near $h_\star$. In this case, we can model π_i as

$$\text{(i)} \ \pi_i \ \propto \ 1/d_{i\star}, \qquad \text{or}$$
$$\text{(ii)} \ \pi_i \ \propto \ \exp\left(-d_{i\star}^2\right).$$

where $d_{i\star}$ denotes the distance between citizen i and $h_\star$. For example, (ii) corresponds to a gaussian decay in π_i as citizens get far from $h_\star$.

We point out that $\mathbf{G}$ does not capture this information. Without taking into account π_i, this model could yield very poor performance. To see this, suppose that citizen i is so far from $h_\star$, that $\mathbf{P}_{i\star}$ is much smaller than $\mathbf{P}_{ij}$ for some houses h_j neighboring citizen i. But this does not mean that citizen i suspects of any of these houses. In fact, citizen i may not suspect criminal activities in *any* house. In this case, citizen i is unlikely to provide a tip, which equates to π_i being small. But if we ignore π_i, and still ask this citizen to provide a tip, it is very likely (because $\mathbf{P}_{i\star}$ is very small) that he suggests some of his neighboring houses, contaminating the information provided to the police.

3 Proofs

3.1 Proof of Theorem 1

Let $N_\star$ and N_j denote the number of suggestions that $h_\star$ and h_j receive. We want to show that with high probability, the criminal lives in $\hat{h}$, the most suggested house. So union bounding over $\mathcal{H} \backslash h_\star$, we have that

$$P(\hat{h} \neq h_\star) = P\left(\bigcup_{j=1}^{n} \{N_\star \leq N_j\}\right) \leq \sum_{j=1}^{n} P(N_\star \leq N_j). \tag{3}$$

Let $Z_j := \frac{1}{m}(N_\star - N_j)$ such that $P(N_\star \leq N_j) = P(Z_j \leq 0)$. Letting

$$Z_{ij} := \begin{cases} 1 & \text{if } i^{th} \text{ citizen suggested house } h_\star \\ -1 & \text{if } i^{th} \text{ citizen suggested house } h_j \\ 0 & \text{otherwise,} \end{cases} \tag{4}$$

it is clear that $Z_j = \frac{1}{m}\sum_{i=1}^{m} Z_{ij}$. Since citizens suggest independently, the Z_{ij}'s are i.i.d. random variables with mean $p_\star - p_j$. Using Hoeffding's inequality [9] we obtain

$$P(Z_j \leq 0) = P(E[Z_j] - Z_j \geq (p_\star - p_j)) \leq e^{-\frac{m}{2}(p_\star - p_j)^2} \leq e^{-\frac{m}{2}(p_\star - p_1)^2},$$

where the last inequality follows because $p_1 \geq p_j \; \forall j$ by assumption. Going back to (3), we have that

$$(3) = \sum_{j=1}^{n} P(Z_j \leq 0) \leq \sum_{j=1}^{n} e^{-\frac{m}{2}(p_\star - p_1)^2} < n e^{-\frac{m}{2}(p_\star - p_1)^2} \leq \epsilon,$$

where the last inequality follows by (1). $\square$

3.2 Proof of Theorem 2

Let $\mathcal{C}_{\gamma_k}$ be a set satisfying the conditions of Theorem 2. We start as before:

$$P(\hat{h} \neq h_\star) = P\left(\bigcup_{j=1}^{n} \{N_\star \leq N_j\}\right) \leq \sum_{j=1}^{n} P(N_\star \leq N_j). \tag{5}$$

In the worst case scenario, all citizens will most likely suggest the same house (other than $h_\star$), which we will assume without loss of generality to be h_1 (equivalently, $\mathbf{P}_{i1} \geq \mathbf{P}_{ij} \; \forall i, j$). It follows that $P(N_\star \leq N_j) \leq P(N_\star \leq N_1) \; \forall j$, which further implies

$$(5) \leq nP(N_\star \leq N_1) = nP(Z_1 \leq 0), \tag{6}$$

where the last inequality follows by letting $Z_1 := \frac{1}{m}(N_\star - N_1)$. Defining Z_{ij} as in (4), we can write

$$Z_1 = \frac{1}{m}\sum_{i=1}^{m} Z_{i1} = \frac{1}{m}\sum_{i \in \mathcal{C}_{\gamma_k}} Z_{i1} + \frac{1}{m}\sum_{i \notin \mathcal{C}_{\gamma_k}} Z_{i1}.$$

In the worst case scenario, all the non-perceptive citizens will suggest h_1, whence $Z_{i1} = -1$ for every $i \notin \mathcal{C}_{\gamma_k}$. Then

$$Z_1 \geq \frac{1}{m} \sum_{i \in \mathcal{C}_{\gamma_k}} Z_{i1} - \frac{m - k}{m},$$

which implies

$$P(Z_1 \leq 0) \leq P\left(\frac{1}{m} \sum_{i \in \mathcal{C}_{\gamma_k}} Z_{i1} \leq \frac{m - k}{m}\right) = P\left(\frac{1}{k} \sum_{i \in \mathcal{C}_{\gamma_k}} Z_{i1} \leq \frac{m - k}{k}\right). \tag{7}$$

Letting $Z_1' := \frac{1}{k} \sum_{i \in \mathcal{C}_{\gamma_k}} Z_{i1}$ we obtain

$$(7) = P\left(Z_1' \leq \frac{m - k}{k}\right) = P\left(E[Z_1'] - Z_1' \geq E[Z_1'] - \frac{m - k}{k}\right), \tag{8}$$

and by Hoeffding's inequality [9],

$$(8) \leq e^{-\frac{k}{2}(E[Z_1'] - \frac{m - k}{k})^2} \leq \frac{\epsilon}{n},$$

where the last inequality follows because $E[Z_1'] = \frac{1}{k} \sum_{i \in \mathcal{C}_{\gamma_k}} (\mathbf{P}_{i\star} - \mathbf{P}_{i1})$; by the definition of $\mathcal{C}_{\gamma_k}$, every term of this sum, is at least γ_k, which implies $E[Z_1']$ is lower bounded by γ_k. We thus conclude that $P(Z_1 \leq 0) \leq \frac{\epsilon}{n}$. Substituting this in (6), we obtain the desired result.   □

4   Experiments

In this section we present a series of experiments to study the behavior of our detection scheme for different geographic dependency matrices $\mathbf{G}$, which together with the inherent *suspiciousness* level of the houses $\mathbf{p}$, determines the likelihood that citizens perceive suspicious activities. We will test the following cases of $\mathbf{G}$:

(a) **Constant.** This is equivalent to the most basic setup described at the beginning of Sect. 2, where each citizen suggests each house independently and identically according to $\mathbf{p}$.
(b) **Difference.** Setup described in Example 1, where the geographic dependency is given by the number of houses in between.
(c) **Euclidian.** Same setup as in Example 1, but with the geographic dependency given by the inverse distance, measured in number of houses, i.e.,

$$\mathbf{G} = \begin{bmatrix} 0 & 1 & 1/2 & 1/3 \\ 1 & 0 & 1 & 1/2 \\ 1/2 & 1 & 0 & 1 \\ 1/3 & 1/2 & 1 & 0 \\ & \vdots & & & \ddots \end{bmatrix} \cdots .$$

In each trial, we first generate a vector with independent entries selected uniformly at random according to the uniform distribution on $(0,1)^{n+1}$. Next we normalize and sort this vector to obtain $p_\star \geq p_1 \geq p_2 \geq \cdots \geq p_n$. The location of $h_\star$ in the street (and the rest of the houses) is selected uniformly at random. In each trial, m citizens will provide a tip. The citizens that provide tips will be distributed independently over the $n+1$ houses (sample with replacement) according to two tipping priors:

(i) $\pi_i := \mathsf{P}(\text{citizen } i \text{ provides a tip}) = \text{same for every } i,$

(ii) $\pi_i := \mathsf{P}(\text{citizen } i \text{ provides a tip}) = \exp\left(-\dfrac{d_{i\star}^2}{400}\right),$

where $d_{i\star}$ denotes the distance (measured in number of houses) between h_i and $h_\star$, and 400 represents a variance of roughly 20 houses before the exponential decay. Setting (i) corresponds to the basic model where all citizens are

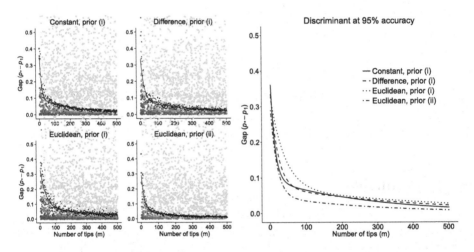

Fig. 2. Left: Phase transition diagram of the success rate at identifying $h_\star$ as a function of the number of tips m and the gap between $p_\star$ and p_1, for four different settings. The gray level at each pair $(m, p_\star - p_1)$ indicates the success rate over $10,000$ replicates: brightest gray represents 100% accuracy; darkest gray represents 0% accuracy. Each $(m, p_\star - p_1)$ pair was selected randomly. For each m, all pairs above the black point have at least 95% accuracy. The curve is the best exponential fit to these points. These curves represent the discriminant between at least 95% accuracy (above curve) and less than 95% accuracy (below curve). Intuitively, if we are above the curve, i.e., if we have enough tips, and enough gap between $p_\star$ and p_1, we will likely identify $h_\star$. **Right:** Comparison of the discriminants at 95% accuracy for the four settings in the left. The lower the curve the better, because then fewer tips and gap are required to identify $h_\star$. The Euclidian setting with prior (i) requires more tips and gap, which means identifying $h_\star$ is more difficult. Prior (i) corresponds to the basic model where all citizens are equally likely to provide tips. Prior (ii) corresponds to the more realistic model where citizens near $h_\star$ are more likely to provide tips. Under this model, identifying $h_\star$ requires fewer tips and gap. This can be appreciated by comparing the two Euclidian settings.

equally likely to provide tips. As discussed in Sect. 2.2, setting (ii) corresponds to the more realistic scenario where citizens near $h_\star$ are more likely to provide tips.

Each citizen that provides a tip will suggest a house suspected to host criminal activities according to $\mathbf{P}$. Recall that a citizen living in house h_i will suggest house h_j with probability $\mathbf{P}_{ij}$. Since $\mathbf{P} = \mathbf{Gp}$, this probability depends on the house where the citizen lives through the geographic dependency matrix $\mathbf{G}$. We will then select the most suggested house, and we will verify whether it corresponds to $h_\star$. We repeat this experiment 10,000 replicates for different values of m and $\{p_1, \ldots, p_n, p_\star\}$. The results are summarized in Fig. 2.

As predicted by our theory, $h_\star$ can be consistently identified as long as there are enough tips, and there is enough gap between $p_\star$ and p_1. Observe that vis-à-vis, under prior (i), the Euclidian setting demands more tips and gap than the rest of the settings. This is because the Euclidian matrix $\mathbf{G}$ has a faster decay with distance. We can interpret this as houses being farther apart from one another. This suggests, in accordance to intuition, that it is easier to find $h_\star$ in denser areas, like highly populated cities, where people are close.

5   Conclusions and Discussion

In this paper we introduce a simple model to identify houses hosting criminals. We prove that under reasonable assumptions, a crowdsourcing strategy will succeed at this task with large probability. Our experiments support our theoretical findings. We now give some simple generalizations to the models described in Sect. 2, along with other settings where our ideas may be easily extended.

Increasing our Odds. Recall that $p_\star$ and p_1 denote the underlying probabilities that a citizen suggests $h_\star$ and h_1, where h_1 is the most suspicious among the innocent houses. As shown by Theorems 1 and 2, the gap between $p_\star$ and p_1, and the number of citizens that provide tips (m) will determine whether our strategy will work. These quantities can be influenced in our favor through media campaigns to promote participation (to increase m), to encourage citizens to be more aware (to increasing $p_\star$) and to avoid unfounded suggestions, bias or prejudice (to restrict p_1).

Organized Crime. It is also possible that the city has not only one, but several criminals. Moreover, these criminals could be organized and determined to collaborate in an optimal way to avoid detection. In this case, it is in the criminals' best interest to suggest the most suspicious innocent house, h_1. This can be modeled by letting the rows of $\mathbf{P}$ corresponding to criminals take the value 1 in the column corresponding to h_1, and zeros elsewhere. In fact, Theorem 2 is shown assuming that all the citizens not in $\mathcal{C}_{\gamma_k}$ will suggest h_1. Recall that $\mathcal{C}_{\gamma_k}$ denotes the set of perceptive citizens that are at least γ_k more likely to suggest $h_\star$ than any other house.

This implies that Theorem 2 follows regardless of whether the citizens not in $\mathcal{C}_{\gamma_k}$ are criminals or not. We thus conclude that as long as there are enough honest citizens (at least k) with sufficient accuracy (at least γ_k), then with high probability we will find a house hosting a criminal. Hence, we can easily generalize our model to include several criminals' houses. The pattern of identified houses can help detect criminal networks.

Observe that one implicit requirement of Theorem 2 is that the set $\mathcal{C}_{\gamma_k}$ contains at least half of the citizens. This can be seen mathematically because if $k \leq \frac{m}{2}$, then $\frac{m-k}{k} \geq 1$, whence (1) requires that $\gamma_k > 1$, which implies $\mathcal{C}_{\gamma_k} = \emptyset$. In other words, Theorem 2 requires that there are more perceptive citizens than not. This is precisely because Theorem 2 is considering this worst-case adversarial scenario. If there are more organized criminals than honest citizens, then with high probability, h_1 will have more suggestions than $h_\star$.

Detecting Corruption. Of course, none of the ideas discussed above will work if the police force is corrupt. Fortunately, the same ideas can be adapted to detect patterns of corruption, or equivalently, to find the most honorable policemen. Consider, for an example, the following scenario. Suppose a citizen runs a light and is caught by a policeman. It is the policeman duty to assign a ticket and report it in the system. But if the policeman is corrupt, he will take a bribe and there will be no record of this transaction.

Suppose instead that citizen i runs a light, and is caught by a policeman. An other citizen i' sees that a policeman (who can be identified by the police car) is interacting with the first citizen, and he reports this to the system (anonymously, through a website, a cell phone app, text message, phone call, etc.). Citizen i' does not know the nature of the interaction between the policeman and citizen i, yet he reports that an interaction occurred.

If many citizens report that there was an interaction between a certain policeman, but there is no report of a fine in the system, this would suggest that the policeman took a bribe. If there are many cases suggesting that a particular policeman took bribes, it is likely he did. This would also allow us to identify the most honorable policemen: the ones whose interactions with citizens (reported by citizens) match the fines in the system (reported by the policeman). We can then analyze the hierarchical structure of the corrupt policemen to determine patterns of corruption in higher levels. This will be the case of future study.

References

1. Brabham, D.C.: Crowdsourcing as a model for problem solving: An introduction and Cases. Convergence: Int. J. Res New Media Technol. **14**, 75–90 (2008)
2. Estellés-Arolas, E., González-Ladrón-de, G.F.: Towards an integrated crowdsourcing definition. J. Inf. Sci. **38**, 189–200 (2012)
3. Doan, A., Ramakrishnan, R., Halevy, A.Y.: Crowdsourcing systems on the World-Wide Web. Commun. ACM **54**, 86 (2011)
4. Sullivan, B., Wood, C., Iliff, M., Bonney, R., Fink, D., Kelling, S.: eBird: A citizen-based bird observation network in the biological sciences. Biol. Conserv. **142**(10), 2282–2292 (2009)
5. Lin, A.Y.-M., Huynh, A., Lanckriet, G., Barrington, L.: Crowdsourcing the unknown: The satellite search for Genghis Khan. PLoS ONE **9**(12), e114046 (2014)

6. Goodchild, M.F., Glennon, J.A.: Crowdsourcing geographic information for disaster response: a research frontier. Int. J. Digit. Earth **3**, 231–241 (2010)
7. Featherstone, C.: Identifying vehicle descriptions in microblogging text with the aim of reducing or predicting crime. In: International Conference on Adaptive Science and Technology (ICAST), pp. 1–8 (2013)
8. Byrne Evans, M., O'Hara, K., Tiropanis, T., Webber, C.: Crime applications and social machines: crowdsourcing sensitive data. In: Proceedings of the 22nd International Conference on World Wide Web, pp. 891–896. ACM (2013)
9. Hoeffding, W.: Probability inequalities for sums of bounded random variables. J. Am. Stat. Assoc. **58**(301), 13–30 (1963)

Signature Recognition: Human Performance Analysis vs. Automatic System and Feature Extraction via Crowdsourcing

Derlin Morocho[1,2(✉)], Mariela Proaño[1], Darwin Alulema[1],
Aythami Morales[2], and Julian Fierrez[2]

[1] Universidad de las Fuerzas Armadas – ESPE, Sangolquí, Ecuador
{dmorocho,mcproano6,doalulema}@espe.edu.ec
[2] ATVS – Biometric Recognition Group,
Universidad Autonoma de Madrid, Madrid, Spain
{aythami.morales,julian.fierez}@uam.es

Abstract. This paper presents discriminative features as a result of comparing the authenticity of signatures, between standardized responses from a group of people with no experience in signature recognition through a manual system based on crowdsourcing, as well as the performance of the human vs. an automatic system with two classifiers. For which an experimental protocol is implemented through interfaces programmed in HTML and published on the platform Amazon Mechanical Turk. This platform allows obtaining responses from 500 workers on the veracity of signatures shown to them. By normalizing the responses, several general features which serve for the extraction of discriminative features are obtained in signature recognition. The comparison analysis in terms of False Acceptance Rate and False Rejection Rate founds the presented features, which will serve as a future study of performance analysis in the implementation of automatic and semiautomatic signature recognition systems that will support financial, legal and security applications.

Keywords: Amazon Mechanical Turk · Workers · FAR · FRR · Crowdsourcing

1 Introduction

Since the beginning of humanity, it has been a need to identify and verify people by various means. One way to check the identity of people is the handwritten signature, which corresponds to a biometric feature. Biometric verification is a very important topic of research and is focused on identity verification applications. Biometric measurements are used for security purposes, since these are not easy to duplicate, and also cannot be stolen or forgotten, so they are safer [1]. Because of the warranty presented by written signatures, it has been seen today the need to build systems that perform biometric verification of signatures' strokes. There are automatic signature verification systems that require a lot of training information for proper performance. At present, many activities of daily life require a signature to identify an individual. This identification is performed by humans with no experience in forensic document analysis in

© Springer International Publishing Switzerland 2016
J.F. Martínez-Trinidad et al. (Eds.): MCPR 2016, LNCS 9703, pp. 324–334, 2016.
DOI: 10.1007/978-3-319-39393-3_32

most applications such as a bank or a public notary, where time and decision-making depend on having agility and ability to observe minor variations.

To massively collect data in signature recognition by humans, crowdsourcing, is the tool that allows such activities, which is implemented through Amazon Mechanical Turk (MTurk). At present, crowdsourcing is a tool that uses a multitude of human beings to solve different types of problems, in the state of the art there are research projects that use this method, such as face recognition [2], gait recognition [3], and biometric security [4] for which this method is an alternative for signature recognition. Face recognition is used in applications such as mobile device authentication [5] and site security verification [6]. To run these applications, crowdsourcing helps determining human performance regarding algorithms and a COTS face comparator (Commercial-off-the-shelf) [2], concluding that humans have a better performance in terms of False Acceptance Rate (FAR) with a low value of False Rejection Rate (FRR) [2]. This is because human beings have the ability to use contextual information known as soft biometrics (ethnicity, gender, hair profile, etc.) for recognition of strokes and faces [3].

Human recognition capabilities were tested to identify people and scenarios in low quality surveillance videos. The recognition is based on human perception to tag subjects in various environments. Humans have proven to be objective, outstanding, reliable and robust to changes in viewing distance, as mentioned in [3].

MTurk is a platform used in several studies. Currently there are security systems based on voice recognition [7]. In [4] a method for identifying participants that execute mimics for different speakers of a target population is proposed.

The performance of people doing signature recognition is studied in [8, 14] and shows promising results in human performance as they are being able to distinguish a genuine signature from a forged one. Due to the positive results of these studies, it is proposed to use the tool of crowdsourcing implemented in MTurk to attack the problem of intravariability to make handwritten signature recognition with human intervention.

The article is distributed as follows: Sect. 1 shows the related works and importance of this paper, Sect. 2 shows state of de art of Automatic and manual recognition systems. The analysis and performance comparison are analyzed in Sect. 3 and finally the conclusions are discussed in Sect. 4.

2   State of the Art of Automatic and Manual Recognition Systems

2.1   Crowdsourcing via Amazon Mechanical Turk

Crowdsourcing is a participatory type of online activity in which an individual, institution, nonprofit organization or Company proposes to a set of people with different kinds of knowledge, diversity and number, through a free call, a task to be carried out voluntarily. The execution of the task, of varying complexity and modularity, and in which the crowd should participate by contributing with their work, money and knowledge, always involves a mutual benefit. The advantages of this tool such as large numbers of participants, maximized performance, fast response rate and low costs of operation, have allowed researchers to have good results that have seen the human potential regarding to automated systems [2–4].

To implement crowdsourcing, it is required the use of MTurk [10]. Due to the features shown by MTurk, this platform has been used in face recognition [2], the obtained results show that humans have a superior performance compared to machines because they are capable of using contextual information of the image.

2.2 Manual Handwritten Signature Recognition System Based on Crowdsourcing

Manual signature recognition is performed by forensic document analyzers which have a method for studying handwriting features as mentioned in [13]. Manual signature recognition system is formed by the requester, HIT, MTurk platform, workers and the obtained results. Figure 1 shows how the manual signature recognition system is formed.

The requester is responsible for designing the HIT according to the experimental protocol to be evaluated. HITs are tasks that require some level of intelligence for its completion and are implemented using HTML. MTurk allows requesters to publish HITs which are solved by workers to obtain the results.

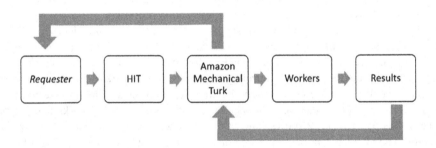

Fig. 1. System based on crowdsourcing

For manual recognition based on crowdsourcing, it is required to implement a GUI that contains instructions, training and test signatures and answer options. The interface shows four training signatures and one test signature, where the worker observes training signatures and qualify the test signature giving a similarity value between 1 and 10. Figure 3 shows the activities the workers should perform to evaluate the signatures (Fig. 2).

The aim of the protocol is to compare the performance of the human in a possible real scenario with more information. The protocol configuration parameters are shown in Table 1 (Fig. 4).

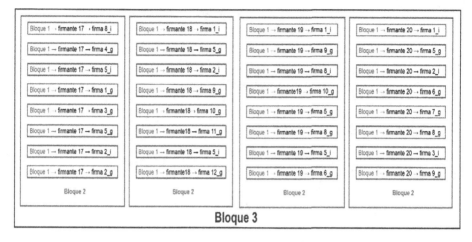

Fig. 2. Protocol interface with 8 test signatures for 4 signers

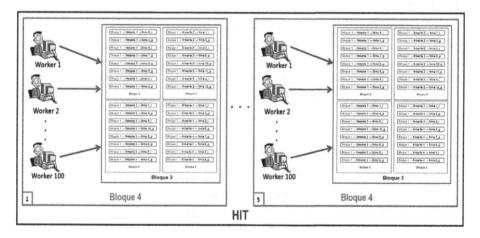

Fig. 3. HIT for 500 workers with 20 different signers

Table 1. Data configuration of the crowdsourcing system

Parameter	Configuration
Database	BiosecurID
Signers	20
Execution time	3 min
Number of workers	500
Number of activities	4
Number of analyzed signatures	240 divided into: - 80 genuine training signatures - 160 genuine and forged test signatures

Fig. 4. Protocol interface with the first test signature

2.3   Automatic Recognition System

The recognition system used for the proposed comparison performance [11], uses global and local features. To classify the local features a Back Propagation Neural Network (BPNN) was used while global features use the probabilistic model. In order to use the recognition benefits from both classifiers, an "AND" combination of the classifiers is done to make the final decision [11] (Table 2).

Table 2. Configuration data of the automatic system

Parameter	Configuration
Database	SVC2004
Signers	40
Number of analyzed signatures	600 divided into: - 200 genuine training signatures - 400 genuine and forged test signatures

The global features' classifier uses the probabilistic model for its development. First, feature values of the training signatures are obtained, then its mean and variance is calculated and thus a threshold value is defined. This threshold is the lowest value of the training signature' features, which is compared with the probability score (PS). If the PS value is greater than or equal to the threshold value, the user is considered genuine, otherwise is rejected as a forgery [11].

The local features' classifier corresponds to a BPNN with a 3-layer MLP and in which the output of the neural network is evaluated based on the desired output. If the response is not satisfactory, the connections (weights) between layers are modified and the process is repeated until the error is low [12]. In this case the threshold value (desired output) is 0.5. If the values obtained from the output of the BPNN is greater than or equal to 0.5 the user is considered genuine, otherwise it is rejected as a forgery [11].

In [11] the logical operation "AND" to combine the two classifiers is explained and its benefits are shown. Thus, a user is considered genuine if it is recognized as genuine in both classifiers, otherwise he/she is considered as a forger.

For performance comparison, the values of FAR and FRR from the combination of both classifiers and the probabilistic classifier are used [11].

3 Performance Comparison and Results' Analysis

Performance comparison is accomplished by using data obtained from online recognition system based on two classifiers and manual system based on crowdsourcing.

The manual system based on crowdsourcing uses the BiosecurID database and the comparison for 20 signers is done. The automatic system uses de SVC2004 database for 40 signers.

The FRR and FAR values used correspond to those obtained from the combined responses of workers and the decision obtained from the combination of the two classifiers. The combination of responses of the manual system based on crowdsourcing is done by obtaining FAR and FRR mean values of several workers taken randomly, as shown in (1). Furthermore, 10 repetitions of this process are made, afterwards the mean value of the obtained results in repetitions is calculated as shown in (2).

$$cr = \frac{1}{n}\left(\sum\nolimits_{i=1}^{n} w_i\right) \tag{1}$$

$i = 1, 2, \ldots, n$	Random number of answers to combine
w	Worker's answer in terms of FAR or FRR
cr	Mean value of n combined answers

$$er = \frac{1}{m}\left(\sum\nolimits_{j=1}^{m} cr_i\right) \tag{2}$$

$j = 1, 2, \ldots, m$	Number of repetitions
cr	Mean value of n combined answers
er	Evolution of the m repetitions of combined answers

In this case the number of combined responses ranges from 1 to 500 in the manual system. Table 3 shows the values of FRR and FAR of the variations of the automatic and manual systems.

Table 3. FFR and FAR response performance in handwritten signature recognition

Method	FRR (%)	FAR (%)
1 Automatic probabilistic (40 signers)	27	9
2 Automatic combined (40 signers)	3	5
3 Manual crowdsourcing (20 signers)	32	38
4 Manual Crowdsourcing Combined (20 signers)	7	24

Table 3 shows that Manual Crowdsourcing Combined System compared to Automatic probabilistic system in terms of FRR is better, thus the Manual Crowdsourcing Combined system rejects less genuine signatures, demonstrating the ability of humans in signature verification, but in terms of FAR, the Manual Crowdsourcing Combined system, admits a greater number of impostors firms, implying that the human is more permissible to accept impostors signatures as genuine.

When analyzing the FRR of Automatic Probabilistic, it can be observed that it is slightly lower than the value of Manual Crowdsourcing system, this shows that both, human and machine, when working without combinations may wrongly reject genuine signatures. However, the value of FAR is different because automatic probabilistic system shows superior performance compared to Manual Crowdsourcing system. This shows that the automatic probabilistic recognition accepts less forged signatures as genuine.

The Manual Crowdsourcing system has a FRR value (32 %) and FAR value (38 %) higher than the automatic combination (3 % FRR and 5 % FAR) because it uses for signature recognition, people that are not experts in forensic document examination. Figures 5 and 6 show the evolution of FAR and FRR respectively when workers' responses are combined in signature recognition.

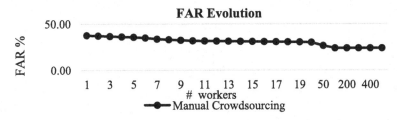

Fig. 5. Performance evolution FAR

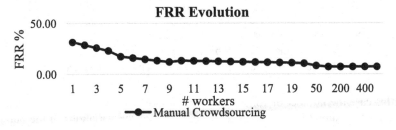

Fig. 6. Performance evolution FRR

Figures 5 and 6, show that FAR = 7 % and FRR = 24 % respectively, stabilize when the answers of 100 workers are combined in handwritten signature recognition. Manual Crowdsourcing System provides information of discriminative features for the identification of a handwritten signature. Figure 7 shows the signature features and the percentage of workers who used each one of them.

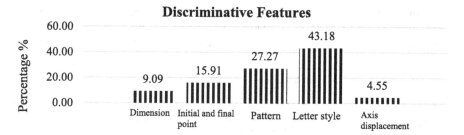

Fig. 7. Features extraction defined by workers

In Fig. 7, it can be seen that **Letter style**, is the characteristic chosen by the workers to determine whether the signature is genuine or forged. Furthermore, the **Axis displacement** characteristic is the less applied in signature recognition.

From the discriminative features of Fig. 7, a subdivision of them is extracted and is shown in Figs. 8, 9 and 10. It is noted that the traits present in **Letter style**, are the most used by workers in handwritten signature recognition, this happens because humans are able to detect minute variations in features due to their perception [3].

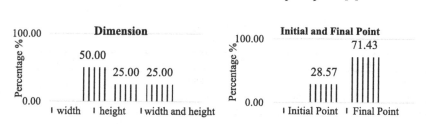

Fig. 8. Features: dimension and initial and final point

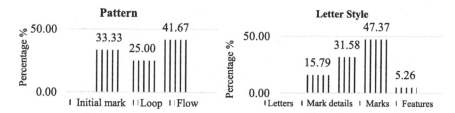

Fig. 9. Features: pattern and letter style

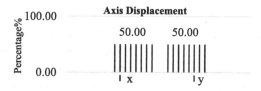

Fig. 10. Features: axis displacement

The discriminative features obtained through crowdsourcing and the results of this classification, generate a set of 7 features discriminating for handwritten signature recognition. Tables 4 and 5 show and briefly describe the 7 discriminative features.

Table 4. Discriminative features 1

Characteristic	Definition	Example
Legibility	It refers to the existence of defined literal characters	
Total area occupied by the signature	The most relevant signatures' points are taken (4 strategic points containing signature) to calculate the area of the signature	
Perimeter	It is proposed to measure the contour of the signature that best represents it.	
Pressure of initial and final points	It is propose the identification of signatures' initial and final points and the measurement of pressure of these points.	

Table 5. Discriminative features 2

Characteristic	Definition	Example
Connection between characters	The signature is observe to determine whether it is legible or illegible. If the signature is illegible then it is identified the existence of relevant connections between characters. If there are connections they should be classified in rounded, sharp or mixed.	
Character Area	Measuring the area of a character through a box is proposed. The box should contain inside the character which desired area is being measured.	
Proportionality	The signature strokes are observed and height levels representative points are defined, with the aim of establishing a dependent dimensionless ratio in the signature	

4   Conclusions

According to research and analysis of responses from the workers, these alone are not competent and able to recognize signatures. Static information displayed to the workers for signature recognition is not sufficient and appropriate to improve performance. However, by combining the responses of each worker, performance has improved by 40 % in FAR and FRR by 80 %.

The criteria from workers' responses allow extracting discriminative features of a signature such as: dimension, initial and final points, pattern, letter style and displacement, permitting the release of new features that will help automatic signature recognition systems and in the future be support for generating semiautomatic signature recognition systems.

Performance results of manual system based on crowdsourcing envision a bright future because, by combining the answers it may give a better performance to automatic and semiautomatic signature recognition systems. It is noted that the results are comparable with an automatic classifier based on a probabilistic model, therefore, merging the benefits of a manual and an automatic system could lead to the development of a high performance automatic system with human oversight. In addition, the combination of workers' responses in HITs, provide future research, which would be the analysis of discriminant features that provide better performance.

This paper is the beginning of a new research that includes HIT generation in signature recognition by improving or establishing new parameters such as: number of signers, number or workers, comparison levels, focusing on new scenarios of human assistance applied in signature recognition through crowdsourcing.

References

1. Toscano Medina, L.: Reconocimiento Dinámico y Estático de Trazos (2009)
2. Best-Rowden, L., Han, H., Otto, C., Klare, B., Jain, A.: Unconstrained face recognition: identifying a person of interest from a media collection. IEEE Trans. Inform. Forensic Secur. **9**(12), 2144–2157 (2014)
3. Martinho-Corbishley, D., Nixon, M., Carter, J.: Soft biometric recognition from comparative crowdsourced annotations. In: 6th International Conference on Imaging for Crime Prevention and Detection (ICDP 2015). IET, pp. 1–6 (2015)
4. Panjwani, S., Prakash, A.: Crowdsourcing attacks on biometric systems. In: Symposium On Usable Privacy and Security (SOUPS 2014), 2014, pp. 257–269 (2014)
5. Sandoval, A. López, E., Martínez, C., Cruz Rivas, L.: Sistema de Autenticación Facial mediante la Implementación del algoritmo PCA modificado en Sistemas embebidos con arquitectura ARM. La Mecatrónica en México, **4**, pp. 53–64, 2015
6. Andrade, C.N.: Autenticación por reconocimiento facial para aplicaciones web, utilizando software libre (2015)
7. Pérez Badillo, O., Poceros Martínez, F., Villalobos Ponce, J.: Sistema de seguridad por reconocimiento de voz (2013)
8. Morocho, D., Morales, A., Fierrez, J., Tolosana, R.: Signature recognition: establishing human baseline performance via crowdsourcing. In: Proceedings of 4th International Workshop on Biometrics and Forensics (IWBF). Limassol, Cyprus (2016)
9. Estelles-Arolas, E., Gonzalez-Ladron-de-Guevara, F.: Towards an integrated crowdsourcing definition. J. Inf. Sci. **38**(2), 189–200 (2012)
10. Amazon Mechanical Turk, [En línea]. Available: http://www.mturk.com. Último acceso: 3 Noviembre 2015
11. Alhaddad, M.: Multiple classifiers to verify the online signature. World Comput. Sci. Inf. Technol. J. (WCSIT) **2**, 46–50 (2012)

12. Cilimkovic, M.: Neural Networks and Back Propagation Algorithm. Institute of Technology Blanchardstown (2008)
13. Harrison, D., Burkes, T., Seiger, D.: Handwriting examination: meeting the challenges of science and the law. Forensic Sci. Commun. **11**(4) (2009). https://www.fbi.gov/hq/lab/fsc
14. Morocho, D., Morales, A., Fierrez, J., Vera-Rodriguez, R.: Towards human-assisted signature recognition: improving biometric systems through attribute-based recognition. In: IEEE International Conference on Identity, Security and Behavior Analysis (ISBA 2016). Sendai, Japan (2016)

Automated Image Registration for Knee Pain Prediction in Osteoarthritis: Data from the OAI

Jorge I. Galván-Tejada[1]([✉]), Carlos E. Galván-Tejada[1],
José M. Celaya-Padilla[1], Juan R. Delgado-Contreras[2], Daniel Cervantes[1],
and Manuel Ortiz[1,2]

[1] Unidad Académica de Ingeniería Eléctrica,
Universidad Autónoma de Zacatecas, Zacatecas, Zacatecas, Mexico
gatejo@uaz.edu.mx
[2] Instituto Tecnológico Superior de Zacetcas Sur, Tlaltenango, Zacatecas, Mexico
http://www.uaz.edu.mx

Abstract. Diagnose Knee osteoarthritis (OA) is a very important task, in this work an automated metrics method is used to predict chronic pain. In early stages of OA, changes into joint structures are shown, some of the most common symptoms are; formation of osteophytes, cartilage degradation and joint space reduction, among others. Using public data from the Osteoarthritis initiative (OAI), a set of X-ray images with different Kellgren Lawrence score (K & L) scores were used to determine a relationship between bilateral asymmetry and the radiological evaluation in K & L score with the chronic knee pain. In order to measure the asymmetry between the knees, the right knee was registered to match the left knee, then a series of similarity metrics; mutual information, correlation, and mean square error were computed to correlate the deformation (mismatch) and K & L score with chronic knee pain. Radiological information was evaluated and scored by OAI radiologist groups, all metric of image registration were obtained in an automated way. The results of the study suggest an association between image registration metrics, radiological K & L score with chronic knee pain. Four GLM models wit AUC 0.6 and 0.7 accuracy random forest classification model was formed with this information to classify the early bony changes with OA chronic knee pain.

Keywords: Osteoarthritis · Knee pain · K & L · Image registration

1 Introduction

Osteoarthritis (OA) is one of the most common diseases in the industrialized world. This incapacitating disease brings poor quality of life to many people, turning in a painful task working or even making daily life activities. This disease affects 1 in 10 adults over 60 years, only in the United States [25], OA is the most common manifestation of arthritis. In Mexico, OA is one of the principal causes of medical attention requests.

© Springer International Publishing Switzerland 2016
J.F. Martínez-Trinidad et al. (Eds.): MCPR 2016, LNCS 9703, pp. 335–345, 2016.
DOI: 10.1007/978-3-319-39393-3_33

X-Ray imaging is the first hand information that Diagnosing OA and determining the patient disease stage is critical, medical imaging is one of the techniques used to primary diagnose and staging.

There are several radiological methods to establish the stage in the OA. Expert radiologists evaluate radiological evidence of x-ray images to determine morphological changes such as appearance of osteophytes, structural changes or joint space narrowing (JSN) [1]. This radiological evidence has not been fully studied or associated with the most common symptoms [12,14,18,24]; pain and stiffness [8]. One of the biggest challenges is clearly associate radiological evidence with early symptoms that commonly have late onset, as it is the most common symptom in OA; pain.

Early diagnosis is very important because the knee pain is one of the most disabling symptoms even at early ages. There is a lack of standards in the form of how images are assessed. To attack this problem, OA community has developed a series of clinical questionnaires and some atlas [2,10,16,17] for image evaluation [3,22]. There are methods such as Kellgren-Lawrence (K & L) [16] that are commonly used by radiologists to evaluate images, but these methods still depend on the judgment of an experienced radiologist, and studies demonstrate that the same image can be evaluated very differently by two or more different radiologists, and even the same radiologist may assess differently the same film at two different times [19,20].

The development of an automatic method for the evaluation of the images is very important, this way the bias induced by the human factor is reduced, and the patient can be evaluated in an objective way. Using computational methods to evaluate x-ray images can be used as a second opinion and thus achieve an automatic patient early diagnosis of OA. For this reason, a large number of associations makes huge efforts to generate information about the disease in search of finding treatments or even a cure that does not exist now.

The Osteoarthritis Initiative (OAI) has been recollecting thousands of clinical data in OA patients, subjects at risk, and control subjects using validated questionnaires and standardized image assessment procedures. The OAI effort has bring very important information that will enable a better understanding of the disease process. Using x-ray images of the knee available in open databases OAI, the goal of this work is to use computational methods to correlate metrics obtained automatically with chronic pain as the main symptom.

Using image registration techniques [26] allows the alignment of both knees to measure the degree of asymmetry between knees, this asymmetry metrics aim to help find a relationship between the deformation of the bone structure and chronic knee pain.

Three different asymmetry metrics were obtained using the registered knees; Mutual information, Correlation, and Mean Square error in order to explore the relation of the asymmetry between the knees as the chronic pain. Using the automated metrics and the assessed K & L score, the association of the bony structure with the chronic pain is analyzed.

2   Materials and Methods

In this work, a series of images were processed by computational and radiological techniques. All the Materials and Methods are described below.

2.1   Data Acquisition

Public data from OAI was used to develop this work. "OAI is a multi-center, longitudinal, prospective observational study of knee OA. The OAI will establish and maintain a natural history database for osteoarthritis that will include clinical evaluation data, radiological (x-ray and magnetic resonance) images, and a bio-specimen repository from 4796 men and women aged between 45 and 79 years old, enrolled between February 2004 and May 2006" (https://oai.epi-ucsf.org). Bilateral knee x-ray images and the Central Reading of Knee X-rays for K-L Grade, chronic pain information, and Individual Radiographic Features of Knee Datasets were used. Subjects in the study were selected from the database using the baseline K & L score information and chronic pain data.

2.2   Subject Selection

A case/control study using K & L scores to predict pain cohort was used. Using a random process, 50 subjects of this cohort were selected, 10 for each K & L score level (0–4). The selection criteria for the cohort was:

Control subjects were selected under the criteria of:

1. Not presenting pain as a symptom since the baseline visit to 60 month visit;
2. Not presenting a symptomatic status since the baseline visit to 60 month visit;
3. Taking no pain medication from the baseline visit to the 60 month visit.

Case subjects were selected under the criteria of:

1. Not presenting pain as a symptom at baseline visit;
2. Not presenting a symptomatic status at baseline visit;
3. Taking no pain medication at the baseline visit;
4. Develop chronic right knee pain in some time point after baseline and up to 60 month visit.

2.3   Image Processing

Due the high dynamism of the knees, a direct comparison of the left and right image cannot be performed straight forward. An alignment of the images must be carried out prior to the similarity analysis. Three main steps were performed:

1. The knees of the patient are segmented in order to remove unwanted information.
2. An alignment (registration) of the left knee to the right is carried out.

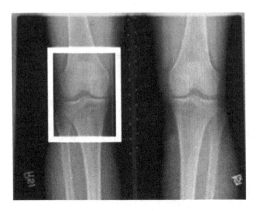

Fig. 1. Input image, ROI to be used is displayed in a yellow rectangle (Color figure online)

3. A similarity metrics are computed in order to evaluate the relationship of the degree of similarity between knees and the stages of the disease.

All images were pre-processed manually. During this process, the area of interest in the images was enclosed to prevent unwanted information i.e. (Tags, noise, marks) into the regions. The region of interest (ROI) in the x-ray image was adjusted, this process generates an individual image for each knee, then, the left knee image is horizontally flipped to enable the image registration process, a logarithmic transformation in each pixel is performed to enhance the low intensity pixels in the image [15]. In Fig. 1 shows an example of the ROI used in this research.

Image Segmentation. All background noise and all image artifacts were removed from the x-ray images using an automated segmentation process. The automatic segmentation of the knee was based on the background noise. This process created a segmentation mask, this mask discards any pixel above a threshold value, this threshold value is unique for each image. Afterwards, the largest connected region of the entire image, went through a hole filling algorithm based on morphological operations to ensure a solid surface extraction. This segmentation process using the mask to extract only the knee bones structure and discard all background.

Image Registration. For the registration process of the left knee into the right knee, the segmentation mask was used as ROI. The left knee image was mirrored and then co-registered with it's corresponding right knee image.

A B-Spline multi resolution algorithm was used to optimize the Mattes mutual information metric in the bilateral image registration [5]. A deformable B-Spline transform is used, this is based on transforming an image modifying control points contained in a mesh based on a maximization of a similarity measure, this approach often avoids local minimal in the parameter search space and reduces computational time [9,23], In this multi resolution approach, all images are registered in the lowest resolution, and in the next steps, the transformation parameters are scaled to the higher resolutions and parameter optimization is computed again. The registration algorithm returns a transformation file: T(x,y) that is used to find a point in the left image for each point (x,y) in the right image [4].

Metric Quantification. To establish the relation between the registered image and the target image, three representative measurements were computed, sum of square differences [7], correlation coefficient, mutual information. This set of measurements is wide used in the literature [13] as a measurement of similarity between two images.

The sum of squared differences (SSD) in Eq. 1 assumes that the images are identical at registration, and the implicit assumption that there are no differences between intensity profiles. This metric is sensitive to outliers: small number of voxels having large intensity differences. Where f(x) is the intensity at a position x in an image, m(t(x)) is the intensity at the corresponding point given by the transformation t(x). N is the number of pixels in the region of overlap.

$$SSD = \frac{1}{N} \sum_X (f(X) - m(t(X)))^2, \tag{1}$$

The correlation coefficient (CC) is a measure of the linear dependence between two variables X and Y or images, in Eq. 2 is described the CC, where f (x) is the intensity at a position x in the fixed image, and m (t (x)) is the intensity at the corresponding point given by the transformation t (x) in the moving image, the number of pixels in overlapped region is represented by N.

$$CC = \frac{\sum_X (f(X).f) * (m(t(X)) - m)}{\sqrt{\sum_X (f(X) - f)^2 * \sum_X (m(t(x)) - m)^2}}, \tag{2}$$

Mutual information (MI) shown in Eq. 3, provides a measure of probabilistic dependence between two intensity distributions. For this research, the Shannon-Wiener entropy measure H in Eq. 3 is used, this measurement is one well-known used measure of information in image processing, this measurement was originally developed as part of communication theory in the 1940 s [17].

$$MI(m, f) = H(m) - H(f|m), \tag{3}$$

In Eq. 4 H(X,Y) represents the joint entropy, and H(X) the individual entropy of random image X, Y, N stands for the number of intensity levels, and Px (Pxy) is the probability of value X (x,y) in the (joint) probability distribution of variable X (X and Y).

$$H(X) = -\sum_{x=0}^{N} P_x log_2(P_x), \tag{4}$$

2.4   Statistical Analysis

In a first step, using only the variables obtained by the method of image registration; MI, CORR, and MS, logistic regression as cost function is used for the analysis. A general linear model is generated and analyzed, odds ratio, and the area under the receiver operating characteristic (ROC) curve (AUC) were calculated. The ROC curve was constructed and the curve is a graphical representation of the sensitivity against 1-specificity for a binary classifier system as the discrimination threshold is varied. The outcome variable in this analysis was the left knee chronic pain.

In a second case, the information obtained from the image registration process and the information evaluated for the K & L score is used with logistic regression to generate a new general linear model, ODD, and AUC were calculated for the model. In this case the outcome variable was chronic left knee pain.

Finally, using the variables that had the highest ODD, a known classifier, it is generated using a Random Forest technique. The classifier was run using chronic left knee pain as outcome variable, and 5000 trees as defaults.

All statistical analyzes were performed using statistical software R and its packages [21].

3   Results

Based on the experimental work, the obtained results are presented. First the results obtained from the image registration metrics are shown in Table 1, results on MI, CORR, and MS obtained for each of the levels of K & L of 50 analyzed patients for this work are presented by OA stage.

After obtaining the metrics, linear models were generated using logistic regression as classification function, statistical results and their AUC are presented in the Table 2. In Fig. 2, ROC curves for each model are shown.

Once the ODD ratios were obtained, the 5000 trees random forest classifications presented the results below:

CORR. Using the variable with the highest ODD, a 0.7143 accuracy classification was obtained.

CORR+K & L. Using the variables in the best GLM models, a 0.6122 accuracy classification was obtained.

Table 1. Metrics by K & L

SUBJECT	K & L score = 0			K & L score = 1			K & L score = 2			K & L score = 3			K & L score = 4		
	MI	CORR	MS	MI	CORR	MS	MI	CORR	MS	MI	CORR	MS	MI	CORR	MS
1	0.98	0.97	6.92E+05	0.51	0.92	1.94E+06	0.6	0.93	8.40E+05	0.6	0.93	2.84E+06	0.64	0.93	5.70E+06
2	0.95	0.95	1.58E+06	0.84	0.97	4.24E+06	1.11	0.95	2.53E+06	0.92	0.96	4.92E+06	0.77	0.96	4.75E+06
3	1.16	0.97	5.59E+05	1.19	0.98	3.10E+06	1.29	0.98	2.76E+06	0.98	0.97	2.34E+06	0.86	0.95	2.11E+06
4	1.04	0.97	4.53E+06	0.74	0.95	1.55E+06	0.88	0.96	4.04E+06	0.85	0.95	1.80E+06	0.7	0.92	2.36E+06
5	0.86	0.96	2.10E+06	1.11	0.98	3.37E+06	0.75	0.93	4.99E+06	1.05	0.97	4.95E+06	0.64	0.97	2.15E+06
6	0.78	0.95	6.73E+05	1.27	0.98	2.39E+06	1.35	0.99	6.07E+06	1.03	0.97	1.62E+06	0.79	0.97	6.42E+05
7	1.25	0.97	3.62E+06	0.86	0.97	3.12E+06	0.85	0.96	5.44E+05	0.6	0.95	7.00E+05	0.72	0.96	4.02E+06
8	1.13	0.98	7.30E+05	0.67	0.97	4.31E+06	0.91	0.97	5.76E+06	0.99	0.96	5.60E+05	0.65	0.97	5.49E+06
9	1.17	0.98	4.08E+06	0.86	0.96	1.87E+06	0.57	0.97	1.27E+06	0.66	0.91	3.58E+06	0.61	0.95	6.55E+06
10	1.02	0.98	7.06E+05	0.56	0.96	4.95E+05	0.63	0.97	2.17E+05	0.67	0.97	3.92E+06	1.12	0.97	2.54E+06

Table 2. Linear models statistical information

Model	Variable	ODD	2.50 %	97.5 %	AUC
GLM1	MI	0.98	0.08	11.96	0.61
	CORR	2.68	0.27	29.82	
	MS	0.65	0.05	7.28	
	K & L	0.83	0.51	1.31	
GLM2	MI	1.11	0.09	13.41	0.58
	CORR	2.67	0.27	28.84	
	MS	0.51	0.04	5.34	
GLM3	CORR	2.67	0.28	28.90	0.624
	K & L	0.81	0.51	1.27	
GLM4	CORR	2.6	0.28	27.04	0.58

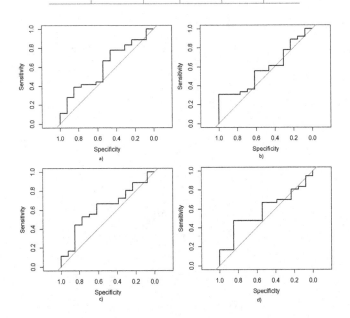

Fig. 2. ROC curves; (a) GLM1, (b) GLM2, (c) GLM3, (d) GLM4

4   Conclusion

This paper presents for the first time in medical diagnosis the use of automatic asymmetry metrics to associate symptoms such as chronic knee pain. The results shown in the Table 2, demonstrate a behavior similar to models based on a search within tens of variables. The association of changes in the bone structure can be detected by the asymmetry that might be associated with early chronic knee pain.

The set of information obtained automatically, and radiological assessment of an expert improves performance classification of a power regression model.

On the other hand, using the variable that determines a greater risk to the subject of developing the symptoms, and use that variable as a classifier on a random forest, delivers us an accuracy of more than 0.7, that within the standard prediction of pain is above those reported in the work of sorting pain [11].

In works like [18], a cohort with all subjects diagnosed with OA, the AUC obtained is 0.64 using radiological evaluations, and 0.70 using evaluations and K & L scores. In a [6] study, the reported the OR are between 1.95 and 2.11.

In this work, the metrics obtained without a radiologist intervention have a similar performance in prediction and OR.

Pain is one of the most complex to explain symptoms, the fact that morphological characteristics can be associated at least as an early risk factor may help develop a treatment by early diagnosis.

This type of association of pain may be the first step in developing a tool that helps to have a second opinion to decrease the workload of the radiologist, and thus deliver more accurate diagnoses.

The limitations of this study are mainly the number of patients enrolled in the study, although medical standards comply with 20 patients minimum.

For future work, the number of patients which metrics are performed will be increased. The association with chronic pain will be complemented by other scales known as WOMAC pain and KOOS. To obtain a robust model is to combine it with some other type of clinical or radiological variables may obtain a biomarker for OA.

Acknowledgment. "The OAI is a public-private partnership comprised of five contracts (N01-AR-2-2258; N01-AR-2-2259; N01-AR-2-2260; N01-AR-2-2261; N01-AR-2-2262) funded by the National Institutes of Health, a branch of the Department of Health and Human Services, and conducted by the OAI Study Investigators. Private funding partners include Merck Research Laboratories; Novartis Pharmaceuticals Corporation, GlaxoSmithKline; and Pfizer, Inc. Private sector funding for the OAI is managed by the Foundation for the National Institutes of Health. This manuscript was prepared using an OAI public use data set and does not necessarily reflect the opinions or views of the OAI investigators, the NIH, or the private funding partners."

References

1. Altman, R., Asch, E., Bloch, D., Bole, G., Borenstein, D., Brandt, K., Christy, W., Cooke, T., Greenwald, R., Hochberg, M., et al.: Development of criteria for the classification and reporting of osteoarthritis: Classification of osteoarthritis of the knee. Arthritis Rheum. **29**(8), 1039–1049 (1986)
2. Altman, R.D., Gold, G.: Atlas of individual radiographic features in osteoarthritis, revised. Osteoarthritis Cartilage **15**, A1–A56 (2007)
3. Bellamy, N.: Validation study of womac: A health status instrument for measuring clinically-important patient-relevant outcomes following total hip or knee arthroplasty in osteoarthritis. J. Orthop. Rheumatol. **1**, 95–108 (1988)

4. Celaya-Padilaa, J.M., Rodriguez-Rojas, J., Trevino, V., Tamez-Pena, J.G.: Local image registration a comparison for bilateral registration mammography. In: IX International Seminar on Medical Information Processing and Analysis, pp. 892210–892210. International Society for Optics and Photonics (2013)

5. Celaya-Padilla, J., Martinez-Torteya, A., Rodriguez-Rojas, J., Galvan-Tejada, J., Treviño, V., Tamez-Peña, J.: Bilateral image subtraction and multivariate models for the automated triaging of screening mammograms. BioMed Res. Int. 2015 (2015)

6. Creamer, P.: Osteoarthritis pain and its treatment. Curr. Opin. Rheumatol. **12**(5), 450–455 (2000)

7. Crum, W.R., Hartkens, T., Hill, D.: Non-rigid image registration: theory and practice. Br. J. Radiol. (2014)

8. Dekker, J., Boot, B., van der Woude, L.H., Bijlsma, J.: Pain and disability in osteoarthritis: a review of biobehavioral mechanisms. J. Behav. Med. **15**(2), 189–214 (1992)

9. Diez, Y., Oliver, A., Lladó, X., Freixenet, J., Martí, J., Vilanova, J.C., Marti, R.: Revisiting intensity-based image registration applied to mammography. IEEE Trans. Inf. Technol. Biomed. **15**(5), 716–725 (2011)

10. Duryea, J., Li, J., Peterfy, C., Gordon, C., Genant, H.: Trainable rule-based algorithm for the measurement of joint space width in digital radiographic images of the knee. Med. phys. **27**(3), 580–591 (2000)

11. Galván-Tejada, J.I., Celaya-Padilla, J.M., Treviño, V., Tamez-Peña, J.G.: Multivariate radiological-based models for the prediction of future knee pain: Data from the OAI. Comput. Math. Methods Med. **2015**, 1–10 (2015)

12. Glass, N., Torner, J., Law, L.F., Wang, K., Yang, T., Nevitt, M., Felson, D., Lewis, C., Segal, N.: The relationship between quadriceps muscle weakness and worsening of knee pain in the most cohort: a 5-year longitudinal study. Osteoarthritis Cartilage **21**(9), 1154–1159 (2013)

13. Guo, Y., Sivaramakrishna, R., Lu, C.C., Suri, J.S., Laxminarayan, S.: Breast image registration techniques: A survey. Med. Biol. Eng. Comput. **44**(1–2), 15–26 (2006)

14. Hochman, J., Davis, A., Elkayam, J., Gagliese, L., Hawker, G.: Neuropathic pain symptoms on the modified paindetect correlate with signs of central sensitization in knee osteoarthritis. Osteoarthritis Cartilage **21**(9), 1236–1242 (2013)

15. Jourlin, M., Pinoli, J.C.: Logarithmic image processing: The mathematical and physical framework for the representation and processing of transmitted images. Adv. Imaging Electron Phys. **115**, 129–196 (2001)

16. Kellgren, J., Lawrence, J.: Radiological assessment of osteo-arthrosis. Ann. Rheum. Dis. **16**(4), 494 (1957)

17. Kellgren, J., Lawrence, J.: Atlas of Standard Radiographs: The Epidemiology of Chronic Rheumatism, vol. 2. Blackwell Scientific, Oxford (1963)

18. Kinds, M.B., Marijnissen, A.C., Bijlsma, J.W., Boers, M., Lafeber, F.P., Welsing, P.M.: Quantitative radiographic features of early knee osteoarthritis: Development over 5 years and relationship with symptoms in the check cohort. J. Rheumatol. **40**(1), 58–65 (2013)

19. Kornaat, P.R., Ceulemans, R.Y., Kroon, H.M., Riyazi, N., Kloppenburg, M., Carter, W.O., Woodworth, T.G., Bloem, J.L.: MRI assessment of knee osteoarthritis: knee osteoarthritis scoring system (koss)inter-observer and intra-observer reproducibility of a compartment-based scoring system. Skeletal Radiol. **34**(2), 95–102 (2005)

20. Pathria, M., Sartoris, D., Resnick, D.: Osteoarthritis of the facet joints: Accuracy of oblique radiographic assessment. Radiology **164**(1), 227–230 (1987)

21. Ripley, B.D.: The R project in statistical computing. MSOR Connections **1**(1), 23–25 (2001)
22. Roos, E.M., Lohmander, L.S.: The knee injury and osteoarthritis outcome score (koos): From joint injury to osteoarthritis. Health Qual. Life Outcomes **1**(1), 64 (2003)
23. Rueckert, D., Sonoda, L.I., Hayes, C., Hill, D.L., Leach, M.O., Hawkes, D.J.: Nonrigid registration using free-form deformations: application to breast MR images. IEEE Trans. Med. Imaging **18**(8), 712–721 (1999)
24. Shimura, Y., Kurosawa, H., Sugawara, Y., Tsuchiya, M., Sawa, M., Kaneko, H., Futami, I., Liu, L., Sadatsuki, R., Hada, S., et al.: The factors associated with pain severity in patients with knee osteoarthritis vary according to the radiographic disease severity: a cross-sectional study. Osteoarthritis Cartilage **21**(9), 1179–1184 (2013)
25. White, D.K., Tudor-Locke, C., Felson, D.T., Gross, K.D., Niu, J., Nevitt, M., Lewis, C.E., Torner, J., Neogi, T.: Do radiographic disease and pain account for why people with or at high risk of knee osteoarthritis do not meet physical activity guidelines? Arthritis Rheum. **65**(1), 139–147 (2013)
26. Zitova, B., Flusser, J.: Image registration methods: A survey. Image Vis. Comput. **21**(11), 977–1000 (2003)

Towards a Supervised Incremental Learning System for Automatic Recognition of the Skeletal Age

Fernando Montoya Manzano[1], Salvador E. Ayala-Raggi[1(✉)],
Susana Sánchez-Urrieta[1], Aldrin Barreto-Flores[1],
José Francisco Portillo-Robledo[1], and Verónica Edith Bautista-López[2]

[1] Facultad de Ciencias de la Electrónica,
Benemérita Universidad Autónoma de Puebla, Avenue San Claudio and 18 Sur,
Col. Jardines de San Manuel, 72570 Puebla, Puebla, Mexico
fmm_ferix@hotmail.com, {saraggi,surrieta,abarreto,portillo}@ece.buap.mx
[2] Facultad de Ciencias de la Computación,
Benemérita Universidad Autónoma de Puebla, Avenue San Claudio and 18 Sur,
Col. Jardines de San Manuel, 72570 Puebla, Puebla, Mexico

Abstract. In this work, we proposed and developed a simple system to estimate skeletal maturity based in using Active Appearance Models in order to create an increasing set of shape-aligned training images which are incrementally stored and used by a $K - NN$ regression classifier. For that purpose, we designed an original layout of landmarks to be located in representative regions of the radiographical image of the hand. Our results show that is possible to use pixels directly as classification features as long as the training and testing images have been previously aligned in shape and pose.

Keywords: Skeletal maturity recognition · Bone age estimation · Active appearance models · $K - NN$ regression

1 Introduction

Skeletal maturity estimation is an important issue in the proper diagnosis of a great set of diseases and growth problems. Estimation is often performed manually by the radiologist who subjectively compares bones and joints between the test image, normally a radiographic image of the left hand, and several template images taken from a standard handbook [1,2]. This procedure is prone to produce wrong or inaccurate age classification mainly caused by two important reasons: human errors due to subjective comparisons, and the fact that there is a

V.E. Bautista-López—The authors wish to thank Imagen Exakta S.A. de C.V., Dr. Patricia Ayala-Raggi MD, and Dr. Juan de Dios Meza-Rojas MD for providing them with a special permission to use a set of anonymous radiological images for academic purposes.

© Springer International Publishing Switzerland 2016
J.F. Martínez-Trinidad et al. (Eds.): MCPR 2016, LNCS 9703, pp. 346–355, 2016.
DOI: 10.1007/978-3-319-39393-3_34

single template image per age in the manual. Even though most utilized manuals have been carefully designed by experts in the field, the fact of using a single prototype hand image could be unsuitable because the great appearance variability in human hands. On the other hand, other approaches commonly used by some radiologists based in calculating different scores for different groups of bones and then carrying out a weighted average is extremely impractical for many physicians. And although it is more accurate than the former methods, it is subjective too [3,4]. Many automatic approaches have been proposed to overcome the problem of subjectivity. Niemeijer et al. [5] proposed an automatic system based on Tanner-Whitehouse method. By using a large number of Regions Of Interest, or shortly $ROIs$, labeled with ages by a radiologist, a mean image for each age is computed. Then, when a new input ROI is entered for classification, an active shape model, or ASM [6], is used to align that to the mean images, and by using correlation the algorithm selects the age label assigned to the mean image that best matches the input ROI. In that approach we observe two drawbacks: first, the mean images used do not contain the variability observed in the original images utilized to construct them. Second, the set of ages is finite, and therefore the estimated age is not a continuous variable. In [3], Hsieh et al. proposed to manually extract geometric features from carpal bones radiographs for ages from one to eight years, and then use artificial neural networks, shortly $ANNs$. Somehow, it is in part of an expert system combined with a supervised learning approach. The purpose of our work was not to automate some of the known clinical methods for bone age assessment [1,2], as done in some previous works [7]. Instead of that, we found it very interesting to investigate and design an automatic bone age learning system that is capable to learn from test examples. As previous knowledge, only the age labels from those examples should be used. Although there have been efforts to achieve bone age estimation algorithms based only on statistically learned features from lots of examples like [8,9], we have seen that no efforts have been done in designing automatic learning systems for estimating bone age with the capability of improving their performance and learning as more images are presented to the system. It would be practical and useful that a system could improve its performance even when it is being used for testing. Typical classification algorithms that are inherently incremental in that sense are precisely those based on the K nearest neighbors search or simply $K-NN$ algorithms. In our work, we have used a regression version of the classical $K-NN$ algorithm which is based on radial basis functions. At first glance, it would seem possible to add new images to the system every time you want to increase learning, and therefore the performance in the classification process too. However, and considering that the orinal images are not aligned, it would be necessary a considerably large number of training images in order to reach an acceptable classification rate. To address this problem, we have implemented a previous alignment algorithm based on a pre-trained Active Appearance Model, or shortly AAM [10], which segments the region of the hand and then warps its texture into a normalized hand shape. Thus, each example to add will become shape-aligned, and important features will be aligned too. This shape-normalized image data and the reduced set of shape parameters returned by the AAM fitting process can be joined together to create a new feature vector which can be added to a knowledge base or training examples set where a $k - NN$ algorithm performs classification.

2   System Overview

In our proposed recognition system, we distinguish two important moments, the first one was the development stage and the second one is the usage stage. During the development an AAM model [11] was created from a set of manual labeled hand images with sufficient variability in shape and age. The purpose of this AAM model is to align or fit the model to a new test hand image during the usage stage. Then, during the usage stage, the user has two options: test a new image, or enter a new image for incremental learning. In both options, the user enters a new image to the system and the AAM alignment is carried out after the user provides an initial location to the model. When the fitting process ends and the model converges and resembles the original hand image, we can segment the hand's region and warp the original texture from the test image into a shape-normalized grid in order to create a shape-normalized hand image. This image is joined to the shape parameters returned by the AAM fitting process creating a feature vector. If the user selected the *test* option, the feature vector can be classified using $K - NN$ regression. On the other hand, if the user selected the *incremental learning* option, the feature vector together with the age label is added to a knowledge base, which is a set of these feature vectors. At the starting point of the usage stage, when no test images have been given to the system, the knowledge base is set up with the feature vectors corresponding to the hand images used to train the AAM model. Therefore the system can learn during the usage stage and it is not required to re-train the AAM model, which would be an expensive operation. Figure 1 shows the whole process.

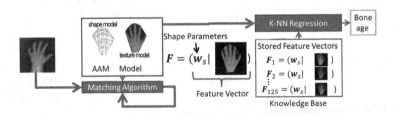

Fig. 1. Overall process diagram

3   Creating an Active Appearance Model of Radiographical Hand Images

We have built an AAM model in order to model the shape and texture variability of hands from radiographical images. $AAMs$ are parametric models based in applying principal component analysis, or PCA, both to textures and shapes from a large set of training hand images. Shape is modeled by manually placing a set of landmarks over distinctive points on each hand image. We call a *shape* or simply $\mathbf{s}_i$ to the set of landmarks (x_j, y_j) of a particular hand image. Thus, and after an aligning procedure, we apply PCA to the set of $\mathbf{s}_i$ in order to obtain a reduced set of *eigenshapes* capable to represent a high percentage of the shape variance showed by the training set. On the other hand, textures contained inside the $\mathbf{s}_i$ are normalized in shape by mapping o warping them to the mean shape before applying PCA to them. Similarly to the shape, we will obtain a reduced set of *eigentextures*, [12], capable to represent a high percentage of the texture variance showed by the training set. This capability of high representativeness using just a reduced set of *eigenshapes* and *eigentextures* is possible thanks to the high similarity among the training images. By computing linear combinations of the *eigenshapes* and *eigentextures* we are able to reconstruct every hand of the training set, or even create or synthesize a novel hand image. The coefficients, more commonly known as *weights*, of those linear combinations are the parameters of the AAM model. Once the AAM model has been created, it can be used for aligning it to an input image. By using a fitting iterative algorithm similar to that implemented in [10], it is possible to recover the shape and texture parameters for the model which best matches the hand portrayed in the input image.

In order to construct the shape model, we designed a proper layout of landmarks and a respective grid of triangles such that the triangles geometrically could never flip due to variations in shape present in the training set. Figure 2 shows our proposed layout of landmarks placed in the corners of the hand bones, and Fig. 3 illustrates our proposed manual triangulation used in our project compared with an automatic *delaunay* triangulation [13] automatically generated by MATLAB. We can see that our triangulation is more robust to variation in landmark locations than the *delaunay* triangulation because the landmarks of some triangles are near to be collinear.

Fig. 2. Set of landmarks placed in distinctive points of the bones structures

Once the images have been manually labeled with the same set of landmarks, an iterative procedure based in Procrustes Analysis [11,14] has to be applied to the set of *shapes* in order to align them in the rigid body sense, i.e. *scale*, *rotation*, and *traslation*. After applying PCA to the that set of aligned shapes, and giving a proper set of shape parameters, every *shape* in the training

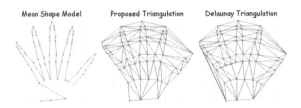

Mean Shape Model   Proposed Triangulation   Delaunay Triangulation

Fig. 3. Mean shape model, proposed triangulation and triangulation calculated by *delaunay* method

set can be reconstructed by the following expression,

$$\mathbf{s} = \bar{\mathbf{s}} + \mathbf{Q}_s\mathbf{w}_s \qquad (1)$$

where $\mathbf{s}$ is the output shape, $\bar{\mathbf{s}}$ is the mean shape, $\mathbf{Q}_s$ is a matrix whose columns are *eigenvectors* that we call *eigenshapes*, and $\mathbf{w}_s$ is the vector of shape parameters. The number of *eigenshapes* and therefore the number of shape parameters can be dramatically reduced if we preserve only the *eigenshapes* corresponding to their largest respective *eigenvalues*. Thus, the number of *eigenshapes* needed to represent a high percentage of the variance in the training set could be much smaller than the number of training examples. By using Eq. 2 and the respective set of *shape* parameters corresponding to each training image, textures of the training images can be warped into the mean shape, creating a set of *shape − normalized* textures. Figure 4 illustrates that process. Thus, we can apply *PCA* to this new set of *shape − normalized* training images. This process is known as *eigenfaces*, see [12].

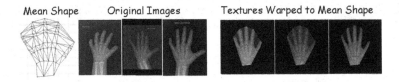

Mean Shape   Original Images   Textures Warped to Mean Shape

Fig. 4. Example of how warped images to mean shape look like.

Any texture from the training set can be approximated using the following expression:

$$\mathbf{g} = \bar{\mathbf{g}} + \mathbf{Q}_g\mathbf{w}_g \qquad (2)$$

where $\mathbf{g}$ is the reconstructed texture, $\bar{\mathbf{g}}$ is the mean texture, $\mathbf{Q}_g$ is a matrix whose columns we call *eigentextures* and $\mathbf{w}_g$ is a vector of texture weights or texture parameters. The number of *eigentextures* and therefore the number of texture parameters can be reduced if we preserve only the *eigentextures* corresponding to their largest respective *eigenvalues*. The *AAM* model parameters are completed when pose parameters (s, Θ, t_x, t_y) (scaling, in-plane rotation angle, translation) are added, $\mathbf{p} = (\mathbf{w}_s^T, \mathbf{w}_g^T | s, \Theta, t_x, t_y)$.

3.1   Fitting the *AAM* Model to an Input Hand Image

We have used the iterative algorithm described in [10] to align the *AAM* model
to an input hand image. Although in essence it is the same alignment algorithm,
some modifications have been necessary in order to work properly with radiographic images instead of face images. In face images the background is independent from face. Faces appear like solid objects over the background. However,
in radiographic images, hands are objects with certain degree of transparency
over a flat background. Therefore, the gray level of the background is present
inside the hand structure.

The alignment process of the model to an input image is computed by using
an iterative algorithm (see [10] for details) which calculates the residual $\mathbf{r}(\mathbf{p})$
(the current difference between the model and image) each iteration. This residual is always measured in a shape-normalized texture frame). By assuming the
relationship between the residual $\mathbf{r}(\mathbf{p} + \Delta\mathbf{p})$ and $\Delta\mathbf{p}$ as approximately constant
(denoted as $\frac{\delta r}{\delta p}$), a proper additive increment to the parameters $\Delta\mathbf{p}$ can be computed in each iteration by

$$\delta\mathbf{p} = -\mathbf{R}\mathbf{r}(\mathbf{p}) \quad where \quad \mathbf{R} = (\frac{\delta r}{\delta \mathbf{p}}^T \frac{\delta r}{\delta \mathbf{p}})^{-1} \frac{\delta r}{\delta \mathbf{p}}^T \tag{3}$$

where $\frac{\delta r}{\delta \mathbf{p}}$ is a Jacobian matrix composed whose number of columns equals the
number of model parameters. The jth column of this Jacobian was computed by
systematically displacing each parameter from its initial value for the synthetic
mean model. In order to avoid numeric errors and for increasing speed, the
pseudo-inverse matrix of Moore-Penrose is used instead of normal matrix inverse
during the calculation of Matrix $\mathbf{R}$.

In [10], the *AAM* alignment algorithm works fine on an input image with
black background if and only if the matrix $\mathbf{R}$ has been computed from synthetic
face images with black background. Similarly, in the case of aligning to radiographical images, a proper alignment will be possible if and only if the matrix
$\mathbf{R}$ has been computed from synthetic hand images with a background which
depends on the gray level inside the hand structure. In order to fulfill the former requirement, a synthetic background gray level has been generated during
the calculation of the matrix $\mathbf{R}$. After multiple tests, we proposed a method to
compute the background gray level by

$$b = \bar{m}_{x,y} - 0.8\sigma_{x,y} \tag{4}$$

where $\bar{m}_{x,y}$ is the mean of the gray level of all pixels (x, y) inside the synthetic
hand region (including inter-fingers spaces), and $\sigma_{x,y}$ is the standard deviation
of the gray level for all same pixels (x, y).

Figure 5 shows two cases in the process of construction of the matrix $\mathbf{R}$.
In both cases mean texture $g(\mathbf{p} = 0)$ was altered by systematically displacing
parameters $\mathbf{p}$ in an increment $\Delta\mathbf{p}$. At the left, the background gray level was set
to zero. At the right, the background gray level was set to the value calculated
by Eq. 4.

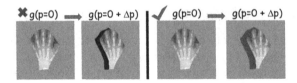

Fig. 5. Two cases in the construction of matrix **R**. Left: In this case the background in synthetic images is set to zero, and the residuals calculated for construction of the matrix **R** are different in nature to those obtained during the fitting process producing a wrong alignment. Right: The background in synthetic images is calculated from the hand internal pixels, and the residuals calculated for construction of the matrix **R** are similar to those computed during the fitting process

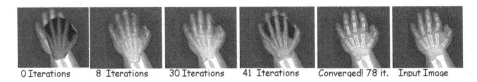

Fig. 6. Iterative alignment of the *AAM* model to an input image.eps

We found that by using Eq. 4 the convergence of the fitting process was reached correctly. Figure 6 illustrates the alignment of fitting process on a baby radiographic hand image.

4   Bone Age Classification

The shape parameters returned from fitting the model to a new image are joined to the shape-normalized texture from the original image in order to create a feature vector useful for $K - NN$ classification.

In theory, we can use the parameters of shape and texture returned by the *AAM* alignment process to create short feature vectors which are suitable for classification. However, proceeding in that way does not allow us to implement an incremental learning process because a new *AAM* model would have to be rebuilt every time a new example image is added to our knowledge base. Of course, this is an expensive operation. Therefore, in order to design a practical incremental learning system, we have chosen to join the shape parameters returned from the fitting process with the shape-normalized texture from the input image, creating new vector of aligned features $\mathbf{F} = (\mathbf{w}_s^T | \mathbf{g}^T)$.

Thus, we do not need to recompute the *AAM* model every time a new image is added for learning. Therefore, we must ensure that *AAM* model is made from enough variability in shape and texture in order to be capable to fit almost every input image.

4.1 Age Classification by $K - NN$ Regression

The estimated feature vector $\mathbf{F}$ can be classified by comparing it with those stored in a set that we call *knowledge base*. In order to carry out the classification we implemented a K-Nearest Neighbor regression algorithm [15,16]. This process consists of finding the k feature vectors stored in the knowledge base which are the nearest in the Euclidean distance sense. We use these distances d_i ($i = 1, ..., k$) to compute respective weight values W_i for each neighbor such that the greater the distance the weight value will be lower.

$$W_i = \exp \frac{-d_i^2}{2\sigma^2} \tag{5}$$

where σ is a constant that can be obtained by trial and error. Therefore, bone age can be computed as a weighted average of the respective age labels of the k neighbors

$$\mathbf{age} = \frac{\sum_{i=1}^{k} W_i Y_i}{\sum_{i=1}^{k} W_i} \tag{6}$$

where Y_i are the age values of the k neighbors.

5 Experimental Results

To test our age recognition method, we used an image set composed by 165 radiographical images all them cropped and resized to 256 × 256. 125 images from that set were utilized for training and learning, and the remaining 40 were reserved for testing. From the training and learning set, 65 images were used for training the AAM model. The shape-normalized textures of these 65 images joined to their respective shape parameters were used as an inicial set of feature vectors for the knowledge base. The remaining 60 were used for incremental learning during the usage stage.

For the AAM model we designed a layout of 71 landmarks located over distinctive corners of hand bones. We decided to preserve 30 *eigenshapes* which provide us 99 % of the variance observed in the training set. Similarly, we used 25 *eigentextures* representing 99 % of the variance observed in the training set. For $K - NN$ regression, by setting $k = 5$ we obtained the best results. Figure 7(B) shows the estimated ages for 40 tests by using a knowledge base containing 125 feature vectors. We obtained a Mean Absolute Error (MAE) of 1.8 years, a Mean Error (ME) of −0.08 years, and a Root Mean Square Error ($RMSE$) of 1.87 years.

In order to evaluate the capability of our system to incrementally learn, we tested age recognition using the set of 40 test images every time the knowledge base was incremented with additional subsets of 10 feature vectors. The process started with a knowledge base containing just the first 65 features vectors, and finished with 125 feature vectors. Figure 7(A) shows a gradual decrease in the error rate ($R - MSE$) every time that the knowledge base is incremented.

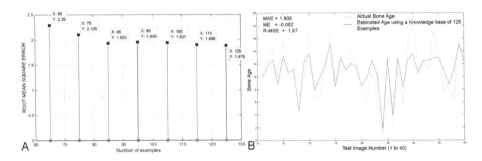

Fig. 7. (A). The root mean square error was measured for 40 test images each time a set of 10 learning images was added to the knowledge base. Learning images have been added from image 65 to image 125. (B). Actual ages vs estimated ages with 125 examples

6   Conclusions and Future Work

In this paper we have proposed and implemented an automatic skeletal maturity recognition system which is capable of estimating bone age with an acceptable accuracy. In contrast to other works, we proposed to investigate the problem of bone age estimation in the context of supervised learning, based in giving only the training examples and their age labels to the computer. Just images and ages should be used as previous knowledge. In addition, we proposed a simple incremental learning methodology which gradually reduces the classification error when more images are presented to the system during the usage stage, without the need to re-train the *AAM* model. This incremental learning skill can not be implemented using only the *AAM* parameters, because the addition of novel training images implies to make a reconstruction of the *AAM* model, and that is not practical. On the other hand, unaligned images can be directly used in classifiers like $k - NN$ or neural networks but the required quantity of them should be really large in order to obtain acceptable results. In our approach, we proposed to use an *AAM* model just for segmenting and aligning the hand region in order to produce a shape-normalized hand image that can be joined to the shape parameters returned by the *AAM* fitting process. This join of vectors can be used as a unique vector which contains aligned features. Therefore, by using a small quantity of images, it is possible to reach acceptable classification rates as we have demonstrated. As a future work, we propose to investigate approaches for reducing the quantity of examples required for incremental learning. Redundant examples should be avoided in such a way that only novel information could enter to the system.

References

1. Greulich, W., Pyle, S.: Radiographic Atlas of Skeletal Development of Hand and Wrist, 2nd edn. Stanford University Press, Stanford (1971)
2. Tanner, J., Whitehouse, R., Cameron, N., Marshall, W., Healy, M., Goldstein, H.: Maturity and Prediction of Adult Height (TW2 Method), 2nd edn. Academic Press, London (1975)
3. Hsieh, C.W., Jong, T.L., Chou, Y.H., Tiu, C.M.: Computerized geometric features of carpal bone for bone age estimation. Chin. Med. J. **120**, 767–770 (2007)
4. Molinari, L., Gasser, T., Largo, R.H.: Tw3 bone age: Rus/cb and gender differences of percentiles for score and score increments. Ann. Hum. Biol. **31**, 421–435 (2004)
5. Niemeijer, M., van Ginneken, B., Maas, C., Beek, F., Viergever, M.: Assessing the skeletal age from a hand radiograph: automating the tanner-whitehouse method. In: Sonka, M., Fitzpatrick, J. (eds.) SPIE Medical Imaging, vol. 5032, pp. 1197–1205. SPIE (2003)
6. Cootes, T.F., Taylor, C.J., Cooper, D.H., Graham, J.: Active shape models-their training and application. Comput. Vis. Image Underst. **61**, 38–59 (1995)
7. Aja-Fernndez, S., de Luis-Garcia, R., Martin-Fernandez, M.A., Alberola-Lpez, C.: A computational TW3 classifier for skeletal maturity assessment. A computing with words approach. J. Biomed. Inf. **37**, 99–107 (2004)
8. Liu, H., Chou, Y., Tiu, C., Lin, C., Chen, C., Hwang, C., Hsieh, C., Jong, T.: Bone age pre-estimation using partial least squares regression analysis with a priori knowledge. In: 2014 IEEE International Symposium on Medical Measurements and Applications, MeMeA 2014, Lisboa, Portugal, pp. 164–167, 11–12 June 2014
9. Adeshina, S.A., Cootes, T.F., Adams, J.E.: Evaluating different structures for predicting skeletal maturity using statistical appearance models. In: Proceedings of MIUA (2009)
10. Cootes, T.F., Edwards, G.J., Taylor, C.J.: Active appearance models. IEEE Trans. Pattern Anal. Mach. Intell. **23**, 681–685 (2001)
11. Cootes, T.F., Edwards, G.J., Taylor, C.J.: Active appearance models. In: Burkhardt, H., Neumann, B. (eds.) ECCV 1998. LNCS, vol. 1407, pp. 484–498. Springer, Heidelberg (1998)
12. Turk, M., Pentland, A.: Eigenfaces for recognition. J. Cogn. Neurosci. **3**, 71–86 (1991)
13. Luo, B., Hancock, E.: Iterative procrustes alignment with the EM algorithm. Image Vis. Comput. **20**, 377–396 (2002)
14. Ross, A.: Procrustes analysis. Technical report, Department of Computer Science and Engineering, University of South Carolina, SC 29208 (2004). www.cse.sc.edu/songwang/CourseProj/proj2004/ross/ross.pdf
15. Passerini, A.: K-nearest neighbour learning. Department of Information Engineering and Computer Science. University of Trento, Italy (2015). http://disi.unitn.it/passerini/teaching/2015-2016/MachineLearning/slides/02b_nearest_neighbour/talk.pdf
16. Altman, N.S.: An introduction to kernel and nearest-neighbor nonparametric regression. Am. Stat. **46**, 175–185 (1992)

Author Index

Printed in the United States
By Bookmasters